SOME PROBLEMS IN CHEMICAL KINETICS AND REACTIVITY

SOME PROBLEMS IN CHEMICAL KINETICS AND REACTIVITY

VOLUME 2

By N. N. SEMENOV

Translated by Michel Boudart

1959

Princeton University Press

Princeton, New Jersey

L. C. Card 57-10321

Printed in the United States of America

TABLE OF CONTENTS: VOLUME I*

Page

CHAPTER I: REACTIVITY OF MONORADICALS.......................... 1
1. Main Types of Radical Reactions....................... 1
2. Activation Energy and its Experimental Determination.. 4
3. Data on Bond Energies................................. 10
4. Relationship Between Activation Energy and Heat of Reaction.. 29
5. Activity of Radicals and Molecules.................... 33
6. Empirical Formulae to Calculate Bond Energies in Organic Molecules............................... 49
7. Addition to the Double Bond........................... 57
8. Decomposition of Radicals and Energy of the π-bond.... 65
9. Isomerization of Radicals............................. 68
10. Additional Remarks on Substitution Reactions.......... 74
11. Role of Polar Factors in Organic Reactions............ 75
12. Role of Polar Factors in Polymerization............... 85

CHAPTER II: COMPETITION BETWEEN MONORADICAL REACTIONS............ 95
1. Competition Between Different Radical Reactions....... 95
2. Effect of Temperature and Pressure on the Competition Between Radical Reactions............. 99
3. Intermediate and Final Products of Chain Reactions.... 108

CHAPTER III: REACTIONS OF DIRADICALS............................ 145
1. Transition of Atoms to a Valence-active State......... 145
2. Reactivity of O_2, S_2, Se_2 Molecules................. 146
3. The Divalent State of Carbon.......................... 148
4. Diradicals of Complex Structure....................... 149
5. Production and Reactions of the Diradicals $-\overset{|}{C}H_2$ and $-O-$ 154
6. Diradical Participation in Chain Reactions............ 159

CHAPTER IV: DISSOCIATION OF MOLECULES AND RECOMBINATION OF RADICALS.. 169
1. Homogeneous Initiation of Chain Reactions............. 169
2. Homogeneous Recombination of Radicals................. 175
3. Dependence of the Overall Kinetics on the Nature of Chain Termination.................................. 181

* Published by Princeton University Press, 1958.

CONTENTS

Page

4. Peculiarities of Radical Generation and Recombination in Liquids........................... 184

5. Inhibition of Chain Reactions by Certain Additives... 186

CHAPTER V: INITIATION OF CHAIN REACTIONS BY IONS OF VARIABLE VALENCE... 197

1. Formation of Radicals and Radical-ions by Electron Transfer................................ 197

2. Thermodynamic and Kinetic Characters of Production of Free Radicals by Ions of Variable Valence...... 200

3. Initiation of Radical Chain Reactions by Ions of Variable Valence................................. 204

CHAPTER VI: WALL INITIATION AND TERMINATION OF CHAIN REACTIONS.. 211

1. Wall Destruction and Generation of Radicals.......... 211

2. The Method of Differential Calorimetry.............. 216

3. Free Valences on Reactor Walls....................... 220

4. Conditions for Wall Activity in the Process of Radical Generation.............................. 225

5. Heterogeneous Radical Initiation with Impurities in the Gas Phase................................. 227

6. Heterogeneous Initiation Due to Molecular Reactions Producing Free Radicals.......................... 228

7. An Attempt to Apply the New Ideas to Heterogeneous Catalysis....................................... 230

TABLE OF CONTENTS: VOLUME II

Page

PART III: KINETICS OF CHAIN REACTIONS 1

CHAPTER VII: COMPETITION BETWEEN CHAIN REACTIONS AND REACTIONS BETWEEN SATURATED MOLECULES 1

§1. Energy Factors in Chain Reactions and Reactions Between Molecules 2

§2. A Comparison Between the Rates of Direct and Chain Reactions 5

Hydrogen Isotope Exchange 10

Reaction Between Methyl Iodide and Hydrogen Iodide 10

Formation of Nitric Oxide 11

Hydrogenation of Ethylene 12

Chlorination of Ethylene 15

Decomposition of Acetaldehyde 17

Decomposition of Dimethyl Ether 20

Decomposition of Di-iodo-ethane 21

Decomposition of Secondary Butyliodide 24

§3. Decomposition of Alkylbromides 25

§4. Pyrolysis of Alkylchlorides 39

§5. Cracking of Hydrocarbons 42

§6. Factors Affecting Chain Length 49

§7. The Elementary Step of Decomposition 51

§8. Generation of Free Radicals by Reactions Between Saturated Molecules 61

§9. Some Examples 63

Oxidation of Hydrogen 63

Oxidation of Hydrocarbons 63

Reactions of Fluorination 65

Chlorination and Bromination of Olefins 66

Chlorination of Dienes 68

Polymerization 70

Some Bimolecular Cyclization Processes 71

Oxidation of Nitric Oxide 73

References 75

PART IV: BRANCHED CHAIN REACTIONS AND THERMAL EXPLOSIONS 81

INTRODUCTION 81

CHAPTER VIII: THERMAL EXPLOSIONS 87

References 109

CONTENTS

Page

CHAPTER IX: CHAIN IGNITION.................................... 111

§1. Generalities on Chain Ignition.................................... 111

§2. Oxidation of Phosphorus.................................... 121

1. Experimental Data on Phosphorus Oxidation....... 122

2. Theory of the Lower Limit.................................... 130

3. Theory of the Upper Limit.................................... 137

§3. Oxidation of Sulfur.................................... 142

§4. Limits in the Oxidation of Phosphine, Silane, Carbon Disulfide, Carbon Monoxide and in the Decomposition of NCl_3.................................... 143

References.................................... 147

CHAPTER X: CHAIN IGNITION IN HYDROGEN-OXYGEN MIXTURES............. 149

§1. Lower Ignition Limit.................................... 156

§2. The Induction Period.................................... 162

§3. Kinetics of the Reaction Above the Lower Limit......... 165

§4. The Upper Explosion Limit.................................... 188

References.................................... 196

CHAPTER XI: CHAIN INTERACTION.................................... 199

§1. Negative Chain Interaction.................................... 199

§2. Positive Chain Interaction.................................... 202

Widening of the Explosion Region in the System $H_2 + O_2$ by Short Wave-Length Illumination or Addition of O atoms.................................... 204

Cold Flame Propagation in Lean Oxygen-Carbon Disulfide Mixtures.................................... 209

References.................................... 214

CHAPTER XII: CHAIN REACTIONS WITH DEGENERATE BRANCHING............ 217

§1. Introduction.................................... 217

§2. Oxidation of Methane.................................... 228

§3. Kinetics of Oxidation of Hydrogen Sulfide.............. 248

§4. Kinetics of Liquid Phase Oxidation of Hydrocarbons..... 256

Rates of Chain Propagation and Termination.......... 257

Chain Initiation and Degenerate Branching........... 261

The Induction Period and the Kinetics of the Reaction.................................... 265

Accumulation of Hydroperoxides in Liquid Phase Oxidation.................................... 273

§5. Some New Phenomena in the Oxidation of Hydrocarbons and Aldehydes.................................... 274

Stages in Oxidation Processes.................................... 276

References.................................... 280

CONTENTS

Page

APPENDIX I: THE METHOD OF THE ACTIVATED COMPLEX.................. 285

APPENDIX II: QUANTUM MECHANICAL CALCULATIONS OF THE ACTIVATION ENERGY.. 299

References.............................. 310

APPENDIX III: ADDITIONS TO VOLUME I.............................. 311

1. On Radical Isomerization.......................... 311
2. On Polar Factors in Organic Reactions........... 314
3. On the Nature of Intermediates in the Oxidation of Hydrocarbons.......................... 322
4. On the Mechanism of Propane Oxidation........... 322
5. On the Kinetic Tracer Method.................... 322
6. On the Mechanism of Butene Cracking............. 324
7. On the Inhibition of Reactions in the Liquid Phase...................................... 325
8. Heterogeneous Catalysis in Biology.............. 327

Author's Acknowledgements.. 331

PART III

KINETICS OF CHAIN REACTIONS

CHAPTER VII: COMPETITION BETWEEN CHAIN REACTIONS AND REACTIONS BETWEEN SATURATED MOLECULES

In every chemical system, there always take place simultaneously, a simple uni- or bi-molecular reaction and a chain process. The only question then is that of their competition and relative rates. In some cases, the rate of the direct reaction is so much larger than that of the chain reaction that the latter may be neglected, and vice versa in other cases. There are also cases where the rates of both types of change are close to each other. The rate of simple uni- and bi-molecular reactions is due to the sum of molecular acts and naturally does not depend on the conditions of the experiment but only on temperature and concentrations. On the contrary, the course of chain reactions strongly depends on the conditions of the experiment, the presence of impurities, of inert diluents, etc. This dependence is due to the action of various factors on the reactions of chain initiation and termination. Therefore, by changing the conditions of an experiment, we may alter the competition between chain and direct reactions. Thus, for instance, ethylbromide decomposes mainly directly into molecular products. However, it is enough to add to the system a very small quantity of molecular bromine (which easily dissociates into atoms and thus facilitates chain initiation) to carry out the decomposition mainly following a chain mechanism.

It must also be noted that, in a system, not one given chain reaction is taking place, but all chain reactions that are possible from the standpoint of structure. All these chain processes then compete with each other. The rates of these various chain mechanisms are usually quite different and then a particular one predominates and determines the observed reaction rate and the composition of intermediate and final products. However, under different conditions, especially at different temperatures, another chain mechanism may take over and give way to different products. In such cases, there will be transition regions within which the rates of the various chain mechanisms are close to each other, and as a result, a large variety of reaction products will be formed.

We will consider in this book a number of examples of this type.

§1. Energy Factors in Chain Reactions and Reactions Between Molecules

Valence-saturated molecules like di-radicals have an even number of valence electrons. However, their reactivity is quite different.

Thirty years ago, before the discovery of chain theory and of the role of free radicals in chemical changes, and before the development of present views on ionic processes, practically all chemical reactions were envisaged as reactions between two valence-saturated molecules (in gases and liquids). Among these were substitution reactions, e.g., $H_2 + I_2 \longrightarrow 2HI$, $CH_3I + HI \longrightarrow CH_4 + I_2$, decompositions such as $CH_3CHO \longrightarrow CH_4 + CO$ and $C_2H_5Br \longrightarrow C_2H_4 + HBr$, hydrocarbon cracking and addition reactions, for instance, $H_2 + C_2H_4 \longrightarrow C_2H_6$ or $I_2 + C_2H_4 \longrightarrow C_2H_4I_2$, $RH + O_2 \longrightarrow RO_2H$ etc. It has been known for a long time that the majority of gas-phase reactions involve considerable activation energies, with activation barriers around 30 - 50 kcal. This is about an order of magnitude higher than the activation energy of the radical reactions considered above. Moreover, activation energies E and activation barriers E_0 seem to be unrelated to heats of reaction q. Let us consider a few examples.

TABLE 37

Heats of Reaction and Activation Energies for Molecular Reactions

	Reaction	q kcal	E kcal	E_0 kcal	Reference
1)	$H_2 + I_2 \longrightarrow 2HI$	+ 2	40	40	[1]
2)	$CH_3I + HI \longrightarrow CH_4 + I_2$	+ 14	33	33	[1]
3)	$H_2 + C_2H_4 \longrightarrow C_2H_6$	+ 32.5	43	43	[1]
4)	$C_2H_5Br \longrightarrow C_2H_4 + HBr$	- 18	50	32	[2]
5)	$C_2H_5C\ell \longrightarrow C_2H_4 + HC\ell$	- 15	59	44	[2]
6)	$CH_3CHO \longrightarrow CH_4 + CO$	+ 5.5	47	47	[3]
7)	$CH_3OCH_3 \longrightarrow CH_4 + HCHO$	0	58	58	[1]
8)	$(CH_2)_2O \longrightarrow CH_4 + CO$	+ 32	42	42	[4]

Table 37 shows that there is no relation between E_0 and $|q_0|$. For all values of $|q|$, the value of E_0 oscillates between 30 and 50 kcal and can be as high as 60 kcal.

It must be pointed out that the values of E collected in Table 37 may be lower than the true values for the direct processes because, for many of these reactions, a chain mechanism involving free radicals accounts for the bulk of the process.

Every molecular decomposition proceeds through the formation of some transition state, and the energy required to form this transition state constitutes the activation energy E for decomposition into molecules. For endothermic reactions $E = E_0 + q_1$ for exothermic reactions $E = E_0$ (E_0: the height of the activation barrier, q: the heat of reaction).

For free radical reactions, as was seen in Chapter I, $\epsilon_0 = A - \alpha|q|$ where A is a quantity close to 11 kcal, slightly dependent on the type of reaction and α is approximately 0.25. For this reason, we were able to assume in first approximation that ϵ_0 is a function of the difference between the energies of the bonds formed and broken during reaction. For decomposition into molecules, it is probable that there exists a similar relationship: $E_0 = A' - \alpha'|q'|$. Here, the quantity q' is of the same order of magnitude as q for radical reactions (not more than 20 or 30 kcal). As to the quantity A', it is several times larger in molecular decompositions that the corresponding quantity in radical reactions. Even if the relative variations

$$\frac{\Delta A'}{A'} \quad \text{and} \quad \frac{\Delta A}{A}$$

were the same for both molecular and radical reactions, variations in rates, determined by the factors $\exp(\Delta A'/RT)$ and $\exp(\Delta A/RT)$ would be much larger for direct decomposition of a molecule into molecular products than for, say, decomposition of a radical into a radical and a molecule. As a result, changes in the quantity A' for various molecular reactions, in contrast with the situation prevailing in the case of radical reactions, exert a decisive influence on rates and the role of the heat of reaction q' becomes a minor one. Large variations $\Delta A'$ in the case of molecular decompositions are due to the fact that at least two bonds of the original molecules (and not one as is the case in radical reactions), take part in the formation of the activated complex. Then, the energy of formation of the activated complex is extremely sensitive to the relative geometric position of the reacting groups and only one single change in a characteristic distance brings about a large modification of A' for molecules of different structure. Moreover, the ease with which the activated complex is formed depends on the strength and energy of the bonds broken (and not on the difference between the energies of bonds formed and broken, which determines q'). Also, the height of the activation barrier depends not only on the properties of bonds broken but also on the properties of bonds formed in the reaction products. All this leads to a strong dependence of E_0 on the chemical structure of reactant and product molecules. The quantum-mechanical calculation of E_0 is very unreliable and more experiments are needed before the problem of the activation energy of molecular reactions can be clarified. The analysis of these experiments should establish convincingly that the experimental activation energy is that of the direct molecular process and not an overall activation energy determined by a radical mechanism. Let us note that, whereas a clear picture of future developments has emerged from studies of free radical reactions, the situation is much less advanced in the case of molecular reactions.

How do we explain the inertness of molecules? One may think that the saturation of chemical forces in molecules makes them relatively stable so that they react only with difficulty, although the absolute stability of the system increases as a result of reaction. It is therefore necessary to stretch bonds in the molecule, in order to overcome this relative stability and this requires high activation energies. A free radical, being a valence-unsaturated particle, exerts a direct influence on a molecule and can react immediately. Qualitative calculations based on the quantum-mechanical equation of London also give high activation energies for molecular reactions of the type (see Appendix II):

$$AB + CD \longrightarrow AC + BD \quad .$$

The London equation gives considerably smaller activation energies for reactions between free radicals and molecules.

As was already pointed out, this conclusion is reached for the end-on attack of a σ-bond: $A + B - C \longrightarrow A - B + C$ where A, B and C are collinear. For a perpendicular attack ${}_{B}\overset{A}{-}{}_{C} \longrightarrow A - B + C$, the quantum-mechanical calculation gives a substantially higher value for the activation barrier, about half of the value for reaction between two molecules:

$$\begin{matrix} A - B \\ C - D \end{matrix} \longrightarrow \begin{matrix} A \\ | \\ C \end{matrix} + \begin{matrix} B \\ | \\ D \end{matrix} \quad .$$

Although this question is not discussed anywhere explicitly, it is apparent that the quantum-mechanical treatment of molecular inertness and radical activity leads to the problem of relative arrangement of reacting species at the moment of attack. A free radical on an atom may attack a given bond under the most favorable angle. Thus the activated complex may have a linear configuration. On the contrary, in the majority of direct reactions between molecules, the geometry of the reacting partners is unfavorable since the activated complex has a trapezoidal or, in general, a quadrangular shape:

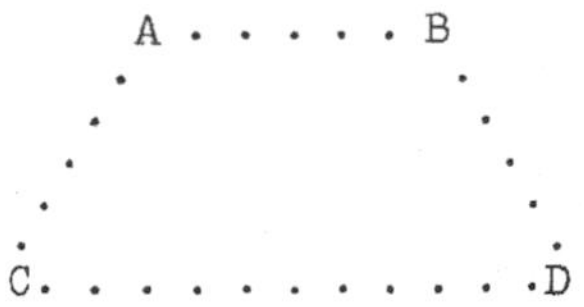

If the reacting molecules AB and CD collide in a favorable way from the viewpoint of activation energy and form a linear complex A ... B ... D ... C, the reaction, of course, does not yield the molecular products AC and BD but a molecule BD and two radicals A and C:

$$AB + DC \longrightarrow A \ldots B \ldots D \ldots C \longrightarrow A + BD + C \quad .$$

To such a process corresponds a low activation barrier but it is energetically unfavorable since the released energy is smaller than for the process: $AB + CD \longrightarrow AC + BD$. The difference is the bond energy Q_{AC} and the process is therefore strongly endothermic. For instance, while the reaction $H_2 + C\ell_2 \longrightarrow 2HC\ell$ is exothermic (44 kcal), the process $H_2 + C\ell_2 \longrightarrow HC\ell + H + C\ell$ requires an expenditure of energy $|q| = 102 - 44 = 58$ kcal. Therefore its activation energy will be $E = E_0 + 58$ kcal, a large quantity even if E_0 is small (5 - 10 kcal). Because of this the process is not a likely one.

To sum up, direct molecular reactions giving molecular products have an unfavorable energy barrier when they are favored energetically and, on the contrary, a favorably small activation barrier goes together with an unfavorable energy factor. In both cases, the activation energy turns out to be quite large. On the other hand, radical reactions $A + BC \longrightarrow AB + C$ are not usually very endothermic since one bond is broken while another is made. This circumstance, as well as the small activation barrier due to a favorable angle of attack make these reactions proceed very readily. For these reasons, a chain mechanism is generally more favorable than a direct reaction path, not only for photochemical or catalytic reactions but frequently also for homogeneous thermal processes.

§2. A Comparison Between the Rates of Direct and Chain Reactions

A direct reaction $AB + CD \longrightarrow AC + BD$ has a rate $w_D = k[AB][CD]$

where k is a bimolecular rate constant, approximately equal to $10^{-10}\exp(-E/RT)$.

Consider now a chain process:

1) $A + CD \longrightarrow AC + D$

2) $D + AB \longrightarrow A + BD$ etc.

If the A radicals are generated artificially, for instance, by light, the rate is now:

$$w_C = k_1[A][CD] = k_2[D][AB] \approx 10^{-10}\exp(-\epsilon_1/RT) \cdot [A][CD]$$

where ϵ_1 is the activation energy of the intermediate step going with the lowest rate. Thus, the ratio of rates of the direct and photochemical processes is equal to:

$$\gamma = \frac{w_D}{w_C} = \frac{[AB]}{[A]}\exp[-(E - \epsilon_1)/RT] \quad .$$

The concentration of A radicals is determined by equating the number I of free radicals formed per second (by photolysis of AB molecules) to the number of A radicals disappearing each second (for instance by recombination). The rate of the latter process is obviously $k_3[A]^2$. Therefore

$$I = k_3[A]^2$$

and

$$[A] = (I/k_3)^{1/2} \quad .$$

Consequently,

$$\gamma = \frac{w_D}{w_C} = [AB](I/k_3)^{-1/2}\exp[-(E - \epsilon_1)/RT] \qquad (1)$$

where

$$w_C \cong 10^{-10}[A][CD]\exp(-\epsilon_1/RT)$$

$$w_D \cong 10^{-10}[AB][CD]\exp(-E/RT) \quad .$$

At atmospheric pressure, $[AB] \approx 10^{19}$. If recombination takes place on each binary collision $k_3 \approx 10^{-10}$. If atoms are involved, recombination takes place on each triple collision between the two atoms and a third particle and $k_3 \approx 10^{-32}[M]$ where $[M]$ is the number of all particles — atoms, molecules, etc. — per cm^3. At atmospheric pressure $[M] \approx 10^{19}$ and $k_3 \approx 10^{-13}$. Since k_3 is raised to the 1/2 power [see equation (1)], the ratios $\gamma = (w_D/w_c)$ differ by a factor of 30 in both cases.

Taking $k_3 \approx 10^{-10}$ and $[AB] \approx 10^{19}$, we get

$$\gamma = \frac{10^{14}}{I^{1/2}} \exp[-(E - \epsilon_1)/RT] \quad .$$

For molecular reactions, E has a large value, say 40 kcal. For radical reactions, ϵ_1 is small, for instance 5 kcal. Then:

$$\gamma = \frac{10^{14}}{I^{1/2}} \exp(-35{,}000/RT) \quad .$$

With strong illumination, $I \approx 10^{15}$ while $I \approx 10^{12}$ with very weak light. Take $I \approx 10^{12}$. Then $\gamma = 10^8 \exp(-35{,}000/RT)$. At 600°K, $\gamma \approx 10^{-5}$ and at 400°K, $\gamma \approx 10^{-11}$.

Under the conditions just discussed, the direct reaction will play a very small part.

The reaction will proceed by a chain mechanism. Our comparison between the photochemical chain process and the direct reaction is, however, not too suggestive. It is more interesting to compare the direct reaction and the chain process without artificial production of primary radicals. Then, radicals are generated by thermal dissociation of the AB molecule into radicals A and B and $I = 10^{13} \exp(-Q/RT)[AB]$. Here I is the number of radicals generated thermally per cm^3. At atmospheric pressure, $I = 10^{32} \exp(-Q/RT)$ where Q is the dissociation energy of AB. Then:

$$[A] = \left(\frac{10^{13} \exp(-Q/RT) \cdot [AB]}{10^{-10}} \right)^{1/2} \approx 10^{21} \exp(-Q/2RT) \quad .$$

Therefore, the rate of the chain process is:

$$w_c \approx 10^{11} \exp[-(\tfrac{Q}{2} + \epsilon_1)/RT] \cdot [CD]$$

while that of the direct process is:

$$w_D \approx 10^9 \exp(-E/RT) \cdot [CD] \quad .$$

According to equation (1):

$$\gamma = \frac{w_D}{w_c} = \frac{[AB]}{(I/k_3)^{1/2}} \exp[-(E - \epsilon_1)/RT]$$

$$= \frac{10^{19}\exp[-(E-\epsilon_1)/RT]}{[10^{32}\exp(-Q/RT)/10^{-10}]^{1/2}} = 10^{-2} \exp[-(E - \epsilon_1 - \tfrac{Q}{2})/RT]$$

For a reaction of the type $H_2 + Cl_2$, the chain is propagated by the steps $Cl + H_2 \longrightarrow HCl + H$; $H + Cl_2 \longrightarrow HCl + Cl$ and recombination occurs by means of triple collisions $Cl + Cl + M \longrightarrow Cl_2 + M$. Decomposition of Cl_2 into radicals is due to the bimolecular process: $Cl_2 + M \longrightarrow Cl + Cl + M$.

Therefore we have:

$$k_3 = 10^{-32}[M]$$

and

$$I = 10^{-10}[Cl_2][M] \exp(-Q/RT) \quad .$$

Then, with $[Cl_2] \approx 10^{19}$:

$$\gamma = \frac{w_D}{w_c} = \frac{10^{19}\exp[-(E-\epsilon_1)/RT}{\left(\frac{10^{-10}\cdot 10^{19}[M]}{10^{-32}[M]}\right)^{1/2}\cdot\exp(-Q/2RT)} \cong 10^{-1}\exp[-(E-\epsilon_1-\frac{Q}{2})/RT] \quad .$$

We see that if $E - \epsilon_1 - \frac{Q}{2} > 0$, the chain reaction always takes over, since then $\gamma < 0.1$ and the rate of the direct reaction is less than 10% of that of the chain process. If, on the other hand, $E - \epsilon_1 - \frac{Q}{2} < 0$, the direct reaction may predominate.

The reaction $H_2 + I_2 \longrightarrow 2HI$ is a simple bimolecular process with an energy of activation $E = 40$ kcal.

The step $H + I_2 \longrightarrow HI + H$ is strongly endothermic. The energy of dissociation of HI into H and $\cdot$I is equal to 70 kcal, so that the step $I + H_2$ requires an expenditure of energy equal to $103 - 70 = 33$ kcal. The activation barrier, ϵ_{01} is very small. Thus we may write $\epsilon_1 = 33$ kcal. The energy of dissociation of I_2 into $I + I$ amounts to 35.5 kcal. Consequently, $E - \epsilon_1 - \frac{Q}{2} = 40 - 33 - 17.8 = -10.8$ kcal. Since $\gamma = 10^{-1}\exp(10{,}8000/RT)$, at $T = 700°K$, $\gamma \approx 2.10^2$. The chain process is two hundred times slower than the direct reaction.

As is well known the reactions $H_2 + Cl_2$ and $H_2 + Br_2$ are chain processes. We do not know the values of the activation energy E for the direct reactions. Remembering, however, that E practically does not depend on the heat of reaction and also that the bonds Br - Br and Cl - Cl are stronger than the bond I - I, we must expect that $E > 40$ kcal for these two processes. Let us adopt the value $E = 40$ kcal.

We know that the activation energy ϵ_1 of the step $Cl + H_2 \longrightarrow HCl + H$ is equal to 6 kcal $(q = 1$ kcal). On the other hand,

$Q_{C\ell-C\ell} = 57$ kcal. From these data, we calculate $E - \epsilon_1 - \frac{Q}{2} = 40 - 6 - 28.5 = 5.5$ kcal and $\gamma = 10^{-1}\exp(-5500/RT)$. At 600°K (i.e., in a temperature range where the $H_2 + C\ell_2$ reaction proceeds at a measurable rate) we have $\gamma = 10^{-3}$. Consequently the direct reaction is one thousand times slower than the chain process.

For the hydrogen-bromine reaction, $Br + H_2 \longrightarrow HBr + H$ is characterized by $q = -16$ kcal, $\epsilon_1 = 17$ kcal, while $Q \approx 46$ kcal. Then, $E - \epsilon_1 - \frac{Q}{2} = 40 - 17 - 23 = 0$, whence,

$$\gamma = \frac{w_D}{w_c} = 0.1$$

and the direct reaction is ten times slower than the chain process. In fact, E, for that reaction, is probably considerably larger than 40 kcal and the ratio of rates will be smaller than the value just calculated. Consequently, both the $H_2 + C\ell_2$ and the $H_2 + Br_2$ processes are chain processes as observed experimentally.

A discussion of competition between chain and direct processes for the formation of hydrogen halogenides, was facilitated first by the fact that we knew the value of E, the activation energy of the direct reaction and also because both propagation steps of the chain are bimolecular substitution reactions. For such steps, the pre-exponential factors are sufficiently similar so that the rate-determining chain propagation step is that with the larger activation energy.

In the majority of cases, we do not know the value E of the activation energy of the direct process. However, if the experimental overall activation energy of the overall process E_0 is found to be equal to $\epsilon_1 + \frac{Q}{2}$, there are good reasons to believe that the process is of the chain type. Here again Q is the expenditure of energy for initiation and ϵ_1 is the activation energy of the slow step. If the primary radicals are generated by dissociation of one of the reactants into atoms, Q is then a dissociation energy which is known in many cases. The value of ϵ_1 must be determined from photochemical experiments or in studies with atoms produced in a discharge. Unfortunately such data are rarely available. When both steps of the propagation cycle are exothermic, $\epsilon = \epsilon_{01}$ (ϵ_{01} is usually less than 11 kcal) and if the inequalities

$$E_0 > \frac{Q}{2}$$

and

$$E_0 - \frac{Q}{2} = \epsilon_{01} < 11 \text{ kcal} \tag{2}$$

are satisfied, there is a good presumption that a chain mechanism is at work. Estimation of ϵ_{01} can be made if the exothermicity of the elementary step is known: $\epsilon_{01} = 11 - 0.25q$ (see Chapter I). When one of the propagation steps is endothermic, we have $\epsilon_1 = \epsilon_{01} + q$ (heat of reaction: $- q$). In this case, a chain process can be presumed when

$$E_0 > \frac{Q}{2} + q \tag{3}$$

and

$$E_0 - (\frac{Q}{2} + q) = \epsilon_{01} < 11 \text{ kcal} \quad . \tag{4}$$

Consider now a few examples where these formulae can be applied.

Hydrogen Isotope Exchange ($H_2 + D_2$) [1]

The reaction $H_2 + D_2 \longrightarrow 2HD$ proceeds homogeneously at temperatures above 600°C. The experimentally determined activation energy is about 57 kcal. A possible chain mechanism is:

0) $H_2 + M \longrightarrow 2H - 103$ kcal, initiation
1) $H + D_2 \longrightarrow HD + D - 1$ kcal } chain
2) $D + H_2 \longrightarrow HD + H - 0.7$ kcal }
3) $D + D + M \longrightarrow D_2 + M$, chain breaking.

With such a mechanism, the activation energy would be:

$$E_c = \frac{103}{2} + 1 + \epsilon_{01} = 52.5 + \epsilon_{01} \quad .$$

Since one of the propagation steps is endothermic, let us use formula 3):

$$E_0 = 57 > \frac{Q}{2} + 1 = 52.5 \text{ kcal}$$

$$E_0 - \frac{Q}{2} - q_1 = 57 - 51.5 - 1 = 4.5 < 11 \text{ kcal} \quad .$$

This result makes a chain mechanism very probable. The directly measured activation energy of the step $H + D_2 \longrightarrow HD + D$ is equal to 6.5 kcal (see [1] p. 377). Then $\epsilon_{01} = 5.5$ kcal and $E_c = 52.5 + 5.5 = 58$ kcal, a value equal, within experimental error, to the observed value $E_0 = 57$ kcal. This now proves that the main part of the reaction takes place by a chain mechanism.

Reaction Between Methyl Iodide and Hydrogen Iodide

$$CH_3I + HI \longrightarrow CH_4 + I_2 + 14 \text{ kcal} \quad .$$

A possible chain mechanism is the following:

0) $CH_3 I \longrightarrow CH_3 + I - 52$ kcal, initiation
1) $I + CH_3I \longrightarrow CH_3 + I_2 - 18$ kcal } chain
2) $CH_3 + HI \longrightarrow CH_4 + I + 32$ kcal } chain
3) $I + I + M \longrightarrow I_2 + M + 35$ kcal, chain breaking.

The activation energy of the chain process is:

$$E_c = \frac{Q}{2} + q_1 + \epsilon_{01} = 26 + 18 + \epsilon_{01} = 44 + \epsilon_{01} .$$

The observed value is $E_0 = 33$ kcal [5]. Let us apply formula 3):

$$E_0 = 33 > \frac{Q}{2} + q = 44 .$$

In this case, condition 3) is not fulfilled and a chain process should be practically excluded. It is possible, however, that initiation takes place not only by dissociation of CH_3I into CH_3 and I with an expenditure of 52 kcal, but also by a bimolecular process $CH_3I + HI \longrightarrow CH_4 + I + I - 21$ kcal.

We will see in the following section that there are good reasons to believe that such processes between molecules, giving two free radicals, have a low activation barrier similar to that for reactions between molecules and radicals, or even less. Then, putting $Q = 21 + \epsilon_{00}$ and keeping the same scheme for propagation and termination, we find for the activation energy of the chain process:

$$E_c = \frac{21}{2} + 18 + \epsilon_{01} + \frac{\epsilon_{00}}{2} = 28.5 + \epsilon_{01} + \frac{\epsilon_{00}}{2} .$$

Applying condition 3), we have:

$$E_0 = 33 > \frac{Q}{2} + q_1 = 28.5 \text{ kcal}$$

and

$$E_0 - \frac{Q}{2} - q_1 = 33 - 28.5 = 4.5 < 11 \text{ kcal} .$$

Conditions 3) and 4) are satisfied and it becomes quite likely that the process considered follows a chain mechanism.

Formation of Nitric Oxide

The reaction $N_2 + O_2 \longrightarrow 2NO - 44$ kcal proceeds at very high temperatures, of the order of 1000°C. A chain mechanism is as follows:

0) $O_2 + M \longrightarrow O + O + M - 118$ kcal, initiation

1) $O + N_2 \longrightarrow NO + N - 75$ kcal
2) $N + O_2 \longrightarrow NO + O + 31$ kcal } propagation
3) $O + O + M \longrightarrow O_2 + M + 118$ kcal, chain breaking.

According to the data of Ya. B. Zel'dovich, P. Ya. Sadovnikov and D. A. Frank-Kamenetskiĭ [6], the experimental value of the activation energy is $E_0 = 129$ kcal. Experiments conducted at such higher temperatures cannot give a value of E_0 better than within $\pm$ 10 kcal.

The scheme just presented gives:

$$E_c = \frac{Q}{2} + q_1 + \epsilon_{01} = 59 + 75 + \epsilon_{01} = 134 + \epsilon_{01} \quad .$$

Reaction 1') is strongly exothermic (75 kcal) and therefore its activation barrier must be very small. Thus $E_c = 134$ kcal, a figure slightly higher than the experimental figure. For the reaction under discussion, one must take into account the error in the determination of E_0 and it may be assumed that the process is of the chain type. This conclusion is supported by the observation of Ya. B. Zel'dovich et al. that the overall reaction rate is proportional to $(N_2)(O_2)^{1/2}$ in agreement with a chain mechanism and not to $(N_2)(O_2)$ as would be the case for a direct bimolecular process.

The simple relationships given above give a solution to the problem of competition between direct and chain processes only when the chain consists of a certain type of bimolecular steps, namely substitution reactions between radicals and molecules.

As we have seen, pre-exponential factors of the rate constants are about the same for reactions of this type. However, if chain propagation involves not only substitution but also addition steps, it is necessary, in order to determine the slow step, to take into account both activation energies and steric factors. It has been pointed out in Chapter I that processes of addition to double bonds are slowed down, as compared to substitutions, by very small steric factors. As a result, additions to double bonds may be rate determining in spite of their having activation energies smaller than those of other steps of the chain.

Consider, as an example, the addition of hydrogen to ethylene.

Hydrogenation of Ethylene [7]

$$C_2H_4 + H_2 \longrightarrow C_2H_6 + 33 \text{ kcal} \quad .$$

The chain is propagated by the following elementary steps:

1) $C_2H_4 + H \longrightarrow C_2H_5 + 38$ kcal

2) $C_2H_5 + H_2 \longrightarrow C_2H_6 + H - 5$ kcal .

We know that the activation energy of reaction 1) is small and equal to ~ 2 kcal. The activation energy of the endothermic step 2) is 11 - 14 kcal, according to several investigations [8]. We will take $\epsilon_2 = 13$ kcal. Then, if the steric factors of both steps were identical, the slower one would of course be step 2) and the ratio of rate constants (k_1/k_2) would be equal to $\exp[(\epsilon_2 - \epsilon_1)/RT] \approx \exp(11{,}000/RT)$. This ratio is about 10^3 at temperatures at which ethylene hydrogenation takes place (circa 800°K). If the steric factors are different, $(k_1/k_2) = (f_1/f_2)\exp[(\epsilon_2 - \epsilon_1)/RT]$.

There is some evidence that steric factors for addition reactions are $10^3 - 10^4$ times smaller than for substitutions. Then $(k_1/k_2) = 10^{-3} \cdot 10^3 = 1$. In this case, therefore, both steps proceed at about the same rate. This leads to an equal concentration of radicals C_2H_5 and H and the main chain breaking step will be the recombination of different radicals:

3) $C_2H_5 + H \longrightarrow C_2H_6$.*

Let us note that the generation of free radicals for ethylene hydrogenation is quite difficult because the dissociation of reactants:

$$H_2 \longrightarrow 2H - 103 \text{ kcal}$$

$$C_2H_6 \longrightarrow C_2H_5 + H - 97 \text{ kcal}$$

requires a sizeable expenditure of energy. One may then assume that the main chain initiation step will be the bimolecular step:

$$H_2 + CH_2 = CH_2 \longrightarrow H + CH_3CH_2 - \sim 65 \text{ kcal} \quad .$$

The complete scheme for ethylene hydrogenation may then be written:

0) $H_2 + C_2H_4 \longrightarrow H + C_2H_5 - 65$ kcal, initiation

1) $H + C_2H_4 \longrightarrow C_2H_5 + 38$ kcal } chain

2) $C_2H_5 + H_2 \longrightarrow C_2H_6 + H - 5$ kcal } chain

3) $C_2H_5 + H \longrightarrow C_2H_6$, chain breaking.

The activation energies of steps 1), 2) and 3) are:

* The reaction $H + H + M \longrightarrow H_2 + M$ is slower because triple collisions are required. The recombination $C_2H_5 + C_2H_5 \longrightarrow C_2H_4 + C_2H_6$ is also slower than the process $C_2H_5 + H \longrightarrow C_2H_6$ because of the large velocity of H atoms and also possibly because of the existence of a steric factor.

$$\epsilon_0 = Q_0 = \epsilon_{00} = 65 + \epsilon_{00} \quad ;$$
$$\epsilon_1 \cong 2 \text{ kcal} \quad ;$$
$$\epsilon_2 \cong 13 \text{ kcal} \quad .$$

As ratio of steric factors, we will take $(f_1/f_2) = 10^3$. Writing down the kinetic equations and solving them, we find for the over-all rate:

$$w = \left(\frac{k_0 k_1 k_2}{k_3} \right)^{1/2} [H_2][C_2H_4] \quad .$$

We see that the chain process, insofar as dependence on concentrations $[H_2]$ and $[C_2H_4]$ is concerned, imitates the behavior of a simple bimolecular law (actually found experimentally). With:

$$k_0 = 10^{-10}\exp[-(65000 + \epsilon_{00})/RT] \quad ;$$
$$k_1 = f_1 10^{-10}\exp(-2000/RT)$$
$$k_2 = f_2 10^{-10}\exp(-13000/RT)$$

and

$$k_3 = 10^{-10} \quad ,$$

we find:

$$w = (f_1 f_2)^{1/2} \cdot 10^{-10}\exp[-(65000 + 2000 + 13000 + \epsilon_{00})/_2RT] \cdot [H_2][C_2H_4] \quad .$$

With $(f_1/f_2) = 10^3$ and $f_1 \cong 1$, we get:

$$w = 3 \cdot 10^{12}\exp[-(40000 + \frac{\epsilon_{00}}{2})/RT] \cdot [H_2][C_2H_4] \quad .$$

The quantity ϵ_{00} hardly exceeds 2 or 3 kcal. Then:

$$w = 3 \cdot 10^{-12}\exp(-41500/RT) \cdot [H_2] \cdot [C_2H_4] \quad .$$

The experimental value [7] is:

$$w = 10^{-11}\exp(-43150/RT) \cdot [H_2][C_2H_4] \quad .$$

In this fashion we obtain very good agreement between theory and experiment. The correctness of our views concerning the chain nature of this process is supported by the fact that the reverse reaction, decomposition of ethane into ethylene and hydrogen, is a chain process as shown directly by a number of facts.

Chlorination of Ethylene

$$C_2H_4 + C\ell_2 \longrightarrow C_2H_4C\ell_2 + Q \quad (43.6 \text{ kcal})$$

In photochemical studies, initiation is due to the reaction $C\ell_2 + h\nu \longrightarrow 2C\ell$, and chain propagation, as generally accepted, is due to the alternation of the two elementary steps:

1) $C\ell + C_2H_4 \longrightarrow C_2H_4C\ell + 26 \text{ kcal}^*$
2) $C_2H_4C\ell + C\ell_2 \longrightarrow C_2H_4C\ell_2 + C\ell + 17.6 \text{ kcal}$.

The chain length at room temperature was determined by Schumacher [9] and is $3 \cdot 10^6$ molecules per quantum. The temperature coefficient of the reaction is very small: 1.09 for 10°. This indicates a value of ~ 1500 cal for the activation energy of the slower of the two steps. A small activation energy accounts for the long chain length at room tempertture.

The photochemical reaction is strongly inhibited by traces of oxygen. The latter is not consumed, so that it probably forms unstable compounds such as $C\ell O_2$ or $C_2H_4C\ell O_2$, which then decompose without regenerating a chlorine atoms. It is known for instance that $C\ell O_2$ decomposes into $C\ell_2$ and O_2:

$$2C\ell O_2 \longrightarrow C\ell_2 + 2O_2 \quad .$$

The thermal chlorination of ethylene may follow two paths: a molecular path with direct addition of a chlorine molecule to ethylene, and a chain path via decomposition of $C\ell_2$ into two chlorine atoms followed by steps 1) and 2).

The main reaction product at 200 - 250°C is dichloroethane and substitution is virtually absent (see Chapter II). Oxygen also strongly inhibits the reaction under these conditions.

Recently, ethylene chlorination has been thoroughly studied by A. M. Chaĭkin at Moscow State University [10]. The reaction was carried out at 227°C and 200 mm Hg. A quantitative measurement of the initial rate as a function of added oxygen led to an equation that fits the data very well:

$$w = w_0 + \frac{a}{b + c(O_2)} \quad .$$

Note that the term $c(O_2)$ is already ten times larger than b for 2.5% oxygen. The rate w_0, a very small quantity, is that of the

* The heat of reaction was calculated indirectly and may be somewhat in error.

completely inhibited reaction. In the absence of oxygen, w_0 represents only about 1% of the total rate.

The decrease of the reaction rate $w - w_0$, as represented by this expression, clearly points to the chain nature of the process. The rate of the direct molecular process may not be larger than w_0 and if a molecular process takes place at all, it must be approximately 100 times slower than the chain process. We believe that, in all likelihood, this molecular process takes place on the surface of the reactor, since its rate depends on the concentration of the reaction product, dichloroethane.

Using the method of differential calorimetry [11], A. M. Chaǐkin studied in detail the space distribution of reaction. As was done before in the case of the hydrogen-chlorine reaction, he showed that chains also start on the surface in ethylene chlorination. Further kinetic studies elucidated the dependence of the reaction rate on surface to volume ratio S/V, for both the uninhibited and the oxygen inhibited reactions. The results were the same as for the $H_2 - Cl_2$ reaction: in the uninhibited process, chains start and finish at the wall; in the inhibited process, chains start at the wall and end within the volume because of the presence of oxygen. The value found for the activation energy (17 kcal) may be explained only on the basis of surface chain initiation. Indeed, in the case of homogeneous initiation $Cl_2 + M \longrightarrow 2Cl + M - 57$ kcal, the activation energy would be considerably higher.

The chain initiation at the wall explains the irreproducibility of the experiments noted by many investigators and due to uncontrollable changes at the surface. The wall reaction may also explain a phenomenon reported by Rust and Vaughn [12]: small quantities of oxygen do not inhibit the reaction but accelerate it at more elevated temperatures.

The accelerating action of small amounts of O_2 has also been observed by V. V. Voevodskiǐ and V. A. Poltorak [13] in propane cracking as well as by Barton and Howlett [13] in the pyrolysis of several chlorine substituted hydrocarbons (1,2-dichloroethane, 1,1,2-trichloroethane, 1,1,2,2-tetrachloroethane).

Chaǐkin also reported a strong inhibition of ethylene chlorination by its reaction products, both with and without oxygen. The reaction rate is at first second order but, as reaction proceeds, the order increases and the rate becomes proportional approximately to the fifth power of pressure.

This self-inhibition is similar to that encountered in hydrocarbon cracking. Its nature is not clear and the question requires further work.

A detailed study of all kinetic facts led to the following expression for the reaction rate:

$$w - w_0 = \frac{a'(C\ell_2)(C_2H_4)}{b + c(O_2)}$$

where $b = b_0 + b_1\Delta X$ and ΔX is the quantity reacted or the extent of conversion or the quantity of product formed.

This rate law can be deduced from the following reaction scheme:

$$
\begin{array}{lll}
0) & C\ell_2 \xrightarrow{k_0} 2C\ell & \text{chain initiation} \\
1) & C\ell + C_2H_4 \xrightarrow{k_1} C_2H_4C\ell & \\
2) & C\ell_2 + C_2H_4C\ell \xrightarrow{k_2} C_2H_4C\ell_2 + C\ell & \text{chain propagation} \\
3) & C\ell + \text{product} \xrightarrow{k_3} & \\
4) & C\ell + \text{wall} \xrightarrow{k_4} & \\
5) & C\ell + O_2 \xrightarrow{k_5} C\ell O_2 & \text{chain termination} .
\end{array}
$$

Then

$$w = \frac{2k_0k_1(C\ell_2)(C_2H_4)}{k_3(\text{product}) + k_4 + k_5(O_2)} .$$

To sum up, gas phase chlorination of ethylene is a chain process and the direct molecular process is virtually absent. In the liquid phase, ethylene chlorination proceeds at much lower temperatures but the mechanism of that process has not been studied thus far.

Let us now discuss decomposition reactions. The chain scheme here consists of two or three elementary steps, one of which is usually unimolecular, the others being bimolecular substitutions. It is then inadequate to determine the slow step by looking only at activation energies since pre-exponential factors are quite different.

Decomposition of Acetaldehyde

$$CH_3C{\overset{O}{\underset{H}{\lessgtr}}} \longrightarrow CH_4 + CO + 5\ \text{kcal} .$$

The direct unimolecular decomposition competes with the following chain mechanism:

0) $CH_3C\begin{smallmatrix}\nearrow O\\ \searrow H\end{smallmatrix} \longrightarrow CH_3 + HCO - 72$ kcal, initiation

1) $CH_3 + CH_3CHO \longrightarrow CH_4 + CH_3CO + 17$ kcal } chain

2) $CH_3CO \longrightarrow CH_3 + CO - 12$ kcal .*

The rate of step 1) is equal to:

$$w_1 = k_1[CH_3][CH_3CHO] = f_1 \cdot 10^{-10}\exp(-\epsilon_{01}/RT)\cdot[CH_3]\cdot[CH_3CHO] \quad .$$

The rate of step 2) is given by:

$$w_2 = k_2[CH_3CO] = 10^{13}\exp[-(|q_2| + \epsilon_{02})/RT]\cdot[CH_3CO] \quad .$$

At the steady state, if chains are sufficiently long $w_1 = w_2$. Whence:

$$\frac{[CH_3]}{[CH_3CO]} = \frac{10^{13}\exp[-(|q_2|+\epsilon_{02})/RT]}{f_1\cdot 10^{-10}\exp(-\epsilon_{01}/RT)\cdot[CH_3CHO]}$$

$$= \frac{10^{23}}{f_1[CH_3CHO]}\exp[-(|q_2| + \epsilon_{02} - \epsilon_{01})/RT] \quad .$$

Since $|q_2| = 12$ kcal and ϵ_{01} and ϵ_{02} may be estimated by means of the formulae: $\epsilon_{01} = 11 - 0.25 \cdot 17 \cong 7$; $\epsilon_{02} = 11 - 0.25 \cdot 12 \cong 8$, we get $\epsilon_2 = \epsilon_{02} + |q_2| = 8 + 12 = 20$ kcal and $|q_2| + \epsilon_{02} - \epsilon_{01} = 13$ kcal. Thus,

$$\frac{[CH_3]}{[CH_3CO]} = \frac{10^{23}}{f_1[CH_3CHO]}\exp(-13000/RT) \quad .$$

---

* According to Szwarc [14], the heat of formation of the radical CH_3CO is 10.8 kcal. Then $Q(CH_3 - CO) = 16$ kcal and

$$Q(CH_3C{\overset{\nearrow O}{}} - H) = 80 \text{ kcal} \quad .$$

Then also $q_1 = 21$ kcal. We have taken $Q(CH_3 - CO) = 12$ kcal, corresponding to

$$Q(CH_3C{\overset{\nearrow O}{}} - H) = 85 \text{ kcal} \quad ,$$

a value accepted earlier by Szwarc [15].

At 800°K,

$$\frac{[CH_3]}{[CH_3CO]} = \frac{10^{23}}{f_1[CH_3CHO]} \cdot 10^{-3.5} \quad .$$

At pressures close to atmospheric, $CH_3CHO \cong 10^{19}$. Then finally:

$$\frac{[CH_3]}{[CH_3CO]} \cong \frac{3}{f_1} \quad .$$

It must be concluded that reaction 1) is rate determining in spite of the fact that the activation energy of the unimolecular step 2) is $\epsilon_2 = 20$ kcal whereas that of the bimolecular step 1) is only $\epsilon_1 = 7$ kcal.

Thus, even if $f_1 = 1$, the concentration of CH_3 radicals is larger than that of CH_3CO radicals. Since, in general $f_1 \approx 0.1$, the concentration of CH_3 is about thirty times larger than that of CH_3CO. The principal chain breaking step is then the recombination of methyl radicals:

3) $CH_3 + CH_3 \longrightarrow C_2H_6$.

The kinetic equations expressing this mechanism give:

$$k_3[CH_3]^2 = k_0[CH_3CHO]$$

or

$$[CH_3] = (k_0/k_3)^{1/2}[CH_3CHO]^{1/2} \quad .$$

The overall rate is given by the expression:

$$W = w_1 = k_1(k_0/k_3)^{1/2}[CH_3CHO]^{3/2} \quad .$$

The activation energy of the chain process is:

$$E_c = \frac{72}{2} + 7 = 43 \text{ kcal} \quad .$$

The experimental value (at pressures close to atmospheric) is $E_0 \cong 48$ kcal and the rate was also found [3] to be proportional to $[CH_3CHO]^{3/2}$. The discrepancy of 5 kcal between calculated and observed activation energies is probably due to errors in the determination of E, Q and ϵ_{01}. Direct photochemical experiments give for ϵ_{01} the value of 10 kcal [16], namely 3 kcal higher than the figure of 7 kcal which we used. This increases the calculated value of the activation energy to 46 kcal, in almost complete agreement with the experimental result $E_0 = 48$ kcal. The chain mechanism of the reaction is supported by various facts such as the inhibiting action of NO [17] and propylene [17], or the accelerating effect of oxygen, biacetyl [3] and ditertiary butyl-peroxide [18].

Decomposition of Dimethyl Ether [19]

$$CH_3 - O - CH_3 \longrightarrow CH_2O + CH_4 \longrightarrow CO + H_2 + CH_4 \quad .$$

The reaction is practically thermoneutral. A possible chain scheme is the following:

$$\begin{aligned} &0)\quad CH_3OCH_3 \longrightarrow CH_3O + CH_3 - Q, \qquad \text{initiation} \\ &\left.\begin{aligned} &1)\quad CH_3 + CH_3OCH_3 \longrightarrow CH_4 + CH_2OCH_3 + |q_1| \\ &2)\quad CH_2OCH_3 \longrightarrow CH_2O + CH_3 - |q_2| \end{aligned}\right\}\text{chain} \quad . \end{aligned}$$

Since the overall process is thermoneutral, $|q_1| = |q_2| = |q|$.
If $q \cong 12$ kcal, the ratio

$$\frac{[CH_3]}{[CH_2OCH_3]}$$

will be the same as in the case of the acetaldehyde decomposition, namely,

$$\frac{[CH_3]}{[CH_2OCH_3]} = \frac{3}{f_1} \quad .$$

But the value of q is apparently larger than 12 kcal and the ratio

$$\frac{[CH_3]}{[CH_2OCH_3]}$$

is then almost equal to unity. This means that all recombination steps are equally likely. Moreover, a similar result would be obtained if it were assumed that chain termination is due to recombination of two different radicals.

$$3)\quad CH_3 + CH_2OCH_3 \longrightarrow CH_2O + C_2H_6 \quad \text{or} \quad \longrightarrow CH_3CH_2OCH_3 \quad .$$

By means of the corresponding kinetic equations, we get, for the velocity of the overall process:

$$w = \left(\frac{k_0k_1k_2}{k_3}\right)^{1/2} [CH_3OCH_3] \quad .$$

Thus the chain process imitates a unimolecular law, as experimentally observed.

The unimolecular constant

$$k = \left(\frac{k_o k_1 k_2}{k_3}\right)^{1/2} \approx \left(\frac{10^{13}\cdot 10^{-10}\cdot 10^{13}}{10^{-10}}\right)^{1/2} \exp[-(Q + |q_2| + \epsilon_{01} + \epsilon_{02})/2RT].$$

Then $k = 10^{13}\exp(-E_c/RT)$ where

$$E_c = \frac{Q}{2} + \frac{\epsilon_{01} + \epsilon_{02} + |q_2|}{2} .$$

Experimentally, one gets [19]: $k = 1.5\cdot 10^{13}\exp(-58000/RT)$.

Using the thermochemical data in Gray's paper [20], we may calculate Q, the heat of the reaction of initiation. The heats of formation of ether, of the methoxy and methyl radicals are equal respectively to -45.6, 0.5 and 31.5 kcal.

$$CH_3OCH_3 \longrightarrow CH_3O + CH_3$$

$$-45.6 = -0.5 + 31.5 - Q .$$

Hence $Q = 76.6$ kcal. The relation $\epsilon_o = 11.5 - 0.25\,|q|$ of Chapter I gives $\epsilon_{01} = 8.5$ kcal since $q = 12$ kcal. In Chapter I, ϵ_{02} was calculated and found to be equal to 7 kcal. Then

$$E_c = 58 = \frac{Q}{2} + \frac{\epsilon_{01} + \epsilon_{02} + q_2}{2} = 38.3 + \frac{8.5 + 7}{2} + \frac{q_2}{2} \cong 46 + \frac{q_2}{2} .$$

But q_2 could hardly be larger than 14 to 16 kcal. Therefore the experimental value of E is a little high (by ~ 4 kcal) as in the preceding case. Although the calculated activation energy for chain decomposition is lower than the experimental value, it is difficult to doubt that the reaction proceeds by a chain mechanism. Apparently, there are some errors involved in the determination of E, Q, ϵ_{01} and ϵ_{02}.

To sum up, the chain character of this reaction is very probable. The marked inhibition by addition of NO supports this conclusion.

Decomposition of Di-iodo-ethane

$$C_2H_4I_2 \longrightarrow C_2H_4 + I_2 - 11 \text{ kcal} .$$

Assume the following chain mechanism:

0) $C_2H_4I_2 \longrightarrow C_2H_4I + I - 48$ kcal, initiation

1) $I + C_2H_4I_2 \longrightarrow I_2 + C_2H_4I - 11$ kcal } chain

2) $C_2H_4I \longrightarrow C_2H_4 + I \pm 0$ kcal } chain

3) $I + I + M \longrightarrow I_2 + M + 35.5$ kcal, chain breaking.

Evidently, the rate determining step of the chain process is the endothermic reaction 1). Its activation energy ϵ_1 is:

$$\epsilon_1 = \epsilon_{01} + |q| \cong \epsilon_{01} + 11 \text{ kcal} \quad .$$

The value of ϵ_{01} for iodine atom reactions is usually quite small, namely a few calories only. In the present case, ϵ_1 has been directly measured by photochemical studies. It is equal to about 12 kcal [21]. Whence, $E_c \approx 24 + 12 = 36$ kcal ($Q = 47 - 48$ kcal). The measured value of E is 36.6 kcal [21].

In the presence of I_2, the decomposition into radicals requires 35.5 kcal. Thus, $E_c = 17.8 + 12 \approx 30$ kcal. Experimentally, one has $E_c = 30.2$ kcal.

Clearly, the process is of the chain type.

Arnold and Kistiakowsky [21], after a study of the decomposition of di-iodo-ethane, came to the conclusion that the reaction is a chain process only when it is catalyzed by iodine. In the absence of iodine, it decomposes directly into the end products, C_2H_4 and I_2. We have shown [2] that this conclusion is incorrect for the following reasons:

1) We have shown that the calculated activation energies of the catalyzed and uncatalyzed reactions (on the assumption that both are of the chain type) coincide with the experimental activation energies.

2) Polanyi has shown [22] that alkyl iodides completely decompose into alkyl radicals and iodine atoms when they pass in a nitrogen stream through a tube heated to a certain temperature.

3) All the data obtained by Kistiakowsky may be interpreted on the assumption that the reaction is of the chain type, with two kinds of initiation:

a) $C_2H_4I_2 \longrightarrow C_2H_4I + I$

and

b) $I_2 + M \longrightarrow I + I + M$.

Writing down and solving the kinetic equations for the scheme of chain decomposition of di-iodo-ethane, we obtain for the rate of the reaction in its initial stage:

$$w = 2k_1\left(\frac{k_0[C_2H_4I_2]}{k_3}\right)^{1/2}[C_2H_4I_2] \quad .$$

Since k_3 - the rate constant for recombination of I atoms in triple collisions I + I + M is $k_3 = x[M]$, where $x \sim 10^{-32}$ and [M] is the initial concentration of $C_2H_4I_2$, we have:

$$k_1 = 10^{-10}\exp(-12000/RT) \quad ,$$

$$k_0 = 10^{13}\exp(-48000/RT)$$

$$w = 10^{12.8}\exp(-36000/RT)[C_2H_4I_2] \quad .$$

Experimentally [21], one has:

$$w = 10^{13}\exp(-36600/RT)[C_2H_4I_2] \quad .$$

Therefore, the chain process imitates unimolecular behavior. Theoretical and experimental rates coincide.

An analogous calculation can be made also for the reaction catalyzed by I_2. Then the initiation step is the dissociation of iodine molecules:

0) $I_2 + M \longrightarrow I + I + M - 35.5$ kcal

with $k_0 = 10^{13}\exp(-35500/RT)$.

The rate of the iodine catalyzed reaction is given by:

$$w^* = 2k_1K^{1/2}[I_2]^{1/2}[C_2H_4I_2]$$

where K is the equilibrium constant of the reaction $I + I \rightleftarrows I_2$. The value of K is given approximately by:

$$K \approx 10^{24}\exp(-35500/RT) \quad .$$

Then,

$$w^* = 2.10^{-10}.10^{12}[I_2]^{1/2}[C_2H_4I_2]\exp(-29750/RT)$$

$$= 200[I_2]^{1/2}[C_2H_4I_2]\exp(-29750/RT) \quad .$$

By comparison, the experimental value is:

$$w^* = 80[I_2]^{1/2}[C_2H_4I_2]\exp(-30200/RT) \quad .$$

Theoretical and experimental activation energies are identical within experimental error. Pre-exponential factors differ by a factor of 2.5. Undoubtedly the process is a chain reaction, with a small chain length of the order of ten.

Decomposition of Secondary Butyliodide

The overall process is: $2C_4H_9I \longrightarrow C_4H_8 + C_4H_{10} + I_2$. The chain mechanism may be written as follows:

0) $C_4H_9I \longrightarrow C_4H_9 + I - Q_{C-I}$ initiation

1) $C_4H_9 + C_4H_9I \longrightarrow C_4H_{10} + C_4H_8I$

2) $C_4H_8I \longrightarrow C_4H_8 + I$ } chain

3) $I + C_4H_9I \longrightarrow I_2 + C_4H_9 - 11.5$ kcal } chain

4) $I + I + M \longrightarrow I_2 + M + 35.5$ kcal chain breaking.

Step 3) is rate controlling. The C - I bond energy in normal C_4H_9I has been measured. It is equal to 45 kcal [22]. We may therefore take 47 kcal as the value of Q_{C-I} in secondary butyl iodide. This should not introduce a large error. Then the calculated activation energy of the chain process is:

$$E_c = \frac{Q_{C-I}}{2} + \epsilon_3; \qquad \epsilon_3 = q_3 + \epsilon_{03}$$

where q_3 is the heat of reaction 3) in which a C - I bond is broken (this requires Q_{C-I} kcal) and a I - I bond is made (a gain of 35.5 kcal). Thus $q_3 = Q_{C-I} - 35.5$ kcal, and

$$E_c = \frac{Q_{C-I}}{2} + Q_{C-I} - 35.5 + \epsilon_{03} = \frac{3}{2} Q_{C-I} - 35.5 + \epsilon_{03} \quad .$$

With $Q_{C-I} = 47$ kcal, we obtain $E_c = 35 + \epsilon_{03}$. In the presence of molecular iodine, the decomposition is accelerated because of the easier chain initiation by means of dissociation of I_2 into I atoms:

$$I_2 + M \longrightarrow I + I + M - 35.5 \text{ kcal} \quad .$$

With sufficient amounts of molecular iodine, initiation takes place almost entirely in this way. The calculated activation energy of the iodine catalyzed chain decomposition of secondary butyliodide is:

$$E_c^* = \frac{Q}{2} + \epsilon_3 = \frac{35.5}{2} + Q_{C-I} - 35.5 + \epsilon_{03}$$

$$= 17.75 + 11.5 + \epsilon_{03} = 29.25 + \epsilon_{03} \quad .$$

The experimental activation energy of the iodine catalyzed reaction was found to be 35.2 kcal [23]. Then, $\epsilon_{03} = 35.2 - 29.95 \approx 6$ kcal. Substituting this value of ϵ_{03} into the expression for E_c, we get:

$$E_c = 35 + \epsilon_{03} = 35 + 6 = 41 \text{ kcal}$$

for the uncatalyzed reaction (in the absence of iodine). The experimental value is 39.4 kcal [23] in agreement with the calculated value, within experimental error. This is good proof that the uncatalyzed reaction proceeds by a chain mechanism.

As in the previous case, the uncatalyzed chain reaction, in its initial stages, is quasi-unimolecular.

The rate of the iodine catalyzed reaction is proportional to $[I_2]^{1/2}$ and $[C_4H_9I]$ in agreement with the observed behavior. The calculated pre-exponential factors of both the catalyzed and uncatalyzed reactions are about 20 times larger than the experimental figures. This may be explained by a steric factor equal to 0.05 for the step: $I + C_4H_9I \longrightarrow I_2 + C_4H_9$.

§3. Decomposition of Alkylbromides

Very interesting peculiarities concerning competition between chain and direct processes are found by an analysis of the decomposition of propyl, iso-propyl and ethyl bromides.

Here for the first time, we see clearly the effect of chemical structure on the competition between direct and chain reactions.

The reverse process, addition of HBr to olefins (see Chapter II, Section 1, p. 96) is a chain reaction with light or peroxides. Addition violates the rule of Markovnikov [24]. At low temperatures and in the absence of initiators in the liquid phase, slow HBr addition takes place, 90% of the product corresponding to Markovnikov's rule [25]. At higher temperatures, as foreseen by Markovnikov [26], the fraction of product violating the rule increases.

The decomposition of alkylbromides taking place at 300°C (the reverse of addition) should, of course, also proceed by a chain mechanism under the action of initiators:

1) $Br + C_nH_{2n+1}Br \longrightarrow HBr + C_nH_{2n}Br - q_1$ (~ 5 kcal)

2) $C_nH_{2n}Br \longrightarrow C_nH_{2n} + Br - q_2$ (~ 13 kcal) .

Here $q_1 + q_2 = U$, where U is the expenditure of energy required for the overall process. For the simpler alkylbromides, this quantity has been measured. It is about 18 kcal [27, 28]. Analysis of the termination mechanism led me to conclude in 1952 [2] that the predominant termination step involves two different radicals:

3) $Br + C_nH_{2n}Br \longrightarrow$ molecular products.

In the purely thermal case, without initiators, chain initiation can only be due to a decomposition:

0) $C_nH_{2n+1}Br \longrightarrow C_nH_{2n+1} + Br - Q_{C-Br}$.

For ethylbromide, Q_{C-Br} is equal to 65 kcal. We shall take $Q_{C-Br} \approx 59$ kcal for isopropylbromide and $Q_{C-Br} \approx 62$ kcal for normal propylbromide.

By means of the kinetic equations, it is easy to obtain an expression for the rate of the chain process:

$$w_c = \left(\frac{k_0k_1k_2}{k_3}\right)^{1/2} \exp[-(Q + U + \epsilon_{01} + \epsilon_{02})/2RT]\cdot[C_nH_{2n+1}Br] .$$

Substituting pre-exponential factors for the elementary steps (10^{-10} for the bimolecular processes and 10^{13} for the unimolecular reactions), we get:

$$w_c = 10^{13}\exp[-(Q + U + \epsilon_{01} + \epsilon_{02})/RT]\cdot[C_nH_{2n+1}Br] .$$

For ethylbromide:

$$\frac{1}{2}(Q + U + \epsilon_{01} + \epsilon_{02}) = \frac{65 + 18.3}{2} + \frac{\epsilon_{01} + \epsilon_{02}}{2}$$

$$= 41.65 + \frac{\epsilon_{01} + \epsilon_{02}}{2} .$$

The experimental rate [2] is:

$$w = 10^{13}\exp(-50000/RT) .$$

This corresponds to the theoretical expression for w_c if we make the reasonable assumption that

$$\frac{\epsilon_{01} + \epsilon_{02}}{2} = 8.35 \text{ kcal} .$$

For isopropylbromide, the experimental activation energy is 47.8 kcal [29]. The theoretical value

$$E_c = 38.5 + \frac{\epsilon_{01} + \epsilon_{02}}{2}$$

agrees with the observed figure if it is assumed that

$$\frac{\epsilon_{01} + \epsilon_{02}}{2} = 8.5 \text{ kcal} \quad .$$

The activation energy E of the corresponding direct unimolecular decomposition had not been found until very recently. But it would be very strange if it were close to the calculated value E_c for the chain process or the experimental value itself. This led us to conclude that the decomposition of alkylbromides proceeds via a chain mechanism just like that of alkyliodides. Calculated chain lengths are very large $(10^4 - 10^5)$, however, as compared to the case of alkyl iodides (3 to 30).

More than six years have passed since I wrote my paper on the subject [2]. In 1951, the experimental material relative to the decomposition of alkylbromides was quite meager. Since then, a series of investigations have appeared in the USSR (Moscow University) and elsewhere, throwing additional light on the decomposition of bromides.

First, Szwarc [30] showed that in contrast to the case of alkyliodides, ethylbromide decomposes faster into HBr and an olefin than into C_2H_5 and Br radicals. This would mean that the activation energy E of the direct unimolecular decomposition should be appreciably smaller than 65 kcal, possibly as low as 50 kcal. Blades and Murphy [31] have investigated the decomposition of ethylbromide, n-propylbromide and isopropylbromide at low pressures in a fast stream in the presence of toluene, which, according to their opinion, inhibits the chain process. They found that the decomposition of the three bromides follows a unimolecular law, activation energies being equal to 52.3, 50.7 and 47.7 kcal respectively.

Careful experiments by G. I. Kapralova and G. B. Sergeev have shown that, in a static system, the activation energy for decomposition of n - C_3H_7Br is 42 kcal [32] whereas it is about 47 kcal for iso - C_3H_7Br, the same value as found in a flow system. Under flow conditions, apparently, what is measured for both isomers is the rate of the direct unimolecular decomposition. The agreement concerning activation energies in flow and static systems for the case of isopropylbromide, suggests that, in both systems, the decomposition of this compound goes directly in true unimolecular fashion. As to the normal compound, a 10 kcal lowering of the activation energy in static systems as compared to that found under flow conditions, indicates that the chain reaction predominates in the static case. Let us note that the kinetic laws for the decomposition of normal- and iso-propylbromide are different when studied in a static system. The reaction order is unity in the case of the iso compound, but 3/2 in the case of the normal isomer.

G. B. Sergeev and G. A. Kapralova [32, 33] have studied the pyrolysis of n- and iso C_3H_7Br in the presence of additives. They found that the pyrolysis of n - C_3H_7Br is accelerated by addition of bromine and oxygen and is somewhat retarded by propylene. It must be noted that this inhibition is effective only at the beginning of reaction.

It is interesting that HBr addition must accelerate slightly the reaction so that simultaneous addition of HBr and C_3H_6 in equimolar amounts has no effect at all on the reaction rate.

On the other hand, propylene has no effect on the pyrolysis of n - C_3H_7Br. These experiments show again the different character of the decompositions of n- and iso - C_3H_7Br.

In an attempt to explain the difference in behavior of these two closely related isomers, we conclude that our previously proposed mechanism [2] is incomplete and that the question of the competition between chain and direct decomposition of these substances must be examined again.

We know that the reverse reaction, HBr addition to propylene, initiated by peroxides, is a chain process always giving a product of normal structure. The addition of HBr violates the Markovnikov rule. If this reaction took place in a similar way in the gas phase at higher temperatures (where no data are available), it would become quite clear that of the two bromides, only the one with normal structure would decompose by a chain mechanism. It is not difficult to see why this should be so.

When in the first step:

$$1)\quad Br + CH_3 - CH_2 - CH_2Br \longrightarrow HBr + CH_3 - \dot{C}H - CH_2Br$$

a bromine atom abstracts an H atom from normal propylbromide, it does so from a CH_2 group in which the H atom is weakly bound. The radical that is formed decomposes into an olefin and a Br atom:

$$2)\quad CH_3 - \dot{C}H - CH_2Br \longrightarrow CH_3 - CH = CH_2 + Br$$

and the Br atom may continue the chain.

However, in a first step, when a Br atom reacts with isopropylbromide, the $CH_3 - \dot{C}Br - CH_3$ radical is formed:

$$1)\quad CH_3 - CHBr - CH_3 + Br \longrightarrow HBr + CH_3 - \dot{C}Br - CH_3 \quad .$$

This radical cannot decompose into an olefin and a Br atom. All it can do is recombine with a similar radical or with a Br atom. Thus, with the iso compound, a chain cannot be propagated and the reaction follows the direct unimolecular path, with an activation energy of 47 kcal.

With n-propylbromide, a chain can be propagated. As a result, the overall activation energy of the chain process (42 kcal) is about 8.7 kcal lower than that of the direct unimolecular path.

It must be emphasized that this discussion of the difference between the isomeric bromides is not rigorously correct and provides only a qualitative explanation of the difference. As is well known, near room temperature, light and peroxides initiate the addition of HBr to propylene, giving the normal bromide. It may be assumed that the higher the temperature, the larger will be the amount of isopropylbromide formed besides the main product, n-propylbromide. If this is so, at high temperature, the reverse chain process is possible not only for the normal compound but also for its isomer. Theoretically, this is perfectly understandable, since the radical $CH_3 - \dot{C}Br - CH_3$ formed in step 1) during the decomposition of isopropylbromide is now capable of isomerization. Isomerization of this radical may take place either intramolecularly or, what appears more likely, by an exchange reaction with isopropylbromide. The radical following isomerization, $CH_3 - CHBr - \dot{C}H_2$ may decompose and a chain decomposition of isopropylbromide is now possible:

1) $BR + CH_3 - CHBr - CH_3 \longrightarrow HBr + CH_3 - \dot{C}Br - CH_3 + q_1$

1') $CH_3 - \dot{C}Br - CH_3 + CH_3 - CHBr - CH_3 \longrightarrow CH_3 - CHBr - CH_3 + CH_3CHBr - \dot{C}H_2 + q_1'$

2) $CH_3 - CHBr - \dot{C}H_2 \longrightarrow CH_3 - CH = CH_2 + Br + q_2$.

It is easy to show that the rate determining step is now reaction 1') and therefore, the $CH_3 - \dot{C}Br - CH_3$ radicals will be the most abundant and their recombination:

3) $2CH_3 - \dot{C}Br - CH_3 \longrightarrow$ molecular products

will be the main chain breaking step.

Suppose that initiation takes place by decomposition of isopropylbromide into radicals:

0) $CH_3 - CHBr - CH_3 \longrightarrow Br + CH_3 - \dot{C}H - CH_3 - 59$ kcal .

This mechanism gives the kinetic law:

$$W_c = k_1'\left(\frac{(k_0)}{k_3}\right)^{1/2} [RBr]^{3/2} , \qquad (5)$$

i.e., the order with respect to isopropylbromide is 3/2.

Before we continue the calculation, we must know the energy required to abstract an H atom from n - C_3H_7Br and iso - C_3H_7Br, but no accurate data are available. The energies required to remove an H atom from the CH_2 and CH_3 groups in propane are known: $Q_{CH-H} = 89$ kcal and $Q_{CH_2-H} = 95$ kcal. Let us assume that substitution of an H atom by a Br atom in the CH_3 and CH_2 groups lowers the C - H bond energies by about 3 kcal. Then $Q_{BrC-H} = 86$ kcal and $Q_{HBrC-H} = 92$ kcal. The dissociation energy of HBr is 86 kcal and U, the endothermicity of the overall decomposition of n - C_3H_7Br and iso - C_3H_7Br, is equal to 18 kcal (according to experimental data).

Then, for the chain mechanism of decomposition of iso - C_3H_7Br, we get $q_1 = \sim 0$ kcal; $q_1' = -9$ kcal; $q_2 = U - q_1 - q_1' = -18 + 0 + 9 = -9$ kcal.

Another reaction scheme is possible: a bromine atom may abstract a hydrogen atom from a CH_3 group.

Then, instead of step 1) we have step 1a):

$$1a)\quad BR + CH_3CHBrCH_3 \longrightarrow HBr + \dot{C}H_2 - CHBr - CH_3 + q_{1a}$$

followed by decomposition of the radical $\dot{C}H_2 - CHBr - CH_3$ into a Br atom and propylene. But $q_{1a} = -95 + 86 = -9$ kcal, so that the ratio of rates of 1a) and 1) will be equal to $\exp(-9000/2.700) \approx 10^{-3}$ at 700°K. Thus, evidently, one may neglect reaction 1a).

Let us calculate the activation energy for the chain decomposition of isopropylbromide: $k_0 = 10^{13}\exp(-Q_{C-Br}/RT)$ where $Q_{C-Br} \sim 59$ kcal for iso - C_3H_7Br; $k_3 \approx 10^{-10}$; $k_1' = 10^{-10}\exp[-(q_1' + \epsilon_{01}')/RT]$ where $q_1' = -9$ kcal. Then

$$E_c = \frac{Q}{2} + q_1' + \epsilon_{01}' = 29.5 + 9 + \epsilon_{01}' = 38.5 + \epsilon_{01}' \quad .$$

According to the formula $\epsilon_{01}' = 11.5 - 0.25q_1'$, we find (Chapter I, p. 29) $\epsilon_{01}' = 9.3$ kcal. Then $E_c = 47.8$ kcal and the rate of decomposition, if $[RBr] \approx 10^{19}$ is equal to:

$$w_c = 10^{11} \cdot 10^{19}\exp(-47800/RT) \quad .$$

As we have seen, flow experiments [31] show that the rate of the unimolecular process is:

$$w_D = 10^{13}[RBr]\exp(-47700/RT)$$

$$= 10^{13} \cdot 10^{19}\exp(-47700/RT) \quad .$$

The activation energies of the simple unimolecular reaction and the complex chain process happen to be identical. But pre-exponential factors are different and $(w_D/w_c) \simeq 100$ i.e., the simple unimolecular reaction is 100 times faster than the chain process.

The situation is different in the case of n - C_3H_7Br. The chain decomposition of this compound may also follow two different paths A and B. An H atom may be abstracted from either a CH_2 or CH_2Br group:

A.

1) $BR + CH_3CH_2CH_2Br \longrightarrow HBr + CH_3 - \dot{C}H - CH_2Br - 4$ kcal .

This is followed by the rapid decomposition of the radical $CH_3\dot{C}HCH_2Br$ (rapid, because unimolecular, in spite of being endothermic).

2) $CH_3 - \dot{C}H - CH_2Br \longrightarrow Br + CH_3 - CH = CH_2 - 14$ kcal .

Other reactions may also take place:

B.

1a) $Br + CH_3CH_2 - CH_2Br \longrightarrow HBr + CH_3 - CH_2 - \dot{C}HBr - 7$ kcal .

The radical $CH_3CH_2 - \dot{C}HBr$ cannot decompose and either recombines or propagates the chain:

1') $CH_3 - CH_2 - \dot{C}HBr + CH_3CH_2CH_2Br \longrightarrow CH_3CH_2CH_2Br$
$+ CH_3 - \dot{C}H - CH_2Br + 3$ kcal

2) $CH_3 - \dot{C}H - CH_2Br \longrightarrow CH_3CH = CH_2 + Br - 14$ kcal .

Reaction 1a), because of its large endothermicity will be about 10 times slower than reaction 1). It may not be overlooked, however, since it produces inactive radicals $CH_3CH_2\dot{C}HBr$ that break the chain:

3) $2CH_3CH_2\dot{C}HBr \longrightarrow$ inactive products .

This reaction mechanism gives a rate law similar to expression (5):

$$w_c = 10\, k_1' \left(\frac{(k_0)}{k_3}\right)^{1/2} [RBr]^{3/2} \quad .*$$

However, the activation energy is different since the values of q_1' and $Q_{C-Br} \cong 62$ kcal are different:

$$E_c = \frac{Q_{C-Br}}{2} + \epsilon_1' \quad .$$

For iso - C_3H_7Br, $\epsilon_1' = q_1' + \epsilon_{01}' = 9 + 9.3 = 18.3$ kcal. For n - C_3H_7Br, reaction 1') is exothermic and $\epsilon_1' = 11.5 - 0.25q_1' \approx 10.5$ kcal. Then

---

* Note that the experimental data of Agius and Maccoll [34], carefully checked by G. I. Kapralova and G. B. Sergeev correspond to a reaction order equal to 3/2 with respect to RBr. Agius and Maccoll have incorrectly interpreted this rate law as being due to chain breaking by recombination between bromine atoms. It can easily be shown that their mechanism leads to an initial reaction rate proportional to [RBr] and not to $[RBr]^{3/2}$ as experimentally observed.

$E_c = 31 + 10.5 = 41.5$ kcal. With $[RBr] \approx 10^{19}$, we get, in agreement with experimental data:

$$w_c = 10^{12} \cdot 10^{19} \exp(-41500/RT) \quad .$$

One year after publication of the first edition of this book, A. Maccoll and co-workers [35-40] published a series of papers on the pyrolysis of several organic bromides: n- and iso - C_3H_7Br, allylbromide etc. The kinetic analysis of the decomposition mechanism presented by these authors is identical to ours. As we have proposed, Maccoll considers the possibility of forming two radicals following attack by a bromine atom: an active radical, capable of continuing the chain (reactions 1 and 2) and an inactive radical (reaction 1'). With n - C_3H_7Br, the inactive radical, according to Maccoll, is $\dot{C}H_2CH_2CH_2Br$, which, it seems to us, is less likely than $CH_3CH_2\dot{C}HBr$. Indeed, the energy required to abstract a H atom from the group CH_2Br is about 5 kcal less than for abstraction from a CH_3 group.

Following Maccoll, chain termination is due to reaction between the inactive radical and a bromine atom. This termination is perfectly likely and its inclusion in the reaction scheme also leads to a 3/2 order of reaction:

3') $Br + CH_3CH_2\dot{C}HBr \longrightarrow$ reaction products

$$w_c = k_1 \left(\frac{k_0 k_1'}{k_3' k_{1a}} \right)^{1/2} (RBr)^{3/2} \quad .$$

The activation energy E_c is then: $E_c = \epsilon_1 + \frac{1}{2} Q_0 + \epsilon'_{01} - \epsilon_{1a}$. The relation of Chapter I: $\epsilon = 11.5 + 0.75|q|$ gives ϵ_1 and ϵ_{1a}: $\epsilon_1 = 11.5 + 0.75 \cdot 4 = 14.5$ kcal and $\epsilon_{1a} = 11.5 + 0.75(7) = 16.7$ kcal. Also: $\epsilon'_{01} = 11.5 - 0.25(3) = 10.5$ kcal.

Hence, in agreement with the experimental value:

$$E_c = 14.5 + \tfrac{1}{2}(63 + 10.5 - 16.7) \sim 42 \text{ kcal} \quad .$$

The unimolecular decomposition of n - C_3H_7Br, as we have seen [31], proceeds at a rate $w_D = 10^{13}\exp(-50700/RT) \cdot [RBr]$. Whence

$$\frac{w_c}{w_D} = 0.1 \exp(9000/RT) \quad .$$

At 700°K, this gives

$$\frac{w_c}{w_D} = 60 \quad .$$

Therefore, in the case of n - C_3H_7Br, the chain reaction is about 60 times faster than the unimolecular decomposition.

It must be noted however that pyrolysis of n - C_3H_7Br in the presence of toluene [31] and cyclohexene [41] may not be treated as a purely molecular process, free from any secondary reaction, e.g., a short chain process. Indeed, if the equation $w_D = 10^{13}\exp(-50700/RT)\cdot(RBr)$ is accepted as correct, one may calculate the reaction rate at the temperature (406°C) at which Sergeev decomposed n - C_3H_7Br in a static system. Then it appears that the molecular reaction rate must be of the same order of magnitude at a pressure of 16 mm Hg as that of the chain process. Since the monomolecular reaction is first order but the chain reaction is 3/2 order, one should observe at 406°C an appreciable departure from 3/2 order. But this is not so. In order to agree with the facts, one should raise the activation energy of the molecular reaction from 50.7 to 54 kcal. Then the rate of the molecular reaction would be only 10% of the rate of the chain reaction and the 3/2 order would be observed. That it is well observed in a static system is shown by Figure 2.

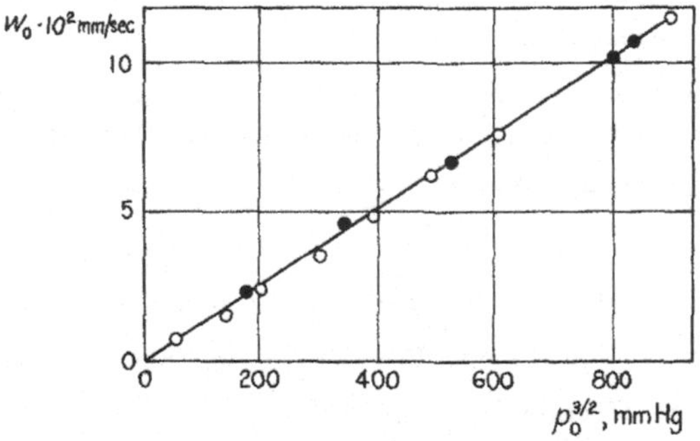

FIGURE 1. Initial rate of pyrolysis of normal propylbromide as a function of initial pressure to the 3/2 power. T = 406°G [33].

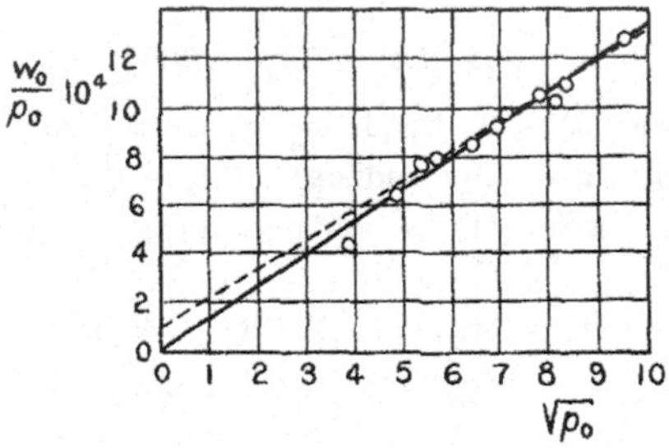

FIGURE 2. Relation between chain and unimolecular reactions [33].

Figure 2 reproduces Figure 1 in the coordinates w_o/p_o and $\sqrt{p_o}$. If, however, the molecular process took place together with the chain process, we would have $w_o = k_1 p_o^{3/2} + k_2 p_o$ or $w_o/p_o = k_1 \sqrt{p_o} + k_2$ where k_1 and k_2 are the rate constants of the chain and molecular reactions. The intercept, at $p_o = 0$, gives k_2. It can be seen that the experimental data gives an intercept not larger than unity. Thus $k_2 = 1.0 \cdot 10^{-4} sec^{-1}$. Such a value for k_2 is appreciably smaller than the rate constant given by the expression $w_c = 10^{13} exp(- 50700/RT) \cdot (RBr)$ which gives at 403°C: $k' = 10^{13} exp(- 50700/2 \cdot 676) = 10^{-3} sec^{-1}$.

In order to obtain a rate constant for the molecular reaction equal to $10^{-4} sec^{-1}$, the activation energy must be equal to 54 kcal. This is the minimum possible value for the activation energy of the pure molecular decomposition. In this case, the chain length may be larger than 60.

The decomposition of ethylbromide proceeds unimolecularly in a static system where a large number of experiments have been performed so that the reactor walls are covered with a thin film of polymeric products.

I have shown by an analysis of experimental data, that the activation energy for decomposition of C_2H_5Br is equal to 49 - 50 kcal. This is the value that was subsequently obtained experimentally by G. I. Kapralova and G. B. Sergeev.

Chain propagation during decomposition of ethylbromide proceeds, as we had assumed it, by alternation of two elementary steps:

1) $Br + C_2H_5Br \longrightarrow HBr + C_2H_4Br$

2) $C_2H_4Br \longrightarrow C_2H_4 + Br$.

If we consider step 1), we see that it can take place in two ways:

1) $Br + CH_3 - CH_2Br \longrightarrow HBr + \dot{C}H_2 - CH_2Br - 5$ kcal .

This is followed by the decomposition step 2):

2) $\dot{C}H_2CH_2Br \longrightarrow C_2H_4 + Br - 13$ kcal .

In this case, chain propagation is easy. Another path is possible, however:

1a) $Br + CH_3 - CH_2Br \longrightarrow HBr + CH_3 - \dot{C}HBr - 2$ kcal .

It is assumed here that the C - H bond energy in a CH_3 group is 3 kcal larger than in a CH_2Br group. The radical $CH_3 - \dot{C}HBr$ cannot decompose with elimination of a Br atom: it recombines or goes through another reaction:

1') $CH_3 - \dot{C}HBr + CH_3 - CH_2Br \longrightarrow CH_3 - CH_2Br$
$+ \dot{C}H_2 - CH_2Br - 3$ kcal .

Only if reaction 1') takes place, is a $\dot{C}H_2 - CH_2Br$ radical formed that may decompose following process 2). Since reaction 1) is more endothermic than reaction 1a), the chain will mainly follow the second path [reactions 1a), 1') and 2)]. Let us note that with respect to competition between direct and chain processes, ethylbromide occupies a position intermediate between those of isopropylbromide and n-propylbromide. The heat of reaction 1') in the case of iso - C_3H_7Br is equal to - 9 kcal while it is + 3 kcal for normal propylbromide and - 3 kcal for ethylbromide. If we assume that the activation barrier for decomposition of C_2H_5Br is 1 or 2 kcal and the energy of dissociation of C_2H_5Br into C_2H_5 and Br is 65 kcal, we obtain 44.5 kcal for the energy of activation of the overall chain process, i.e., about 5 or 6 kcal less than the activation energy for the simple unimolecular process. At 700°K, such a difference between activation energies would favor the chain process by a factor of about 34. The pre-exponential factor is 100 times smaller for the chain process than for the direct unimolecular reaction.

To sum up, the rate of unimolecular decomposition of ethylbromide is slightly faster than the corresponding chain process at pressures near atmospheric. At pressures around 0.1 atm, the rate of the chain process would be smaller by a factor of the order of 10.

It is therefore clear that all factors facilitating even to a small extent the generation of free radicals (decreasing the energy required to form Br atoms) will favor the chain process of ethylbromide decomposition. Such a factor, for instance, is addition of bromine. It is likely that this will also explain the strong effect of reactor walls, oxygen addition and other factors that accelerate the decomposition of ethylbromide.

Sergeev [42] has studied the pyrolysis of the isomeric bromides: n - C_4H_9Br, iso - C_4H_9Br, sec - C_4H_9Br and tert. C_4H_9Br. Table 38 contains the main data of Sergeev, relative to the decomposition of propyl and butylbromide.

Two among the four bromides, namely n- and iso - C_4H_9Br exhibit all the characteristics of chain processes: acceleration by addition of bromine and oxygen, inhibition by propylene and cyclohexene, poor reproducibility, 3/2 reaction order.

Secondary and tertiary butyl bromides on the other hand decompose following a first order rate law and are not affected by bromine and propylene.

The clearest symptoms of chain decomposition are shown by isobutylbromide. This is quite natural since with

$$\begin{array}{c} CH_3 \\ \diagdown \\ \diagup \quad CH - CH_2Br \\ CH_3 \end{array}$$

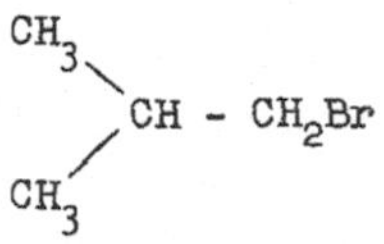

the bromine atom attack will be directed mainly against the H atom of
the C - H group, giving the radical

$$\begin{array}{c} CH_3 \\ \diagdown \\ \quad \dot{C} - CH_2Br \\ \diagup \\ CH_3 \end{array}$$

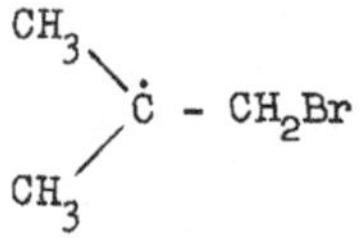

that decomposes readily into isobutylene and a bromine atom involved in
chain propagation. As we have seen in the case of n - C_3H_7Br, H abstraction is also possible, though less likely, from the CH_2Br group.
The same relations are then expected as in the case of n - C_3H_7Br.

        The activation energy for decomposition of isobutylbromide may
be somewhat lower than in the case of n - C_3H_7Br since the bond
dissociation energy Q_{C-Br} in isobutylbromide is likely to be less than
in n - C_3H_7Br. The observed activation energy is, however, only 30
kcal, which is too low and may be explained only by chain initiation at
the wall. Poor reproducibility supports this view.

        Normal butyl bromide, although similar in structure to
n - C_3H_7Br, differs from the latter because there are now two possible
ways of forming an inactive radical: from the CH_2 and CH_2Br groups.
Kinetic analysis of the data gives the same relations as in the case of
n - C_3H_7Br. The only point difficult to understand is that the activation
energy is 6 kcal higher than for pyrolysis of n - C_3H_7Br. But this
may well be due to the possibility just mentioned of forming two inactive
radicals.

        With secondary butyl bromide $CH_3CHBr - CH_2CH_3$ the hydrogen
abstraction is easiest from the CHBr group. This yields an inactive
radical which cannot decompose any further. An active radical can be
formed by hydrogen abstraction from the CH_2 group. But this is energetically harder than to form the inactive radical since there is a
difference of about 5 kcal between the C - H bond energies involved
in both cases. Just as we have seen with iso - C_3H_7Br, a chain decomposition of sec - C_4H_9Br is possible but considerably slower than
the molecular process. The difference between the rate of the molecular
reaction and that of the chain decomposition must be less than in the
case of iso - C_3H_7Br. With the latter, attack of a bromine atom removes

a H atom from a CH_3 group while it removes it from a more reactive CH_2 group in the case of sec. C_4H_9Br.

A similar reason (a smaller C - H bond energy) probably explains also the lowering of the activation energy, from 47 to 45.5 kcal, when one goes from iso - C_3H_7Br to sec. C_4H_9Br.

Finally, with tertiary butyl bromide, H atom abstraction gives a radical that decomposes quite readily into an olefin and a bromine atom. Chain termination is due to recombination of a bromine atom with the active radical, and, as a result, the reaction rate is now first order. The activation energy for decomposition calculated from the elementary steps of the chain reaction, is equal to 42 kcal, in good agreement with the experimental value (41 kcal). All these arguments tend to favor a chain mechanism in the pyrolysis of tert. C_4H_9Br. However, there are various facts refuting this interpretation. Thus, the pyrolysis of tert. C_4H_9Br is not sensitive to bromine addition, a fact supporting a molecular mechanism. On the other hand, a molecular mechanism does not seem compatible with the low activation energy since H abstraction from a CH_3 group requires more energy than is required in the case of sec. C_4H_9Br and iso - C_3H_7Br.

While all facts indicate a molecular decomposition of iso-C_3H_7Br and sec. C_4H_9Br, all bromides, with the exception of these two, show self-inhibited rates and the reaction stops practically completely after a certain time.

As Sergeev has shown [33], this is not due to thermodynamic equilibrium (an explanation advanced by Maccoll [29]) since the concentrations of products when the reaction stops are still far from their equilibrium values. To prove this point, G. B. Sergeev ran a decomposition of n - C_3H_7Br until no pressure change was detected and this state of pseudo-equilibrium was reached. But after introduction of the mixture obtained in this way into a reactor at 400°C, reaction starts again as shown by a pressure change. The interruption of reaction is therefore not due to thermodynamic reasons.

Self-inhibition by small quantities of impurity is a phenomenon characteristic of chain reactions. Oftentimes, it is not due to the main reaction products. An example of this kind is the cracking of hydrocarbons. Inhibition of this type is impossible in unimolecular reactions. If experiments similar to those just described are repeated with iso - C_3H_7Br and sec. C_4H_9Br and the result is the same as with n - C_3H_7Br (this seems highly probable), the molecular character of their decomposition will be established without any doubt.

To sum up, these studies have established for the first time a relation between reaction mechanism and structure of reacting molecules.

TABLE 38
Decomposition of Bromides

	Compounds and Conditions	E kcal	k
1	n - C_3H_7Br P = 10-100 mm; T = 347-397°	42	$k = 1.21 \cdot T^{1/2} e^{-42000/RT}$ $(cm^3/mole.)^{1/2} sec^{-1}$
2	iso-C_3H_7Br P = 10-150 mm; T = 347-497°	47	$k = 5.5 \cdot 10^{12} e^{-47000/RT}$ sec^{-1}
3	n-C_4H_9Br P = 20-150 mm; T = 340-450°	48	$k = 0.6 \cdot 10^2 \cdot T^{1/2} e^{-48000/RT}$ $(cm^3/mole.)^{1/2} sec^{-1}$
4	iso-C_4H_9Br P = 10-150 mm; T = 300-390°	30	$k = 0.58 \cdot 10^{-3} \cdot T^{1/2} e^{-30000/RT}$ $(cm^3/mole.)^{1/2} sec^{-1}$
5	sec-C_4H_9Br P = 10-200 mm; T = 330-400°	45.5	$k = 1.1 \cdot 10^{13} e^{-45500/RT}$ sec^{-1}
6	tert.C_4H_9Br P = 10-150 mm; T = 265-325°	41	$k = 1.7 \cdot 10^{13} e^{-41000/RT}$ sec^{-1}

	Effect of Additives	Character of decomposition	Data of Maccoll [41] on decomposition with excess of cyclohexene
1	Acceleration by bromine and oxygen; weak inhibition by C_3H_6; weak acceleration by HBr	Chain	$k = 8.10^{12} e^{-50700/RT}$ sec^{-1}
2	No inhibition by C_3H_6	Molecular	$k = 4.2 \cdot 10^{13} e^{-47800/RT}$ sec^{-1}
3	Acceleration by bromine; weak inhibition by C_3H_6	Chain	$k = 1.5 \cdot 10^{13} e^{-50900/RT}$ sec^{-1}
4	Acceleration by bromine; weak inhibition by C_3H_6; poor reproducibility	Chain	$k = 1.12 \cdot 10^{13} e^{-50400/RT}$ sec^{-1}
5	no inhibition by C_3H_6	Molecular	$k = 4.25 \cdot 10^{12} e^{-42200/RT}$ sec^{-1}
6	no inhibition by C_3H_6	Molecular?	$k = 10^{14} e^{-42200/RT}$ sec^{-1}

§4. Pyrolysis of Alkylchlorides

Alkylchlorides decompose into hydrogen chloride and olefins (or chlorinated olefins in the case of chlorides with more than one chlorine atom). The characteristic feature of the reaction is the sharp dependence of the rate on the state of reactor walls.

In a fresh quartz reactor, the reaction rate is faster by a factor of the order of ten than in the same reactor after a hundred or so experiments. Obviously, some unsaturated carbon residues are formed besides the main products and they deposit as a very thin layer on the walls of the reactor. This layer is not removed by evacuation and therefore it accumulates after each run. After ten or twenty runs, this surface film becomes quite visible. Its color is dark grey. As the film builds up, the reaction rate decreases and reaches a limiting value after a few hundred experiments. Then, no further change in rate takes place.

The majority of investigators (Barton, Howlett, Williams etc.) have confined their attention to chloride pyrolysis under these stabilized conditions. It was found that all chlorides follow a first order decomposition and the first order rate constant does not change until 30 to 50% conversion. At higher conversions, the rate constant becomes somewhat smaller. Increasing the surface to volume ratio has little effect on the reaction rate.

It was also found that various chlorides behave differently under these stabilized conditions and they may be divided into two groups. In the first group are chlorides the decomposition of which is inhibited by addition of propylene [43], n-hexane [43], acetaldehyde [44] and accelerated by traces of oxygen or chlorine [45]. In some cases, especially with inhibition by propylene, an induction period is observed during the first stages of reaction. To this group belong 1,2-dichloroethane [43], 1,1,1-trichloroethane [46], 1,1,2-trichloroethane [47], 1,1,1,2-tetrachloroethane [48], 1,1,2,2-tetrachloroethane [48] and 1,4-dichlorobutane [47].

Barton [43, 44, 46], Howlett [43, 44] and Williams [47, 48] believe that all these chlorides decompose in chain fashion, e.g., following the scheme:

0) $CH_2C\ell - CH_2C\ell \longrightarrow \dot{C}\ell + \dot{C}H_2 - CH_2C\ell - Q$ Initiation

1) $\dot{C}\ell + CH_2C\ell - CH_2C\ell \longrightarrow HC\ell + \dot{C}HC\ell - CH_2C\ell + q_1 (q_1 \sim 3 - 5 \text{ kcal})$ Chain

2) $\dot{C}HC\ell - CH_2C\ell \longrightarrow \dot{C}\ell + CHC\ell = CH_2 - q_2 (q_2 \sim 24 \text{ kcal})$ Chain

3) $\dot{C}\ell + \dot{C}HC\ell - CH_2C\ell \longrightarrow CHC\ell_2 - CH_2C\ell$ Termination

To the second group belong chlorides which (at least under stabilized conditions) are insensitive to addition of propylene, oxygen etc. Among them are: ethyl chloride [49], 1,1-dichloroethane [49], 1,2-dichloropropane [50], 2-chloropropane [50], n-propyl chloride [51], n-butyl chloride [51] and isobutyl chloride [52]. The investigators who studied the pyrolysis of these chlorides assume that it corresponds to a true direct molecular decomposition into $HC\ell$ and olefin.

This viewpoint is supported by some purely chemical arguments. As was already discussed in connection with the pyrolysis of bromides, the propagation of the chain is arrested if a reaction between a chlorine atom and a chloride molecule is likely to form a radical which cannot yield an olefin by elimination of a chlorine atom. Then the chain decomposition becomes difficult and a direct decomposition into $HC\ell$ and an olefin will predominate. Thus, in the decomposition of ethylchloride and 1,1-dichloroethane, a chlorine atom can react with these molecules (reaction 1) and abstract a hydrogen atom from the group containing the largest number of chlorine atoms: the radicals $\dot{C}HC\ell - CH_3$ and $\dot{C}C\ell_2 - CH_3$ are then formed and these cannot eliminate a chlorine atom.

On the contrary, in the decomposition of the chlorides of the first group (1,2-dichloroethane, 1,1,1-trichloroethane, 1,1,1,2-tetrachloroethane, 1,1,2-trichloroethane, 1,1,2,2-tetrachloroethane, etc.), a reaction with chlorine atoms yields the radicals: $\dot{C}HC\ell - CH_2C\ell$, $CC\ell_3 - \dot{C}H_2$, $CC\ell_3 - \dot{C}HC\ell$, $CC\ell_2 - CH_2C\ell$, $\dot{C}C\ell_2 - CHC\ell_2$. Each one can decompose into a chlorine atom and an olefin and the reaction chain can be propagated.

Most authors who have studied the pyrolysis of chlorides assume that chain initiation and termination are homogeneous and that the inhibition by propylene and other additives is due to a homogeneous reaction of these molecules with chain radicals.

This conclusion is not binding and does not explain why the pyrolysis of chlorides of both groups I and II is considerably more rapid in fresh reactors. The kinetics of the reaction in fresh reactors has been studied quantitatively by Barton [45] in a flow system before a film was formed on the surface of the reactor. Then the reaction is still first order but its rate constant is substantially higher than in aged reactors. In the pyrolysis of 1,2-dichloroethane, Barton [45] finds the following rate constants (all in sec^{-1}): in aged reactors $k_1 = 6.4 \cdot 10^{10} \exp(-47000/RT)$; in fresh reactors (flow system): $k_1' = 1.6 \cdot 10^6 \exp(-27100/RT)$. Thus, at 362°C, $k_1 = 6.4 \cdot 10^{-6}$, $k_1' = 8.8 \cdot 10^{-4}$ and at 485°C, $k_1 = 2 \cdot 10^{-3}$ and $k_1' = 2.8 \cdot 10^{-2}$. The data of G. A. Kapralova, taken in 1954-1955 at the Institute of Chemical Physics, are: $k_1 = 3 \cdot 10^{11} \exp(-48000/RT)$ and $k_1' = 10^8 \exp(-32000/RT)$, which gives at 480°C,

$k_1 = 4.8 \cdot 10^{-3}$, $k_1' = 5.6 \cdot 10^{-2}$ and at 560°C, $k_1 = 9.4 \cdot 10^{-2}$, $k_1' = 4.5 \cdot 10^{-1}$.

The reaction rate in a clean reactor may also be measured in a static system. But precise data are difficult to obtain since the surface gets poisoned and the reation rate decreases from run to run.

Two hypotheses can be made concerning the reaction in fresh reactors: 1) the reaction proceeds directly in heterogeneous catalytic fashion on the reactor surface and it stops after a sufficiently thick film of condensed reaction products has deposited on the walls; 2) most of the pyrolytic reaction takes place homogeneously as a chain process but the chains start on the fresh wall surface.

The first hypothesis implies that the heterogeneous catalytic reaction must have a rate proportional to the reactor surface. Barton [53], in a study of 1,2-dichloroethane decomposition in a flow system at 500°C, noted that doubling the surface increased the rate only by 17 to 19%. This suffices to conclude that the pyrolysis in fresh reactors is not a simple heterogeneous catalytic reaction.

Using the method of differential calorimetry, G. A. Kapralova has shown directly that in fresh reactors, pyrolysis of 1,2-dichloroethane takes place homogeneously. In sufficiently wide reactors (~ 4 cm), some non-uniformity of the homogeneous reaction becomes noticeable. It tends to be stronger near the wall and to proceed in a zone about 1 cm thick around the wall.

The experiments of G. A. Kaprolova provide conclusive proof that the reaction in fresh reactors starts at the wall and propagates within the volume. Chain termination occurs mainly at the surface but also within the volume. This explains the formation of the reaction zone as observed by Kapralova in wide reactors.

Similar results were obtained by Kapralova in aged reactors. Here also the reaction is homogeneous and in a reactor of ~ 3 cm diameter, she observed some more intense reaction in a zone near the wall, about one cm thick. It is possible also that in aged reactors, the decomposition chains start and end principally at the wall. We must note that in the absence of homogeneous termination, the chain decomposition will be first order if chain termination consists in the wall destruction of a radical ACl (where A is an olefinic diradical), such as $\dot{C}HCl - CH_2$ in the case of 1,2-dichloroethane, and if chain initiation at the wall is due to the reaction:

$$ClH_2C - CH_2Cl + \text{wall} \longrightarrow Cl(\text{adsorbed}) + \dot{C}H_2 - CH_2Cl \ .$$

Kapralova also obtained very interesting results in a study of the propylene inhibited decomposition of 1,2-dichloroethane. As was found by Barton and Hawlett [43], propylene additions reduces the rate in an aged reactor to a fraction of its value. Thus, at 555°C, Kapralova finds that 2.7% C_3H_6 reduces the rate by a factor of ~ 2.7; 10%C_3H_6 reduces the initial rate by a factor of 3 and 20%C_3H_6 by a factor of 3.36.

Kapralova first studied the effect of propylene addition on chloride decomposition in fresh reactors. The effect is much stronger than in aged reactors. Thus, at 430°C, ~ 0.5%C_3H_6 reduces the rate by a factor of about twenty. This contrast is in itself a strong argument against the idea that the inhibition is due to homogeneous chain termination. A direct refutation was given by Kapralova by means of the method of differential calorimetry. She found that the reaction zone first contracts somewhat after addition of a small quantity of propylene (0.01%). But further addition, up to 0.5% does not change the width of the reaction zone any further. By contrast, 0.01%C_3H_6 reduces the rate by a factor of only 1.5 to 2 while 0.5%C_3H_6 reduces it by a factor of 20.* Therefore, propylene decreases the rate principally because it changes a surface reaction and not a volume reaction.

Apparently, propylene is adsorbed on the wall, gradually covering its free valences and reducing its rate of chain initiation. The nature of the fully inhibited reaction taking place after addition of a definite amount of propylene, such that further addition has no effect on the rate, has not been elucidated as yet. It may be a purely molecular reaction of the type suggested by Hinshelwood to explain the cracking of hydrocarbons fully inhibited by NO.

It is also possible that the fully inhibited reaction is a chain process with an equilibrium concentration of radicals, similar to the case of hydrogen-chlorine reaction discussed in Chapter VI.

§5. Cracking of Hydrocarbons

The cracking of paraffinic hydrocarbons, especially the lower members of the homologous series, has been particularly well studied. The initial rate at medium pressures (50 - 500 mm Hg) is described by a first order rate law with respect to initial pressure.**

---

* It must be remarked that the inhibition is stronger at the beginning of reaction. Thus at 4% conversion, 0.5%C_3H_6 reduces the rate fiftyfold. At higher conversions, in all cases of inhibited reactions, the reaction accelerates noticeably.

** Even in this pressure range and more so at higher pressures, this is not so for sufficiently long $(n > 6)$ non-branched hydrocarbons. It is necessary, in such cases to use a mixed rate law of first and second order $Ap_0 + Bp_0^2$. With branched hydrocarbons, the first order law is obeyed in a much wider interval.

As was shown earlier, F. O. Rice and K. F. Herzfeld [55] proposed a chain mechanism to explain several features of these reactions. This mechanism has been widely accepted since then (see Chapter II).

In 1933, A. V. Frost and A. I. Dintses [56] observed inhibition of the initial decomposition of hydrocarbons by reaction products or the addition of propylene.* Hinshelwood [54] showed that nitric oxide exerts a similar action. Studying quantitatively the decrease of the rate constant for cracking as a function of the amount of NO added, Hinshelwood showed that, above a certain quantity of NO, the constant does not change any more upon further additions of NO but reaches a limiting value: the rate constant for the 'fully inhibited reaction'. Later it was shown that the action of propylene is similar to that of NO, and what is more important, the limiting inhibited rates turned out to be the same in both cases. The difference between the two gases consists in the fact that a considerably larger quantity of propylene must be used to decrease the rate to a given level. For instance, to inhibit the cracking of n-pentane [54] at 530°, the addition of propylene must be 12.5 times larger than that of NO. Figure 3 shows Hinshelwood's data on the inhibition of n - C_5H_{12} cracking by NO and propylene. A. D. Stepukhovich [57] has compared quantitatively the action of iso-butylene and propylene on the decomposition of various hydrocarbons and has found that the limiting rate reached in these cases is also the same.

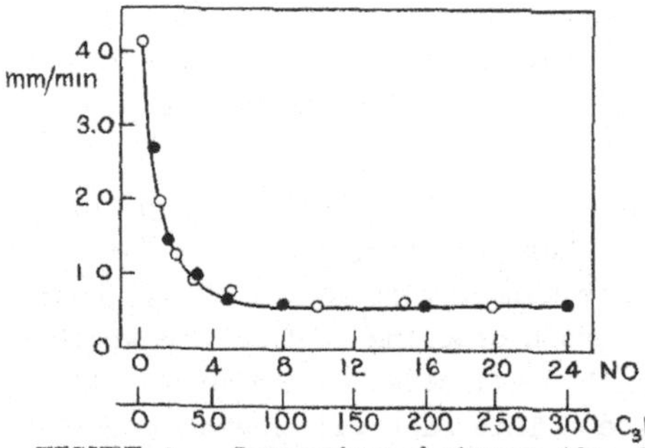

FIGURE 3. Comparison between the inhibiting effect of nitric oxide and propylene on the initial rate of cracking of n-petane [54]. T = 530°C; P = 100 mm Hg.

In several investigations, the effect of NO on the rate of cracking as well as the self-inhibition in the absence of NO were studied in parallel. Thus, for instance, V. A. Poltorak and V. V. Voevodskiǐ [58, 59] have shown that the rate of cracking of pure propane at a sufficiently high percentage of conversion, coincided with the limiting inhibited rate due to NO at the same value of conversion. This means that the limiting inhibitions by NO and by reaction products are the same. Results of these experiments are shown in Figure 4, on the following page.

It appears, therefore, that one must agree with Hinshelwood: for

* Ethylene is an exception. It does not inhibit the decomposition of ethane. However the latter is strongly inhibited during the course of the reaction. This phenomenon is still unexplained.

each hydrocarbon, under specified conditions, there exists a definite limiting inhibited cracking rate depending only on the addition of a sufficient amount of inhibitor and not on the nature of the latter.

The usual explanation of inhibitor action in chain processes is that inhibitor molecules react readily with chain radicals, thus forming less active radicals. With this in mind, Hinshelwood [54] analyzed the studies just mentioned and came to the conclusion that at high concentrations of NO or C_3H_6, the chain process disappears completely. The residual reaction which is not affected by the inhibitor, is then, according to him, a molecular reaction. This also explains the identical rates observed with various inhibitors. Although this uniquely important experimental fact cannot readily be explained in any other way, it must be noted that it appears to be the only proof of Hinshelwood's hypothesis concerning the molecular nature of the fully inhibited process.

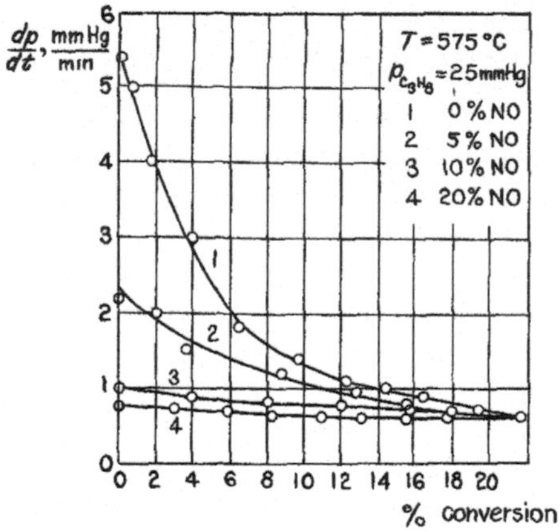

FIGURE 4. Effect of addition of NO on the rate of cracking of propane [58, 59]. T = 575°C; p = 25 mm Hg. 1: without NO; 2: with 5% NO; 3: with 10% NO; 4: with 20% NO.

On the other hand, there are many facts that cast doubt on this simple and seemingly natural hypothesis. Thus, for instance, from the studies of Hinshelwood himself and of other workers, it is well known that the composition of the products of hydrocarbon cracking is the same whether the reaction is fully inhibited or not. This coincidence is true at all temperatures at which cracking takes place. This fact was once more verified recently by V. A. Poltorak [58, 59] who studied propane cracking. The results of one of these experiments are shown in Table 39.

TABLE 39

Composition of the Products of Propane Cracking With and Without NO

T = 590°C $p_{C_3H_8}$ = 25 mm Hg

% Conversion		product composition in % CH_4	H_2	$C_2H_4+C_2H_6$	C_3H_6
5	without NO	23.5	29	16.5	31
	with 20% NO	22.6	29.1	22	26
20	without NO	27.3	21.7	24.7	26.4
	with 20% NO	28	20	28	24

From Hinshelwood's standpoint, it cannot be understood how two completely different reaction mechanisms (chain and molecular) lead to completely identical products. Also difficult to explain is the fact that, for a large number of compounds (hydrocarbons of various structures, ethers etc.) there is generally only a small difference between the rate of the chain process and that of Hinshelwood's molecular reaction. Although the activation energy for decomposition of various organic compounds, under the inhibiting effect of NO, varies appreciably from molecule to molecule (50 to 75 kcal), the activation energies of the inhibited and uninhibited reactions always differ very little and this difference amounts only to a few kilocalories.

Rice and Polley [60], as well as Gol'danskiĭ [61] later and in more detailed fashion, have proposed another explanation for the effect of full inhibition. It is proposed that all molecules capable of ending chains are at the same time capable of starting chains. This can explain the full inhibition caused by sufficient amounts of inhibitor. The residual reaction is then naturally a chain process and this removes the objections formulated above against a molecular mechanism for that process. Nevertheless this interpretation is not satisfactory since it fails to explain this amazing and therefore most important fact: the fully inhibited rates are the same for different inhibitors.

It was therefore necessary to establish by a direct experimental technique, the molecular or radical chain nature of the mechanism of the fully inhibited process. The first attempt in this direction was made by Wall and Moore [62]. They submitted to cracking a mixture of C_2H_6 and C_2D_6, at 610°C, in the absence of NO and with 2.5% NO. It was found that the quantities of H_2, HD and D_2 in the decomposition products, are identical in both cases. It was shown by special experiments that the exchange is due to the process of decomposition itself since at the same temperature but in the absence of ethane, the mixtures $H_2 + D_2$, $H_2 + CD_4$, $D_2 + CH_4$ and $CD_4 + CH_4$ do not exhibit exchange to the same extent.

The isotopic exchange observed by Wall and Moore leads to the conclusion that NO does not exert any noticeable effect on the radical chain mechanism of the process since a purely molecular process in $C_2H_6 - C_2D_6$ mixtures would yield only H_2 and D_2 but not HD. In an analysis of the work of Wall and Moore, Hinshelwood [63] remarks that the work was carried out with insufficient amounts of NO to insure full inhibition. It seems to us, however, that 2.5% NO lowers the rate of what Hinshelwood calls the initial chain mechanism to such an extent that the extent of isotopic exchange should also be lowered appreciably.

In order to resolve the question of the reaction fully inhibited by NO, Poltorak and Voevodskiĭ [64] studied the pyrolysis of propane in the presence of D_2 and measured the deuterium content of the ethylene formed. In the case of a radical mechanism, exchange between the free radicals of the chain and D_2 ought to give products containing deuterium. A molecular decomposition is by itself incapable of giving exchanged products. On the other hand, it was found that exchange always takes place, without NO or with enough NO to reach full inhibition. Moreover, at a given conversion level identical exchange takes place both with or without NO. This result shows that the fully inhibited reaction is a chain process and not a molecular process. The objection cannot be raised that deuteration takes place by means of an exchange process independent of the cracking reaction (e.g., surface reactions), since then this independent process should not be related to the conversion level but only to time. In fact, the exchange corresponds completely to the extent of conversion and not to the time required to reach a given degree of conversion (reaction times differed by a factor of seven in the runs with or without NO). The weak point of this investigation is that the degree of conversion was the same in all runs although runs were made at different temperatures and $C_3H_8 : D_2$ ratios. Yet, it would be essential to show that the exchange is truly proportional to the extent of conversion by studying the exchange at different conversion levels. This, however, was done by Rice and Varnerin [65] who cracked C_2D_6 in the presence of CH_4. The ratio CH_3D/CH_4 was found to be proportional to the extent of reaction and to be identical, at a given conversion level, both in the absence of NO and in the presence of sufficient quantities of NO.

These experiments seem to prove that the fully inhibited process is not molecular but of radical chain nature. But Hinshelwood does not think that this conclusion is fully warranted. In our opinion, the studies of Hinshelwood [66, 67] on the decomposition of CH_4 and C_4H_{10} in the presence of deuterium and of propane containing C^{14}, provide ample proof of the uniqueness and the chain character of the mechanism for the uninhibited and inhibited processes. The objections against these conclusions, raised by Hinshelwood himself, appear to us to be vague. Thus, in the course of his reasoning, Hinshelwood assumes that the pyrolysis of butylene does not proceed by a chain mechanism. Therefore, the exchange in methane taking place during butylene pyrolysis, is explained by him in terms of some purely molecular process. In our opinion, as was already pointed out, butylene decomposes following a chain mechanism and the results of Hinshelwood must be rather considered as a good proof of this view. Besides, Kebarle and Bryce [68] have recently studied the decomposition of propylene in the presence of $Hg(CH_3)_2$. The accelerating effect due to the methyl radicals formed by decomposition of $Hg(CH_3)_2$ proves convincingly that butylene pyrolysis is not a molecular process but involves free radicals.

Although Hinshelwood's hypothesis explains simply and nicely the coincidence of the fully inhibited rates with various inhibitors, it cannot explain the facts if the residual reaction is a chain process.

Recently, in a study of propane cracking (p = 25 to 50 mm Hg, T = 500 to 600°C), Voevodskiĭ and Poltorak [58, 59] have proposed the view that the inhibition of the chain decomposition of paraffins is related to the heterogeneous character of chain initiation and termination. This idea is close to that presented earlier in our explanation of the pyrolysis of chlorides.

In examining this problem, Voevodskiĭ [59] postulates the existence of irreversible decomposition processes taking place at the wall with ejection of a free radical, similar to the reaction:

$$H_2 + C\ell_2 \xrightarrow{\text{wall}} HC\ell + C\ell + [H]$$

considered earlier in our discussion of the thermal $H_2 + C\ell_2$ reaction. These irreversible processes are promoted by free valences at the surface and bring forth, at the beginning of reaction, a higher concentration of radicals within the volume. This in turn gives rise to a higher initial rate in the absence of inhibitors. But these processes also progressively destroy the free valences so that the wall is capable only to dissociate and recombine in reversible fashion. Then some 'equilibrium' concentration of radicals is established within the volume, as was explained in detail in connection with the $H_2 + C\ell_2$ reaction. This equilibrium concentration is always the same since it is determined by thermodynamic conditions. Therefore, if the irreversible wall processes are completely suppressed, the reaction rate is lowered to a definite level, always the same whatever be the state of the surface. In this way, it is possible to explain the limiting rate of self-inhibited cracking of a pure hydrocarbon and the identical limiting rate due to a variety of inhibitors.

From this viewpoint, the fully inhibited reaction proceeds at the true rate of the chain cracking. Then, the concentration of free radicals and also the reaction rate do not depend on the location of chain initiation and termination: within the volume or at the surface or at both places simultaneously. If this is so, the reaction should not depend on the surface to volume ration S/V.

As to the initial stage of the reaction, it appears to be 'irreversible'. Everything happens as if an initiator of the chain reaction had been introduced in the system, the action of the initiator disappearing progressively as it is consumed. In our case, according to Voevodskiĭ, the free valences at the wall play the role of initiator and they are destroyed as reaction proceeds. Then the initial reaction rate should depend on the state of the surface. Voevodskiǐ and Poltorak [58, 59] were the first to observe such a relation between cracking rate and the state of the wall. They showed that the initial rate in the presence of small quantities of O_2, strongly depends on the state of the surface. For instance, a treatment of a quartz reactor by a mixture of HF and NH_4F or hot HF gave rise to a four- or five-fold increase in the initial rate of propane cracking (25 mm Hg C_3H_8 + 1 - 2 mm Hg O_2). It must be noted that after a series of runs, the surface activated with HF gradually lost its properties and ultimately the rate became equal to that taking place in an untreated reactor in the presence of identical quantities of O_2. An addition of NO in a vessel treated with HF immediately lowers the rate to the value characteristic of an untreated vessel. After removal of NO, the lowered rate is still observed.

If NO is introduced in propane that contains oxygen, NO does not inhibit cracking but accelerates it somewhat (see Figure 5).

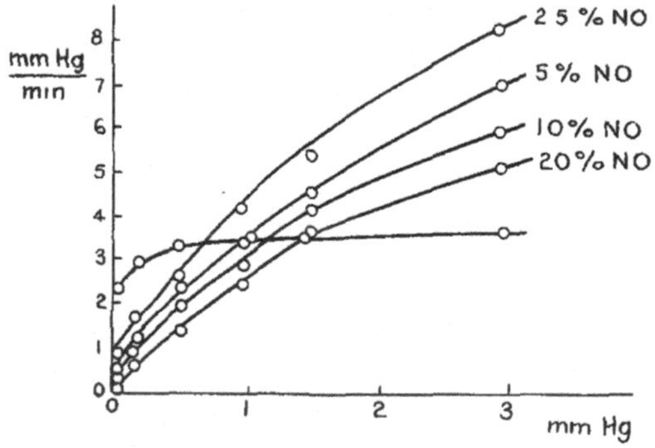

FIGURE 5. Effect on the rate of reaction of the simultaneous addition of nitric oxide and oxygen T = 575°C; $p_{C_3H_8}$ = 25 mm Hg; 1: without NO; 2: with 2.5% NO; 3: with 5% NO; 4: with 10% NO; 5: with 20% NO.

Furthermore, Voevodskiǐ and Poltorak have found that NO has a very weak inhibiting effect in the absence of oxygen if the surface is covered with MoO_3. Thus the reaction in the presence of 20% NO is normally fully inhibited. This corresponds to an eight- to ten-fold decrease. But if the surface is covered with MoO_3, the rate is only 20% lower in the presence of 20% NO. L. Ya. Leǐtis and V. A. Poltorak have also observed that a special treatment of reactor walls may increase (treatment with $Mg(ClO_4)_2$ or decrease (treatment with H_2S + NO) the initial reaction rate. Yet the latter does not depend on the magnitude of the S/V ratio and the effect may not be explained by heterogeneous catalysis. Similarly, in reactors treated with HF, the effect is insensitive to changes in S/V.

It may be concluded that, at least in a variety of cases, chain initiation in cracking occurs on reactor walls. Many authors have noted the fact that the initial rate is practically independent of the S/V ratio. This shows that the walls are active not only in chain initiation but also in chain termination.*

§6. Factors Affecting the Chain Length

As we have already shown, even very small quantities of common impurities (oxygen, peroxides, ions of variable valence) are able to generate primary radicals with relative ease and enhance reaction rates of chain processes considerably above the values just discussed.

The expression for the reaction rate contains the rate of radical initiation which may be increased by these impurities. The overall rate of the chain process is thereby increased. This circumstance is very favorable to chain reactions.

We know many chain reactions, in particular in the liquid phase (e.g., oxidations) that proceed at quite low temperatures (e.g., oxidation of aldehydes) because of the presence of various additives or the formation of peroxides. The chain length of such processes is of the order of ten thousand. The reaction is thus relatively quite rapid in spite of the small rate of generation of free radicals, i.e., the small amount of a catalyst such as ions with variable valence.

Consider the factors that influence chain length.

Chain length is the ratio of the rate of the chain reaction w_c to the rate I_0 of dissociation of the original substance into radicals. Thus, for instance, for the reaction $Hal_2 + H_2$,

$$I_0 = 10^{-10}\exp(-Q/RT)[Hal_2][M] .$$

Since

$$w_c = 10^{-10}\exp(-\epsilon_1/RT)[Hal]\cdot[H_2]$$

$$= 10^{-10}\exp(-\epsilon_1/RT)\cdot(I_0/k_3)^{1/2}[H_2]$$

(with $k_3 = 10^{-32}[M]$ and $[M] \approx 10^{19}$ at atmospheric pressure), the chain length is:

---

* In their study, L. Ya. Leĭtis and V. A. Voevodskiĭ found that at $p = 25$ mm Hg in an untreated quartz reactor, the initial rate decreases by a factor of two or three when the S/V ratio is increased 10 or 15 times. This result could not be explained as yet by the concepts just presented.

$$\nu = \frac{w_c}{I_0} = \frac{10^{-10}\exp(-\epsilon_1/RT)\cdot 10^{19}}{(k_3 I_0)^{1/2}} = \frac{10^9\exp(-\epsilon_1/RT)}{\{10^{-32}[M]\cdot 10^{-10}\exp(-Q/RT)[Hal_2][M]\}^{1/2}}$$

$$\nu = [10^9 \cdot \exp(\tfrac{Q}{2} - \epsilon_1)/RT]/10^{19} \cdot (10^{-23})^{1/2} \approx 10 \cdot \exp(\tfrac{Q}{2} - \epsilon_1)/RT \quad .$$

For the reaction $H_2 + C\ell_2$ with $Q = 57$ kcal and $\epsilon_1 = 6$ kcal, at 600°K, the chain length is about $\nu \approx 10^9$. For the reaction $H_2 + Br_2$ with $Q = 46$ kcal, $\epsilon_1 = 17$ kcal, at 700°K, we obtain $\nu \approx 10^3$. Since $\frac{Q}{2}$ is usually larger than ϵ_1, the chain length ν decreases as the temperature goes up.

At first glance, this seems a little unexpected. It is quite understandable since the rate of recombination increases with the concentration of free radicals i.e., with rate of initiation I_0. Since chain length decreases as the recombination rate goes up, it also decreases as the temperature becomes higher. As the rate of initiation increases, the concentration of free radicals gets larger and the chain becomes shorter.

In photochemical processes, the chain becomes shorter as the light intensity increases, being inversely proportional to the square root of the intensity. Chain length, of course, depends on the magnitude of ϵ_1. If the latter is large, the chain is short, since the ratio of propagation rate to recombination rate (which also determines the chain length ν) decreases when ϵ_1 increases. For photochemical reactions, the chain length

$$\nu = \frac{w_c}{I_0} = \frac{10^{-10}\exp(-\epsilon_1/RT)[CD]}{(I_0 k_3)^{1/2}}$$

The quantity I_0 depends on the number of light quanta absorbed and not on temperature. The chain length of photochemical reactions, at constant I_0, increases with temperature because of the factor $\exp(-\epsilon_1/RT)$.

For photochemical reactions carried out at ordinary temperatures, the chain length depends on the quantity ϵ_1 (the activation energy of the hardest step of the chain). If ϵ_1 is equal to 6 kcal, the chain is very long, but if ϵ_1 is larger than 10 kcal, it is very short. Then the chain length is often unity and radicals simply recombine. Under such conditions, therefore, the products of recombination constitute an important fraction of the reaction products. With long chains, recombination products represent only a small percentage of the main products due to chain propagation.

When the dissociation into free radicals occurs rapidly and the chains are short, an important part of the reaction products consists of the products of various types of radical recombination. This takes place frequently in the case of liquid phase reactions at low temperatures.

Let us return to the problem of the competition between chain and direct processes.

We may express the rate of a chain reaction as the product of chain length by the number of primary radicals formed per second: $w_c = I_0 \nu$. Each strongly endothermic act of dissociation of original molecules into free radicals, requiring an expenditure of energy Q, leads to reaction of more than one decomposing molecule since it starts a long chain of easy processes. In the case of a direct process, each reaction act, even if it is exothermic, requires a large activation energy.

If the chain length is, say, 1000, then even if the number I_0 of primary radicals formed from the original molecules is 1000 times smaller than the number w_D of direct processes, the rates of the chain and direct processes will be equal.

Thus, for example, if $\nu = 1000$ ($T = 800°K$) and the energy of dissociation of a molecule into free radicals is 65 kcal/mole, the rates of the chain and molecular processes will be equal for an activation energy E determined by:

$$\exp -(Q - E)/RT = 10^{-3} = e^{-6.9}$$

if, in first approximation the pre-exponential factors for decomposition into radicals and molecules are about the same, 10^{13}sec^{-1}.

Then $Q - E = 6.9 \cdot RT = 6.9 \cdot 2 \cdot 800 = 11$ kcal. Hence $E = Q - 11 = 65 - 11 = 54$ kcal. If $Q - E < 11$ kcal, the chain process will predominate; if $Q - E > 11$ kcal, the molecular reaction takes over. A chain reaction may thus be faster than the direct process even if the energy of dissociation into radicals Q is larger than the activation energy E of the direct reaction between molecules.

§7. The Elementary Step of Decomposition

The study of the elementary step of decomposition unencumbered by complicating chain propagation and wall reaction, is very interesting indeed. In many cases, Q is smaller than E and molecules rather decompose into radicals than molecular products. Then both a chain reaction and a radical reaction ($\nu = 1$) are more favorable than a molecular process. In other cases, E is smaller than Q and a molecular elementary decomposition is more favorable. Then the primary decomposition step can be stúdied, once it has been shown that it is not a pseudo molecular process consisting of a chain reaction with an overall first order rate. Unfortunately, techniques to make such studies have not been developed sufficiently so that the nature of the primary act has not been investigated in a satisfactory way. Rice [69], using the mirror technique and Szwarc [15] by means of the toluene method (Chapter I) have succeeded in showing that many organic molecules decompose into radicals in a primary step. Yet, in other cases, decomposition into molecules as a result of bond rearrangement is energetically more favorable than a scission into radicals. Conceivably there are cases where both paths are very similar energetically so that for a given molecule, molecular decomposition or radical scission will predominate depending upon the circumstances. Undoubtedly, the nature of the decomposition mechanism depends on molecular structure. It is only recently that this question has received attention.

Szwarc and several other investigators have studied the pyrolysis of various organic compounds and found that a considerable number of molecules decompose primarily into radicals. These experiments were conducted in a rapid flow system, the gas under study being sent through a heated tube at a very low partial pressure (a fraction of a mm Hg) in a stream of inert diluent. The presence or absence of free radicals was detected at the tube exit. This technique has already been described in Chapter I. It has been shown for example that ethylbenzene [70] decomposes practically exclusively into CH_3 and benzyl radicals, and not into styrene and hydrogen as would be the path of a direct unimolecular reaction. Allyl bromide [71] apparently gives only allyl radicals and bromine atoms and not HBr plus propadiene. Ethylene oxide [4] gives CH_3 and HCO radicals and not CH_4 plus CO.

Peroxides also give primarily two radicals by scission of the O - O bond and not molecular products. For example, tertiary butyl peroxide decomposes into two butoxy radicals [72]:

$$(CH_3)_3COOC(CH_3)_3 \longrightarrow 2(CH_3)_3\dot{C}O \quad .$$

Table 40, on the following page, contains data on various molecules for which decomposition into radicals has been proved by Szwarc and other investigators.

Among all the compounds assembled in Table 40, there are quite a few that cannot decompose directly into molecular products for structural reasons. Examples are bromobenzene, bromonaphthalene, bromophenanthrene, bromoanthracene, bromine derivatives of metane etc. . These compounds are

TABLE 40

Compound	Radicals Formed	$E = Q_{(c-x)}$ where x is an atom or a radical	Reference
CH_3Br	CH_3 + Br	67.5	[73]
CH_2Br_2	CH_2Br + Br	62.5	[73]
$CHBr_3$	$CHBr_2$ + Br	55.5	[73]
CBr_4	CBr_3 + Br	49	[73]
CH_2ClBr	CH_2Cl + Br	53.5	[73]
CCl_3Br	CCl_3 + Br	49.0	[73]
CF_3Br	CF_3 + Br	64.5	[73]
C_6H_5Br	C_6H_5 + Br	70.9	[74]
$C_6H_5CH_2Br$	$C_6H_5CH_2$ + Br	50.5	[71]
$CH_2 = CHCH_2Br$	$CH_2 = CHCH_2$ + Br	47.5	[71]
C_6H_5COBr	C_6H_5CO + Br	57.0	[75]
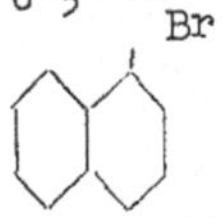	 + Br	70.9	[76]
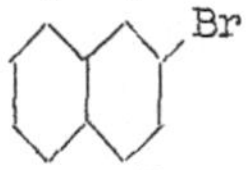	 + Br	70.0	[76]
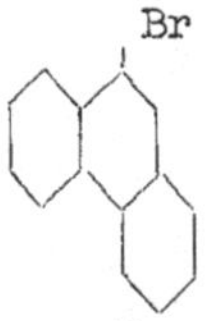	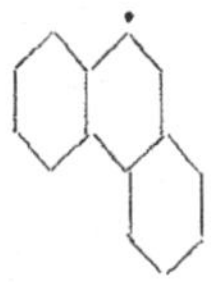 + Br	67.7	[76]
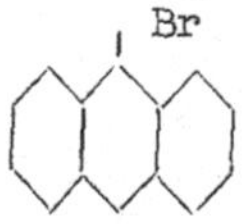	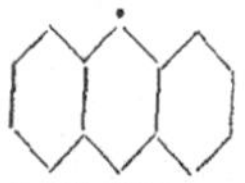 + Br	65.6	[76]
$C_6H_5CH_2Cl$	$C_6H_5CH_2$ + Cl	68.0	[77]
C_6H_5COCl	C_6H_5CO + Cl	73.6	[77]
$CH_3C(=O)$-O-O-$C(=O)CH_3$	$2CH_3C(=O)$-O	29.5	[78]
$(CH_3)_3COOC(CH_3)_3$	$2(CH_3)_3CO$	36.0	[72]
NH_2NH_2	$2NH_2$	60.0	[79]
$C_6H_5CH_2NH_2$	$C_6H_5CH_2$ + NH_2	59.0	[80]
$C_6H_5CH_2CH_3$	$C_6H_5CH_2$ + CH_3	63.2	[70]
$C_6H_5CH_2CH_2CH_3$	$C_6H_5CH_2$ + CH_2CH_3	57.5	[30]
$C_6H_5CH_2CH_2CH_2CH_3$	$C_6H_5CH_2$ + $CH_2CH_2CH_3$	65.0	[81]
$C_6H_5 - CH(CH_3)CH_3$	$C_6H_5CH(CH_3)$ + CH_3	61.0	[82]

TABLE 4c (cont.)

Compound	Radicals Formed	$E = Q_{(c-x)}$ where x is an atom or a radical	Reference
$CH_3C_6H_4CH(CH_3)_2$	$CH_3C_6H_4CHCH_3$ + CH_3	~ 60.0	[82]
$C_6H_5C(CH_3)_3$	$C_6H_5 - C(CH_3)_2$ + CH_3	~ 59.5	[82]
$CH_2 = CHCH_2CH_3$	$CH_2 = CHCH_2$ + CH_3	61.5	[83]
CH_3COCH_3	CH_3CO + CH_3	72.0	[14]
CH_2-CH_2 (with O bridge, ethylene oxide)	HCO + CH_3	44.0	[4]
CH_3I	CH_3 + I	54.0	[22]
C_2H_5I	C_2H_5 + I	52.2	[22]
n - C_3H_7I	n - C_3H_7 + I	50.0	[22]
iso - C_3H_7I	iso - C_3H_7 + I	46.1	[22]
n - C_4H_9I	n - C_4H_9 + I	49.0	[22]
tert.C_4H_9I	tert.C_4H_9 + I	45.1	[22]
$CH_2 = CHCH_2I$	$CH_2 = CHCH_2$ + I	39.0	[22]
$CH_2 = CHI$	$CH_2 = CH$ + I	55.0	[22]
$C_6H_5CH_2I$	$C_6H_5CH_2$ + I	43.7	[22]
C_6H_5I	C_6H_5 + I	54.0	[22]
CH_3COI	CH_3CO + I	50.7	[22]
C_6H_5COI	C_6H_5CO + I	43.9	[22]
CH_3COCH_2I	CH_3COCH_2 + I	45.0	[22]

therefore expected to give radicals upon decomposition.

Among the large number of compounds investigated by Szwarc, ethyl bromide [30], acetyl bromide [84] and acetic anhydride [84] were found to decompose into molecular products.

In ethyl bromide pyrolysis, HBr and C_2H_4 were found but dibenzyl - the indicator of radical decomposition - was detected only in trace amounts.*

Decomposition of CH_3COBr takes place as follows: $CH_3COBr \longrightarrow CH_3Br + CO$ and $CH_3COBr \longrightarrow CH_2CO + HBr$. Dibenzyl again was not found. Nor was it found in the products of acetic anhydride pyrolysis. The products were: $CH_3COOCOCH_3 \longrightarrow CH_2CO + CH_3COOH$ in equal amounts.

* In this section, we deal only with the primary act of decomposition. The term 'decomposition' refers here to that primary act.

Using the toluene method, Blades and Murphy [31] showed that three different alkyl bromides decompose in molecular fashion, namely C_2H_5Br, n - C_3H_7Br and iso - C_3H_7Br. As was shown earlier, the work of Sergeev and Maccoll also establishes that C_2H_5Br and iso - C_3H_7Br decompose molecularly in a static system. In the case of n - C_3H_7Br, the study in a static system revealed a chain decomposition mechanism. The different behavior of these closely related compounds, n- and iso - C_3H_7Br has been explained earlier in this Chapter on the basis of structural differences.

Recently, A. E. Shilov [85, 86] at the Institute of Chemical Physics, has studied the mechanism of the primary act of decomposition for a series of compounds from the viewpoint of a relation between this mechanism and molecular structure. The toluene method was used and the following compounds investigated: C_2H_5Br, C_2H_3Br, $C_2H_2Br_2$, C_3H_5Cl, CCl_4, $CHCl_3$, CH_2Cl_2 and CH_3Cl.

In agreement with earlier data of Szwarc [30] and Blades and Murphy [31], it was found that ethyl bromide decomposes into molecular products C_2H_4 and HBr with a rate constant $k_1 = 1.78 \cdot 10^{13} \exp(-53760/RT)$ sec^{-1}.

In order to elucidate the effect of a double bond on the mechanism of decomposition, the pyrolysis of vinyl bromide was studied. Here also in the absence of dibenzyl in the decomposition products indicates the molecular nature of the process. The activation energy for decomposition, 65 kcal, is close to the bond dissociation energy C - Br in vinyl bromide.*

Introduction of a second bromine atom in vinyl bromide changes the character of decomposition. Dibromoethylene decomposes, at least in part, into free readicals, with an activation energy $E = 62.8$ kcal.

Allyl chloride (and also allyl bromide) has been studied by Szwarc [7]: it splits into radicals, giving a chlorine atom and the allyl radical. The activation energy is 59.3 kcal, in close agreement with the C - Cl bond dissociation energy in C_3H_5Cl determined by electron impact, namely 60 ± 4 kcal [87].

Of special interest are the results obtained with the chlorine derivatives of methane. In contrast with the bromine derivatives of

* Within the range of pressures covered in this study (7 to 50 mm Hg), the decomposition follows a second order rate law: $w = k'(C_2H_3Br)(M)$ where (M) is the total number of molecules (vinyl bromide and toluene). Apparently this is an example of the lowering of the first order rate constant which is found at sufficiently low pressures in many unimolecular decompositions. Then the rate determining step of the unimolecular reaction is the activation of the molecules by collisions.

methane which according to Szwarc [73] (see Table 40), decompose into radicals, the chlorine compounds give primarily diradicals. Thus, according to Shilov, $CHC\ell_3$ and $CH_2C\ell_2$ decompose to form $HC\ell$ and diradicals:

$$CHC\ell_3 \longrightarrow HC\ell + \ddot{C}C\ell_2; \qquad CH_2C\ell_2 \longrightarrow HC\ell + \ddot{C}HC\ell \quad .$$

On the other hand, $CC\ell_4$ and $CH_3C\ell$ decompose into monoradicals, e.g., $CC\ell_4 \longrightarrow CC\ell_3 + C\ell$.

The experiments show that $CHC\ell_3$ decomposes faster than $CC\ell_4$, in spite of the fact that the $C - C\ell$ bond energy in $CC\ell_4$ is less than that in $CHC\ell_3$. No dibenzyl is formed in the decomposition products of $CHC\ell_3$ and $C_2C\ell_6$, a molecule formed by recombination of $CC\ell_3$ radicals is found only in small amounts. All these facts indicate that $CHC\ell_3$ does not decompose into radicals. Since it cannot decompose directly into molecular products, it must apparently form a diradical. Additional proof of the diradical nature of the pyrolysis of $CHC\ell_3$ is provided by experiments with deuterochloroform. If D replaces H in chloroform, an isotope effect is found, which would be absent if $CHC\ell_3$ decomposed into $C\ell$ and $CHC\ell_2$.* No dibenzyl was found in the decomposition products of $CH_2C\ell_2$. In Table 41 are assembled the main data of Shilov relative to the decomposition of halogenated compounds.

TABLE 41

Activation Energies and Bond Dissociation Energies for Decomposition of Halogenated Hydrocarbons

Compound	Mechanism	E	Q(c - Hal)
C_2H_5Br	molec.	53.7	65
C_2H_3Br	molec.	65.5	70 ?
$C_2H_2Br_2$	rad.	62.8	62.8
$C_3H_5C\ell$	rad.	59.3	60 ± 4
$CC\ell_4$	rad.	55.1	68 - 70
$CHC\ell_3$	dirad.	47.0	73.5 ?
$CH_2C\ell_2$	dirad.	66.5	78.5 ?
$CH_3C\ell$	rad.	85.0	83.0

It must be noted that the radical character of the primary

* In 1954, the pyrolysis of $CHC\ell_3$ was investigated by Semelyuk and Bernshteǐn [88]. In spite of the fact that the main decomposition products were $HC\ell$ and $C_2C\ell_4$ (formed by recombination of $CC\ell_2$ radicals), these authors believed that $CHC\ell_3$ decomposition follows a radical mechanism.

decomposition of CH_3Cl has not been established unequivocally, since the temperature was so high (800°C) that toluene started to decompose at an appreciable rate and the toluene method was not applicable. Conclusions as to the radical decomposition were reached on the basis of the kinetics of the overall reaction. The ratio of the main reaction products: HCl, CH_4 and C_2H_2 was calculated from the postulated radical scheme. The analysis of the products revealed that this ratio is actually found. Moreover, the experimental activation energy, 85 kcal, is practically equal to the C - Cl bond energy in CH_3Cl.

There is some doubt concerning the activation energy of the decomposition of CCl_4 (55.1 kcal). If the decomposition takes place following a radical mechanism, as indicated by the presence of dibenzyl in the decomposition products, the activation energy for decomposition E and the C - Cl bond energy in CCl_4 must be equal. But, in this case, E was about 15 kcal smaller than the value of $Q(C - Cl) = 68 - 70$ kcal [89], as determined by electron impact and also calculated indirectly (Chapter I).

Shilov's experimental data permitted him to draw some preliminary general conclusions on the relation between decomposition mechanism and molecular structure.

Consider for instance the elementary decomposition steps of a series of bromide: ethylbromide, vinylbromide, dibromoethylene, allyl bromide and bromobenzene. The energies required for radical and molecular decomposition are called Q_r and Q_m respectively.

1) $CH_3 - CH_2Br \longrightarrow CH_3 - \dot{C}H_2 + \dot{B}r - Q_r$

1') $CH_3CH_2Br \longrightarrow \begin{array}{ccc} CH_2 & \because & CH_2 \\ \vdots & & \vdots \\ H & \ldots & Br \end{array} \longrightarrow CH_2 = CH_2 + HBr - Q_m$

2) $CH_2 = CHBr \longrightarrow CH_2\dot{C}H + \dot{B}r - Q_r$

2') $CH_2 = CHBr \longrightarrow \begin{array}{ccc} HC & = & CH \\ \vdots & & \vdots \\ H & \ldots & Br \end{array} \longrightarrow HC \equiv CH + HBr - Q_m$

3) $CHBr = CHBr \longrightarrow CHBr = \dot{C}H + \dot{B}r - Q_r$

3') $CHBr = CHBr \longrightarrow \begin{array}{ccc} BrC & \overset{=}{\ldots} & CH \\ \vdots & & \vdots \\ H & \ldots & Br \end{array} \longrightarrow BrC \equiv CH + HBr - Q_m$

4) $CH_2Br - CH = CH_2 \longrightarrow \dot{C}H_2 - CH = CH_2 + \dot{B}r - Q_r$

4') $CH_2Br - CH = CH_2 \longrightarrow H_2C \; \cdots \; C = CH_2 \longrightarrow H_2C = C = CH_2$

$$+ HBr - Q_m$$

$$H \; \cdots \; Br$$

5)

5) HC$\diagup$ $\underset{CH - CH}{\overset{Br \; H}{C = C}}$ $\diagdown$ CH $\longrightarrow$ HC $\diagup$ $\underset{CH - CH}{\overset{C \cdot \doteq \cdot C}{}}$ $\diagdown$ CH $\longrightarrow$ HBr $+$ $\underset{CH}{\overset{C \equiv C}{}}$ $\diagdown$ CH $- Q_m$.

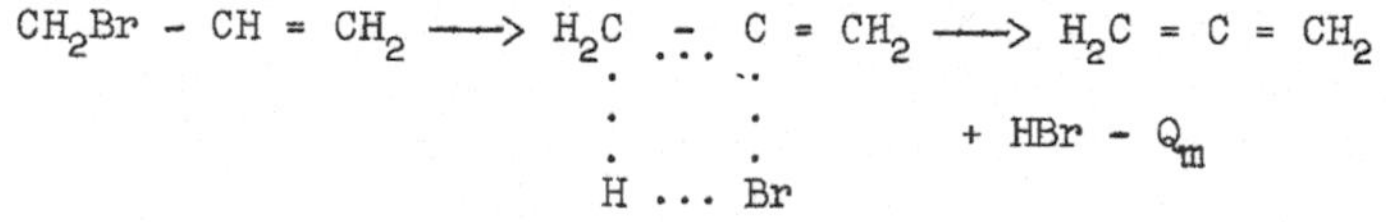

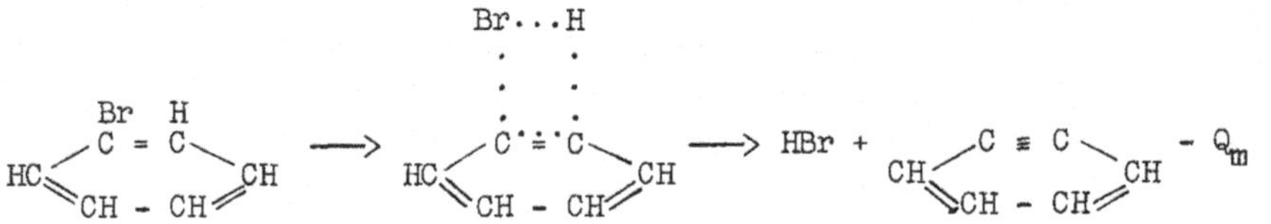

The alkyl halide will be called AHX where X is a halogen atom
and A an olefinic group. As we have seen already, Q_r is equal to the
bond dissociation energy of C - Br in the corresponding bromide. The
quantity Q_m is the difference between the bond energy C - Br in the
bromide molecule (Q(AH - BR)) plus the bond energy of C - H in the radical AH, Q(A - H) with formation of the olefin, and the dissociation energy
of HBr, Q(H - Br):

$$Q_m = Q(AH - Br) + Q(A - H) - Q(H - Br).$$

It is known that splitting of a molecule into radicals requires
practically no activation barrier ϵ_0 while molecular decomposition is
associated with a sizeable activation barrier E_0. Thus, in the case of a
radical decomposition $E_r = Q_r$ while in the molecular case $E_m = E_0 + Q_m$.

Assuming that the pre-exponential factors of molecular and radi-
cal decompositions are identical $((\sim 10^{13} \text{sec}^{-1})$, we get for the ratio of
rates of molecular and radical decompositions:

$$\frac{\exp(- E_m/RT)}{\exp(- E_r/RT)} = \exp -(E_m - E_r)/RT \quad .$$

Thus, if the difference $E_r - E_m$ is positive, the molecular
process predominates while if it is negative, the radical process takes over.

This difference can be expressed more explicitly:

$$E_r - E_m = Q(AH - X) - Q(AH - X) - Q(A - H) + Q(H - X) - E_0 =$$

$$- Q(A - H) + Q(H - X) - E_0.$$

Consider the decomposition of bromides: Q(H - X) = Q(H - Br) =
86 kcal. The values of E_0 change of course from one bromide to another.

The relation between E_0 and molecular structure is still unknown but these changes are not too large, within 25 to 35 kcal. Therefore E_m cannot be calculated exactly. Yet it may be asserted that Q(A - H) exerts a profound, if not decisive, influence on the competition between the two processes. As Q(A - H) becomes larger, the radical mechanism becomes more important.

It is interesting to note that the bond energy Q(AH - X) has no importance in deciding the character of decomposition since it does not enter into the expression for $E_r - E_m$.

A. E. Shilov has tried to estimate the pertinent quantities for the five bromides. The value of Q(A - H) is known for the ethyl radical: $Q(CH_2CH_2 - H) = 38.5$ kcal (see Chapter I). In the case of vinyl bromide, Q(A - H) is not known but may be calculated if it is assumed that it requires 104 kcal to abstract a H atom from ethylene. A thermochemical calculation gives Q(CH = CH - H) = 45 kcal. For dibromoethylene, it is impossible to calculate Q(CBr = CBr - H) since there are no thermochemical data. It appears that Q(CBr = CBr - H) must be larger than Q(A - H) in vinyl bromide since substitution of a bromine atom in the bromide molecule lowers the energy of the π bond in bromoacetylene as compared to acetylene. Thus Q(A - H) must be larger than 45 kcal. For allyl bromide,

$$Q\left(CH_2\overset{H}{C} = CH_2\right)$$

may be obtained from the heat of formation of the allyl radical $\Delta H_{C_3H_5} =$ 30 kcal (see Chapter I). Then Q(A - H) in the allyl radical can be found:

$$CH_2 = CH - CH_2 \longrightarrow CH_2 = C = CH_2 + H - Q$$

$$\Delta H_{C_3H_5} = \Delta H_{C_3H_4} + \Delta H_H - Q \quad .$$

Since $\Delta H_{C_3H_4} = 46$ kcal, $Q = Q(CH_2 - \underset{\displaystyle H}{\overset{|}{C}} - CH_2) = 67.9 \sim 68$ kcal.

In bromobenzene, the bond energy $Q(C_6H_4 - H)$ must even be considerably larger since the decomposition of the phenyl radical must lead to the formation of a compound with a triple bond in the ring. This compound is so energy-rich that it cannot exist.

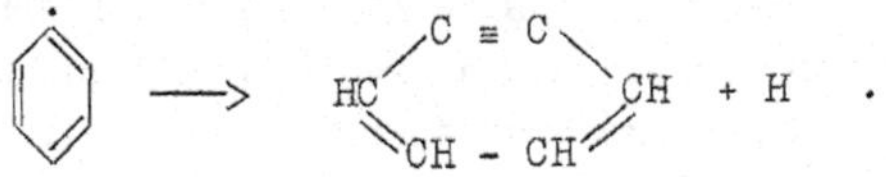

To sum up, in the series: C_2H_5Br, C_2H_3Br, $C_2H_2Br_2$, C_3H_5Br, C_6H_5Br, the energy Q(A - H) must increase and correspondingly, the radical mechanism will become more important than the molecular splitting. Already $C_2H_2Br_2$ decomposes partially via radicals. Allyl bromide and bromobenzene with the highest values of Q(A - H) decompose entirely via radicals.

When Q(H - X) is smaller, the radical mechanism is again favored over the molecular process. Consider, as was done by Shilov, three alkyl halides: $C_2H_5C\ell$, C_2H_5Br and C_2H_5I, with identical values of Q(A - H) = $Q(CH_2CH_2 - H)$. Assume that E_0 will be about the same for all three halides. The values of Q(H - X) are respectively 103, 85 and 69 kcal.

The experiments of Polanyi [22], Szwarc [30] and Shilov [85, 86] show that C_2H_5Br decomposes molecularly and C_2H_5I via radicals. The character of the pyrolysis of $C_2H_5C\ell$ is yet unknown but it appears that it must be molecular.

We have considered so far two types of primary decomposition steps: radical and molecular. It is possible, however, that the decomposition be of the diradical type, although it is energetically always less favorable than the molecular variety since the diradical formed is an energy-rich particle. But in certain cases, a primary molecular step is ruled out.

As was already mentioned, A. E. Shilov has studied the pyrolysis of two compounds of this kind: $CHC\ell_3$ and $CH_2C\ell_2$. He showed that they decompose primarily not into radicals but into diradicals.

Competition between radical and diradical mechanisms is determined by the same factors deciding between molecular and radical mechanisms.

Here $E_r - E_d = -Q(D - H) + Q(H - X) - \epsilon_0'$ where D is a diradical. An exact calculation of this difference is impossible since the activation barrier ϵ_0' is not known for the diradical process, and Q(D - H) is known only in a few cases.

The difference in decomposition mechanism for the halogen derivatives of methane may be illustrated by means of $CC\ell_4$ and $CHC\ell_3$. Let us assume that $CC\ell_4$ also decomposes into diradicals:

1) $CC\ell_4 \longrightarrow C\ell_2 + CC\ell_2 - \Delta H_1$

2) $CHC\ell_3 \longrightarrow CC\ell_2 + HC\ell - \Delta H_2$.

The difference between the heats of reaction can be calculated from thermochemical data: $\Delta H_1 - \Delta H_2 = \Delta H = 24$ kcal. This means that a diradical mechanism for $CHC\ell_3$ has a 24 kcal advantage over a similar mechanism for $CC\ell_4$. Since the bond energy C - Cℓ is less in $CC\ell$ than in $CHC\ell_3$, the radical mechanism is easier for $CC\ell_4$ than for $CHC\ell_3$.

The experimental data concerning the pyrolysis of four halogen derivatives of methane show that for two of them, $CHCl_3$ and CH_2Cl_2, the activation energy for decomposition (47 and 66.5 kcal respectively) is less than the corresponding C - Cl bond energy as calculated by V. V. Voevodskiĭ (73.5 and 78.5 kcal). As a consequence, these two compounds cannot decompose into radicals, as was established experimentally. For $CHCl_3$ and CH_2Cl_2, a diradical split is energetically more favorable.

For two other alkyl halides CH_3Cl and CCl_4, we may calculate what should be the activation energy E^d for diradical decomposition. For CCl_4, assuming that the activation barriers for reactions 1) and 2) are close to each other, we get: $E^d_{CCl_4} = E^d_{CCl_3} + \Delta H = 47 + 24 = 71$ kcal. The C - Cl bond energy in CCl_4 is 68 - 70 kcal; it is less than E^d. Thus, although diradical and radical mechanisms are practically equally probable, the radical mechanism is the one that takes over.

The value of E^d for CH_3Cl may be calculated from the equations:

1) $CH_3Cl \longrightarrow H_2 + CHCl - \Delta H_1$

2) $CH_2Cl_2 \longrightarrow CHCl + HCl - \Delta H_2$.

The difference between the heats of reaction $\Delta H_1 - \Delta H_2 = \Delta H =$ 20 kcal. Then, assuming once more that ϵ_{01} and ϵ_{02} are not too different, we get:

$$E^d_{CH_3Cl} = E^d_{CH_2Cl_2} + 20 = 66.5 + 20 = 86.5 \quad .$$

The bond dissociation energy of C - Cl in CH_3Cl is 83 kcal and the experimental figure for the radical decomposition is 85 kcal. Thus, in this case, the energy required for diradical decomposition is slightly higher than the energy required for a radical split. The latter is also the one that is found experimentally.

To conclude, there is a relation between the character of the primary decomposition step of a molecule and its structure. The investigation discussed above represents an attempt to obtain a qualitative expression of this relation.

§8. Generation of Free Radicals by Reactions Between Saturated Molecules

We have already recalled that theory predicts the possible formation of radicals by reaction between valence-saturated molecules, with small activation barriers. For instance, $H_2 + Cl_2 \longrightarrow H + HCl + Cl - 58$

kcal. There are reasons to believe that when all four atoms are collinear:

$$H - H + C\ell - C\ell \longrightarrow H \ldots H \ldots C\ell \ldots C\ell \longrightarrow H + HC\ell + C\ell \quad ,$$

the activation barrier for this type of reaction is as small as or even smaller than that for reaction between a free radical and a molecule (see Appendix II). Let us assume that this barrier amounts to a few large calories, say 5 kcal, so that the activation energy for the reaction considered is equal to $5 + 58 = 63$ kcal. If this is correct, generation of atoms in the system $H_2 + C\ell_2$ will also be due to this reaction although it is not as fast as the dissociation step: $C\ell_2 + M \longrightarrow C\ell + C\ell + M - 57$ kcal.

The possibility of such a process becomes evident if we consider the reverse step: $H + HC\ell + C\ell \longrightarrow H_2 + C\ell_2$. Indeed, if two valence-unsaturated particles (an H atom and a $C\ell$ atom) approach an $HC\ell$ molecule from both sides, they will tend, of course, to 'attract' one electron each from the $H - C\ell$ bond and their actions will not interfere negatively with each other in the configuration we have chosen. Therefore, if the reaction between an atom and a molecule (e.g., $H + HC\ell \longrightarrow H_2 + C\ell$) is accompanied by a small activation barrier, this will also be the case for the reaction we are now considering (Appendix II). But if this is so, the reverse process: $H - H + C\ell - C\ell \longrightarrow H + HC\ell + C\ell$ will also have a small (or even negligible) activation barrier.

Experimental confirmation of the possibility of reactions between molecules with formation of two free radicals is offered by the disproportionation taking place upon recombination of radicals:

$$C_2H_5 + C_2H_5 \longrightarrow C_2H_4 + C_2H_6 + \sim 60 \text{ kcal} \quad .$$

Such processes apparently occur with very small activation energies. But then, the reverse reaction: $C_2H_4 + C_2H_6 \longrightarrow C_2H_5 + C_2H_5$ also has a small activation barrier and an activation energy $\epsilon \simeq 60$ kcal.

Quite recently, work has been done at the Institute of Chemical Physics that confirms our idea concerning the possibility of forming radicals by reaction between molecules. A. E. Shilov, F. S. D'yachkovskiĭ and N. N. Bubnov have studied the reaction between ethyl lithium and triphenylchloromethane. Using the method of paramagnetic resonance, they have shown the formation of free radicals:

$$LiC_2H_5 + (C_6H_5)_3CC\ell \longrightarrow LiC\ell + \dot{C}_2H_5 + (C_6H_5)_3\dot{C} \quad .$$

At first the quantity of radicals is large, Then, as a result of recombination, their concentration decreases to an equilibrium value. The existence of radicals $(C_6H_5)_3\dot{C}$ in concentrations above equilibrium in the early stages, proves that the radicals are formed in a primary process and are not due to decomposition of hexaphenylethane.

§9. Some Examples

The possibility of free radical formation by reactions between molecules explains a series of facts that have not been understood heretofore.

Oxidation of Hydrogen

The process $H - H + O = O \longrightarrow H + HOO$ is apparently possible. The energy of formation of HO_2 from H and O_2 has been estimated above (47 kcal). The reaction considered is consequently endothermic: the heat of reaction is $47 - 103 = -56$ kcal. This will also be approximately the value of the activation energy $\epsilon \sim 60$ kcal. In the case of hydrogen oxidation, such a way of generating radicals will be incomparably more favorable than the dissociations $H_2 \longrightarrow 2H$ or $O_2 \longrightarrow 2O$ for which the bond energies are 103 and 118 respectively.

It is of interest to point out that in an analysis [90] done in 1943 of our results on the kinetics of the hydrogen-oxygen reaction, we reached the conclusion that the activation energy for the primary act of radical generation was approximately equal to 50 kcal, i.e., close to the figure of 60 kcal that we obtain here.

The process considered may occur even more easily on walls where the adsorption of H atoms makes the formation of HO_2 considerably less endothermic.

Oxidation of Hydrocarbons

It is possible that the relative ease of oxidation of hydrocarbons and other organic compounds (especially in the liquid phase) at temperatures around 100 - 150°C, is due to similar processes of chain initiation: $RH + O = O \longrightarrow R + HO_2$.* Since the abstraction of an H atom from complex hydrocarbons requires considerably less energy than in the case of hydrogen (103 kcal) and can be as low as 80 - 90 kcal, the activation energy of such processes amounts to ca. 40 kcal. With such an activation energy,

* Such a process serves as initiation step in various mechanisms of hydrocarbon oxidation. The first one to use a step of this kind was Hinshelwood [91] in the oxidation of hexane. Yet, in all these cases, the reaction $RH + O_2 \longrightarrow R + HO_2$ was introduced formally, i.e., without an analysis of its activation energy.

a process can proceed at a small rate at temperatures close to room temperature.

Since a peroxide is the primary oxidation product, the chain process will be autocatalytic and only a limited number of primary acts of chain initiation is required before the reaction leaves its induction period and advances at an appreciable rate. A similar mechanism probably explains the initiating effect of traces of oxygen in the addition of HBr to olefins, in polymerization, cracking, etc.

Consider, for instance, the decomposition of acetaldehyde initiated by oxygen, as studied by Niclause and Letort [92]. It was shown that small quantities of oxygen strongly accelerate the reaction which is apparently homogeneous. The authors have proposed the following probable mechanism of the process. Initiation is due to reaction between aldehyde and oxygen:

$$\left.\begin{array}{ll} 0) & CH_3CHO + O_2 \longrightarrow HO_2 + CH_3CO \\ & HO_2 + CH_3CHO \longrightarrow H_2O_2 + CH_3CO \end{array}\right\} \text{chain initiation}$$

$$\left.\begin{array}{ll} 1) & CH_3CO \longrightarrow CH_3 + CO \\ 2) & CH_3 + CH_3CHO \longrightarrow CH_4 + CH_3CO \end{array}\right\} \text{chain propagation}$$

$$\left.\begin{array}{ll} 3) & 2CH_3 \longrightarrow C_2H_6 \\ 4) & CH_3 + CH_3CO \longrightarrow CH_3COCH_3 \end{array}\right\} \text{chain termination}$$

The reaction rate is then given by:

$$w_0 = k_2 \left(\frac{k_0}{k_3} \right)^{1/2} [O_2]_0^{1/2} \frac{[CH_3CHO]^{3/2}}{\left(1 + \frac{k_4 k_2}{k_1 k_3} [CH_3CHO] \right)^{1/2}}$$

where w_0 is the initial rate and $[O_2]_0$ the initial concentration of oxygen (oxygen consumption is not taken into account).

It was shown experimentally that the reaction rate is proportional to the square root of the oxygen concentration and to the 3/2 power of the aldehyde concentration. The overall activation energy E is 22 kcal.

The latter gives a way to calculate E_0:

$$E = E_2 + \frac{E_0}{2} - \frac{E_3}{2} = 22 \text{ kcal} \quad .$$

But E_2 and E_3 are known. They are, respectively 6 kcal and ~ 0 kcal. Hence:

$$E_0 = 2(22 - 6) = 32 \text{ kcal} \quad .$$

The heat of the reaction $O_2 + CH_3CHO$ is $Q_{(CH_3CO-H)} - Q_{(H-O_2)} = 33$ kcal. Therefore, the reaction between oxygen and acetaldehyde proceeds apparently without any activation barrier and its endothermicity is equal to the activation energy.

Reactions of Fluorination

Fluorine is known for its exceptional reactivity. Even at low temperatures (0° and below), it reacts rapidly, often explosively, with many organic compounds. This activity of fluorine as well as the variety of products obtained in the fluorination of halogenated olefins cannot be explained by a direct molecular addition to the double bond, giving molecular products. Many facts indicate that free radicals are intermediates. Radical generation by dissociation of fluorine into atoms is impossible since the dissociation energy of F_2 amounts to 37 kcal. Indeed, around 0°C, less than one fluorine atom is produced per second in this way.

Apparently, the only explanation of fluorine activity, is our proposed reaction in which radicals are formed by reaction between two molecules.

I. $F_2 + RH \text{ (saturated hydrocarbon)} \longrightarrow \dot{F} + HF + \dot{R}$.

Because of the great strength of the HF bond (134 kcal), this reaction must proceed without any expenditure of energy. Thus, with methane, it is practically thermoneutral; with all the other hydrocarbons, it is exothermic:

$$F_2 + CCl_3CHCl_2 \longrightarrow CCl_3 - \dot{C}Cl_2 + HF + \dot{F} + \sim 13 \text{ kcal} \quad .$$

Reaction between F_2 and olefins must be even more exothermic:

II. $F_2 + C_nH_{2n} \longrightarrow \dot{F} + \dot{C}_nH_{2n}F$.

With ethylene, the heat released is about 20 kcal. About the same value obtains with halogenated olefins:

$$F_2 + CCl_2 = CCl_2 \longrightarrow CCl_2F - \dot{C}Cl_2 + \dot{F} + \sim 19 \text{ kcal} \quad .$$

As we have seen, the activation barrier of these processes must be very small (a few kcal). In the majority of cases, a reaction between two molecules to give two radicals is strongly endothermic. With fluorine, these reactions are exothermic and therefore very rapid even at low temperatures.

The activity of fluorine is due to its relatively low dissociation energy (37 kcal) and to the strength of the C - F and H - F bonds that are formed, ~ 115 kcal [93] and 134 kcal respectively. The radicals that are formed in the primary process, react further with the original molecules. The rate of the primary processes I and II is apparently so large that the chain cannot propagate, especially below 0°C (e.g., at - 70°C). Indeed, the activation barrier of reactions I and II is comparable to, and sometimes less than the activation barrier of the chain propagation step.

In 1956, Miller and co-workers [94 - 96] studied the fluorination of halogenated olefins and proposed also steps I and II although he gave no theoretical explanation for the low activation barrier. These remarks suggest that small quantities of fluorine will initiate various hydrocarbon reactions, because of the easy formation of primary radicals.

This effect has in fact been observed by Miller and co-workers [94] in chlorination and oxidation of tetrachloroethylene and pentachloroethane. At low temperatures and in the dark, these reactions do not proceed in the absence of fluorine while they proceed even at low temperatures in the presence of fluorine. The same reactions with chlorine are possible only under illumination which is known to dissociate chlorine into atoms. Schumacher [97] has shown that these reactions, photosensitized by chlorine, are chain processes. There is no doubt that the sole function of fluorine is to provide an alternate path to primary radicals and that the reaction follows the same reaction mechanism as with photosensitization by chlorine. Indeed, in both cases, oxidation of pentachloroethane gives almost equal amounts of CCl_3COCl and $COCl_2$.

Chlorination and Bromination of Olefins

Olefin chlorination in the dark often takes place even at room temperature. Reproducibility is usually then very poor and the rate of reaction depends on the state and nature of the walls of the reaction vessel. Special treatments may reduce the activity of the walls and conditions may be obtained such that the dark reaction will proceed very slowly below 100°C. Schumacher [9] has investigated the photochemical chlorination of ethylene under such conditions. The light dissociates Cl_2 molecules into Cl atoms that start the chain of olefin chlorination:

$$1)\quad \dot{C}l + CH_2 = CH_2 \longrightarrow \dot{C}_2H_4Cl + 26\ \text{kcal}$$

$$2)\quad \dot{C}_2H_4Cl + Cl_2 \longrightarrow C_2H_4Cl_2 + Cl + 17.6\ \text{kcal}\ .$$

Chains are very long, with a length of the order of 10^7. This length is due to the small activation energy of the two exothermic steps written above, through which the chain is propagated.

There are reasons to believe that the dark reaction is also of a chain type, the primary radicals being produced thermally. Homogeneous initiation by means of thermal dissociation ($C\ell_2 \longrightarrow C\ell + C\ell - 57$ kcal) is much too slow and even with a chain length of 10^7, cannot explain the occurrence of ethylene chlorination at 100°C, at a measurable rate in vessels with inactive walls. It seems natural to assume that chain initiation is due to the reaction:

$$C\ell_2 + CH_2 = CH_2 \longrightarrow C\ell + CH_2C\ell - \dot{C}H_2 - q_1$$

where $q_1 = 34.4$ kcal.

As was pointed out above, such a formation of two radicals by interaction of two molecules may have a sufficiently low activation barrier, in this case probably around 6 - 8 kcal. Then the activation energy of this reaction will be equal to:

$$\epsilon = \epsilon_0 + 34.4 \simeq 41 \text{ kcal} \quad ,$$

i.e., about 16 kcal less than the activation energy for dissociation of chlorine molecules into atoms.

This type of initiation is able to account for a chain chlorination of ethylene at 100°C.

Let us calculate q_1, i.e., the energy required for the type of initiation considered. The overall heat of reaction of ethylene chlorination is known:

$$C_2H_4 + C\ell_2 \longrightarrow C_2H_4C\ell_2 + U \quad .$$

$$U = \Delta H_{C_2H_4} + \Delta H_{C\ell_2} - \Delta H_{C_2H_4C\ell_2} = 12.5 + 0 - (-31.1) = 43.6 \text{ kcal}$$

(following Kistiakowsky's data [27]: $U = 43.653$ kcal). The overall addition of chlorine may be divided into two steps:

1) $C\ell_2 + CH_2 = CH_2 \longrightarrow CH_2C\ell - \dot{C}H_2 + C\ell - q_1$

2) $CH_2C\ell - \dot{C}H_2 + C\ell \longrightarrow CH_2C\ell - CH_2C\ell + q_2$.

By means of a cycle we find $-q_1 + q_2 = U$ or $q_1 = q_2 - U$.

The $C - C\ell$ bond energy in step 2) is not known. The $C - C\ell$ bond energy corresponding to the addition of $C\ell$ to the radical $CH_3 - \dot{C}H_2$ is given in Table 6: $Q = 80$ kcal. It is also known that the presence of two $C\ell$ atoms on a single C atom decreases the $C - C\ell$ bond energy by 4 kcal as a result of the presence of the second $C\ell$ atom. In our case, the second $C\ell$ atom is attached to a neighboring C atom. Therefore, $80 > q_2 > 76$ kcal. Let us take $q_2 = 78$ kcal. Then $q_1 = q_2 - U = 78 - 43.6 = 34.4$ kcal.

Between 1930 and 1935, Stewart [98] studied the chlorination of benzene in solution, in the presence of ethylene. He observed the formation of C_6Cl_6 and $C_2H_4Cl_2$. In the absence of ethylene, C_6Cl_6 is formed only when the chlorination of benzene is initiated by light or other factors. In the presence of ethylene, the reaction takes place in the dark.

A similar phenomenon has been observed by E. A. Shilov and I. V. Smirnov-Zamkov [99] in the reaction: $Br_2 + R - C \equiv C - R \longrightarrow R - CBr = CBr - R$. This process, in the dark, exhibits typical radical chain behavior.

In the presence of added toluene, bromination took place with formation of benzyl bromide. As is well known, bromination of toluene in the side chain is initiated by light, i.e., by bromine atoms. Thus, in the system considered here, bromine atoms are also produced in the dark. They may come only from a reaction between bromine and $R - C \equiv C - R$.

In both examples, it may be assumed that halogen atoms are formed following our proposed mechanism:

$$(Hal)_2 + C_2H_4 \longrightarrow Hal + CH_2Hal - CH_2$$

Chlorination of Dienes

Chlorination of dienes, e.g., butadiene, already proceeds at room temperature. The overall reaction is:

$$Cl_2 + CH_2 = CH - CH = CH_2 \longrightarrow CH_2Cl - CH = CH - CH_2Cl \quad .$$

Preferential addition of two Cl atoms in 1,4 positions is direct proof of the chain mechanism of the reaction:

$$1)\quad \dot{C}l + CH_2 = CH - CH = CH_2 \longrightarrow CH_2Cl - CH = CH - \dot{C}H_2$$

$$2)\quad CH_2Cl - CH = CH - \dot{C}H_2 + Cl_2 \longrightarrow CH_2Cl - CH = CH - CH_2Cl + \dot{C}l.$$

If a Cl_2 molecule could add directly to the double bond, Cl atoms ought to be found mostly in 1,2 positions. This takes place to only a slight extent (for bromine addition to butadiene, 1,2 addition represents 20% of the reaction).* Chain initiation is apparently due to the step:

---

* It must be pointed out that 20% of addition in 1,2 positions does not mean at all that direct addition of Cl_2 molecules takes place. It only means that a Cl atom may attack not only a CH_2 group, but also a CH group. The chain is then:

$$1)\quad \dot{C}l + CH_2 = CH - CH = CH_2 \longrightarrow \dot{C}H_2 - CHCl - CH = CH_2$$

$$2)\quad \dot{C}H_2 - CHCl - CH = CH_2 \longrightarrow CH_2Cl - CHCl - CH = CH_2 - \dot{C}l \quad .$$

$$C\ell_2 + CH_2 = CH - CH = CH_2 \longrightarrow C\ell + CH_2C\ell - CH = CH - \dot{C}H_2 - q_1 \quad .$$

The value of q_1, as will be shown presently, is about 18 kcal. If we take $\epsilon_0 \approx 7$ kcal, we get for the activation energy of initiation, the relatively small value $\epsilon \approx 25$ kcal. This explains the ease of chlorination of butadiene at low temperatures.

The value $q_1 \approx 18$ kcal is obtained as follows: Just as we did for ethylene chlorination, let us divide the overall chlorination of butadiene into two steps:

$$C_4H_6 + C\ell_2 \longrightarrow C_4H_6C\ell_2 + U$$

1) $CH_2 = CH - CH = CH_2 + C\ell_2 \longrightarrow CH_2C\ell - CH = CH - \dot{C}H_2 + C\ell - q_1$

2) $CH_2C\ell - CH = CH - \dot{C}H_2 + C\ell \longrightarrow CH_2C\ell - CH = CH - CH_2C\ell + q_2$.

Hence:

$$q_1 = q_2 - U \quad .$$

Within 2 - 3 kcal, q_2 is equal to the energy q_2' of addition of a $C\ell$ atom to the allyl radical: $CH_2 = CH - \dot{C}H_2 + C\ell \longrightarrow CH_2 = CH - CH_2C\ell + q_2'$. Table 6 gives $q_2' = 58$ kcal. Therefore, $q_2 \cong q_2' = 58$ kcal. The overall heat of reaction U is not known but it cannot differ much from the heat of chlorination of olefins, i.e., $U' \approx 40$ kcal, since in both cases a π-bond is destroyed and two $C - C\ell$ bonds are formed. Let us try to calculate U by means of a cycle, subdividing the overall process into four steps:

1) Localize the electrons in a butadiene molecule:

$$CH_2 = CH - CH = CH_2 \longrightarrow (CH_2 = CH - CH = CH_2)^* \quad .$$

The asterisk means that the electrons of the molecule are localized. This process requires about 6 kcal. This figure is the difference between $\Delta H'_{C_4H_6}$ calculated by means of average bond energies and $\Delta H_{C_4H_6}$ obtained from thermochemical data.

2) Form the diradical $(\dot{C}H_2 - CH = CH - \dot{C}H_2)^*$ with localized valence electrons. This requires the destruction of a π-bond, i.e., 57 kcal. Thus: $(CH_2 = CH - CH = CH_2)^* \longrightarrow (\dot{C}H_2 - CH = CH - \dot{C}H_2)^* - 57$ kcal.

3) Dissociate a chlorine molecule:

$$C\ell_2 \longrightarrow C\ell + C\ell - 57 \text{ kcal} \quad .$$

4) Add two $C\ell$ atoms to the diradical:

$$(\dot{C}H_2 - CH = CH - \dot{C}H_2)^* + 2C\ell \longrightarrow CH_2C\ell - CH = CH - CH_2C\ell \quad .$$

Since the dichloride molecule formed has its electrons practically localized, the energy evolved by the step 4) is equal to twice the energy of formation of a $C - C\ell$ bond by addition of a $C\ell$ atom to a radical with localized valence electrons, for instance, C_2H_5. This value is known. It is 80 kcal. Since two $C\ell$ atoms are added in process 4), the energy released is equal to $2 \cdot 80 = 160$ kcal. Then, our cycle gives:

$$U = -6 - 57 - 57 + 160 = 40 \text{ kcal} \quad .$$

The value of q_1 required for initiation of the chain chlorination of butadiene is consequently:

$$q_1 = q_2 - U = 58 - 40 = 18 \text{ kcal} \quad .$$

Polymerization

Dolgoplosk and co-workers [100] have shown that several polymerization reactions at low temperatures (+ 5°) are initiated by addition of two substances: the first one oxidizing and the other one reducing. Dolgoplosk believes that these substances yield free radicals. For example, reaction between hydroperoxide and mercaptans or sulfur dioxide initiates polymerization. It is probable that radicals are formed in the process:

$$1)\quad ROOH + R_1SH \longrightarrow RO + H_2O + R_1S \quad .$$

The energy required can be estimated. The bond dissociation energy in a peroxide of this kind is about 40 kcal. The $H - S$ bond energy in mercaptans is unknown but in H_2S it is ~ 90 kcal. It is likely that, in RSH, the $S - H$ bond energy is somewhat lower, ~ 80 kcal. The $H - OH$ bond energy is 116 kcal. Therefore, reaction 1) is practically thermoneutral or weakly exothermic. Consequently, its activation energy is rather low and it may proceed sufficiently rapidly at 5°C to initiate polymerization.

Of special interest is the mode of initiation of polymerization proposed recently by Ziegler [101] and further extended by Natta [101] and others. The initiator is here a mixture of trialkyl aluminum with a titanium derivative ($TiC\ell_4$ or $TiC\ell_3$). With such initiators, Ziegler succeeded in polymerizing ethylene, propylene and other olefins at much lower temperatures and pressures than formerly.

Here also, one may assume the formation of radicals by reaction between $A\ell R_3$ and $TiC\ell_4$, e.g.:

$$A\ell(C_2H_5)_3 + TiC\ell_4 \longrightarrow A\ell C\ell(C_2H_5)_2 + TiC\ell_3 + C_2H_5 \quad .$$

This reaction may be facilitated by complex formation of the products and adsorption of the radical on the catalyst surface.

However, this polymerization takes place in a different way than in usual radical polymerization. The particular feature of the polymers formed is their stereospecificity, as was shown by Natta. Apparently the reaction is heterogeneous. The mechanisms of such a polymerization and of its initiation are still unknown. Various explanations have been proposed, representing conflicting viewpoints. Many believe that the polymerization follows an ionic mechanism. Yet, the role of radicals in the initiation process cannot be ignored.

Quite recently, A. E. Shilov and N. N. Bubnov [103] at the Institute of Chemical Physics, have used paramagnetic resonance to show surface radicals in the products of reaction between $A\ell(iso - C_4H_9)_3$ and $TiC\ell_4$ or $A\ell(C_2H_5)_3$ and $TiC\ell_3$.

It may be assumed that these radicals take some part in the polymerization process. For example, it is possible that the free valence is not located at the polymer end but binds the polymer chain to the surface.

Some Bimolecular Cyclization Processes

As is well known, dienes readily add to substance having conjugated double bonds with formation of cyclic compounds at relatively low temperatures. Such reactions take place in the liquid phase and in the gas phase. They are bimolecular. In the gas phase where the concentration of reactant molecules is lower than in the liquid phase, these reactions proceed more slowly and are observed at higher temperatures (usually around 150 - 225°C). As examples of reactions of this type, we may quote the dimerization of butadiene:

$$2CH_2 = CH - CH = CH_2 \longrightarrow$$

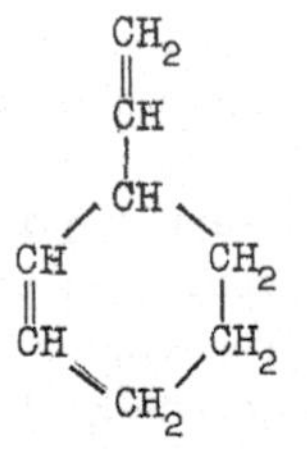

or the addition of acrolein to butadiene:

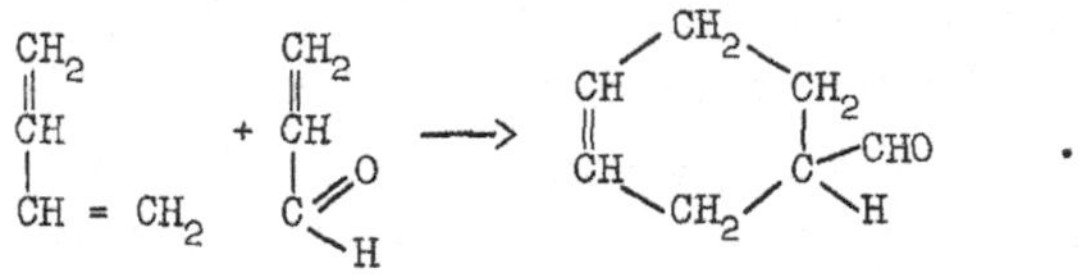

Activation energies are not large, usually of the order of 15 - 25 kcal. Pre-exponential factors are usually smaller ($10^5 - 10^6$) than the value calculated by collision frequencies, i.e., instead of 10^{-10}, values around $10^{-15} - 10^{-16}$ are usually found.

There is some doubt that these processes involve simple direct addition of two molecules.

Maybe the primary process consists in the formation of a diradical similar to $\dot{C}H_2 - CH_2 - CH_2 - \dot{C}H_2$ formed by the interaction of two ethylene molecules. In the case of butadiene, the diradical may be obtained in the following way:

$$CH_2 = CH - CH = CH_2 + CH_2 = CH - CH = CH_2 \longrightarrow$$
$$\dot{C}H_2 - CH = CH - CH_2 - CH_2 - \dot{C}H - CH = CH_2 \quad .$$

To such a process corresponds a decrease of entropy. This leads to a small steric factor $f = 10^{-4} - 10^{-6}$. The heat of reaction is almost zero and therefore the activation energy is $\epsilon = \epsilon_0 + q = \epsilon_0$. The rate of formation of diradicals is given by the expression:

$$w \doteq f \cdot 10^{-10} \exp(-\epsilon_0/RT)[C_4H_6]^2 \quad .$$

The diradical, once formed, can cyclize readily with a small activation energy. The cyclization steps are exothermic and the entire heat of the overall reaction is then liberated. In this fashion, the rate of cyclization is determined by the rate of the primary process of diradical formation. For this reason, dimerization, taking place appreciably faster than polymerization requiring a large activation energy, prevents the formation of polymers. In the presence of several polymerization accelerators such as peroxides, radicals formed by decomposition of the latter react with butadiene giving monoradicals that cannot cyclize. Then polymerization takes place at low temperatures. There are good reasons to believe (Appendix II) that the activation barrier for diradical formation ϵ_0 will be somewhat higher (by a factor of 1.5 or 2) than the activation energy ϵ_0' for reaction between a molecule and a diradical. The latter as we have seen, depends on the heat of reaction q. For thermoneutral reactions ($q \simeq 0$), the activation barrier ϵ_0' is usually of the order of 11 - 14 kcal. Then for thermoneutral processes where two radicals or one diradical are formed from two molecules, we expect an activation energy ϵ_0 of the order of 15 to 22 kcal. This is the value found experimentally for dimerization processes.

It remains to be shown that diradical formation from two butadiene molecules is almost thermoneutral ($q \sim 0$). This reaction involves the disappearance of two π-bonds and the creation of one σ-bond. If there were no resonance energy associated with butadiene or the diradical $\dot{C}_8H_{12}$, the reaction leading to the formation of this diradical would require the same expenditure of energy as the formation of the diradical $\dot{C}H_2 - CH_2 - CH_2 - \dot{C}H_2$ from two ethylene molecules, namely about 30 kcal. But the resonance energy in butadiene amounts to ca. 12 kcal. The resonance energy of the diradical $\dot{C}H_2 - CH = CH - CH_2 - CH_2 - \dot{C}H - CH = CH_2$ is approximately equal to twice the resonance energy of the allyl radicals. This quantity was calculated earlier (see Vol. I) and it is about 20 kcal. Consequently, 40 kcal are evolved by the formation of the diradical.

The heat of reaction of this process is then equal to $- 30 - 12 + 40 \simeq 0$.

Oxidation of Nitric Oxide

Nitric oxide is an inactive radical and the reaction $NO + O_2 \longrightarrow NO_2 + O$ requires a large activation energy since it is strongly endothermic. Here we may also assume that when two NO molecules and one O_2 molecule are collinear, the following exothermic reaction may take place:

$$\dot{N}O + O = O + \dot{N}O \longrightarrow 2NO_2 + 27 \text{ kcal} \quad .$$

Moreover, the activation energy of this process may be small. Experimentally, this reaction is indeed termolecular. To be sure, it has a slightly negative temperature coefficient which our scheme cannot explain.*

What has been said suggests the following question: is it not favorable for reactions to go through a radical mechanism, even when chains are not propagated?

We know that the reaction $H_2 + I_2 \longrightarrow 2HI$ is not a chain process, but does it occur directly by reaction between the molecules? Another reaction path is possible:

---

* It is often postulated that the reaction takes place in two steps: $NO + NO \rightleftarrows (NO)_2$ and $(NO)_2 + O_2 \longrightarrow 2NO_2$.

1) $I_2 + M \longrightarrow I + I + M - 35.5$ kcal

2) $.I + H - H + I \longrightarrow 2HI$

while for the reverse process $2HI \longrightarrow I_2 + H_2$, we may have:

$$I - H + H - I \longrightarrow I + H_2 + I \quad .$$

Consider the rate expression corresponding to this hypothetical mechanism. The concentration of iodine atoms is determined by the formula:

$$\frac{[I]^2}{[I_2]} = 10^{24}\exp(-\ 35500/RT) \quad .$$

Hence: $[I] = 10^{12}\exp(17750/RT)[I_2]^{1/2}$. Triple collisions $I + H_2 + I$ will take place at the frequency:

$$f \cdot 10^{-32}[I]^2[H_2] = f \cdot 10^{-32} \cdot 10^{24}\exp(-\ 35500/RT) \cdot [I_2][H_2] \quad .$$

If the activation energy of step 2) is ϵ_2, then:

$$w = f \cdot 10^{-8}\exp[-\ (35500 + \epsilon_2)/RT] \cdot [I_2][H_2] \quad .$$

The experimental result is:

$$w \simeq 10^{-10}\exp(-\ 39000/RT)[I_2][H_2] \quad .$$

Putting $\epsilon_2 \simeq 4$ kcal and $f = 10^{-2}$, we obtain agreement between experimental and calculated values. The assumed values of ϵ_2 and f are reasonable, since only a small fraction of triple collisions will correspond to an approximately collinear arrangement I ... H ... H ... I. Of course, there are no data to show that this mechanism is correct rather than the universally accepted one where direct collisions between H_2 and I_2 are responsible for HI formation. However, this new viewpoint should not be rejected without doing some specific experiments.

Usually, it is assumed that the molecule dissociates into radicals as a result of the rupture of a bond (the weakest). But another case is possible: two radicals and a molecule may be formed by decomposition, i.e., in a single bond two bonds are broken and one is formed. It is known that azomethane decomposes into two CH_3 radicals and a nitrogen molecule and no chain reaction is observed. The decomposition of azomethane may take place in two ways:

1) $CH_3 - N = N - CH_3 \longrightarrow CH_3 + - N = N - CH_3$

with the $-N=N-CH_3$ radical decomposing in a subsequent reaction act into $N=N$ and CH_3;

2) $CH_3-N=N-CH_3 \longrightarrow CH_3+N=N+CH_3$.

In this case, two CH_3 radicals and a nitrogen molecule are produced in a single reaction act.

The second path appears more likely to us. Recently published experimental data confirm this viewpoint. Indeed, with the first mechanism, the activation energy is equal to the $C-N$ bond dissociation energy, namely, about 53 kcal, whereas the activation energy determined in recent experiments on the decomposition of azomethane is 46 kcal, i.e., 7 kcal less. If decomposition following the second path is actually feasible with a small activation barrier ϵ_0, then the reverse process must also be possible, i.e., the formation of azomethane by simultaneous two-sides attack of a nitrogen molecule by two CH_3 radicals, with formation of a linear complex. This, in last analysis, is perfectly natural, since the activation barrier for the attack of a molecule by a radical is small and there are no reasons to believe that it should be any higher for the attack of a molecule by two radicals approaching it from opposite sides.

One of the most important problems consists in the relative importance of radical and chain reactions on the one hand, and of unimolecular and bimolecular processes on the other hand. We adhere to the view that an important number of reactions proceed via chain and radical mechanisms, and that this type of mechanism, besides the ionic type, represents one of the important kinds of chemical transformations.

REFERENCES

[1] H. J. Schumacher, Chemische Gasreaktionen, Leipzig, 1938.

[2] N. N. Semenov, Usp. Khim., 21, 641 (1952).

[3] M. Letort, J. Chim Phys., 34, 265 (1937); A. Boyer, M. Niclause and M. Letort, J Chim. Phys., 49, 345 (1952).

[4] F. P. Lossing, K. U. Ingold and A. W. Tickner, Disc. Far. Soc., No. 14, 34 (1953).

[5] R. A. Ogg, J. Am. Chem. Soc., 56, 526 (1934).

[6] Ya. B. Zel'dovich, P. Ya. Sadovnikov and D. A. Frank-Kamenetskiĭ, 'Nitrogen Oxidation in Combustion', Moscow, 1947.

[7] R. N. Pease, J. Am. Chem. Soc., 54, 1876 (1932).

[8] E. W. R. Steacie, 'Atomic and Free Radical Reactions', New York, 1946.

[9] H. Schmitz, H. I. Schumacher and A. Jäger, Z. phys. Chem., B 51, 281 (1942).

[10] A. M. Chaĭkin, Thesis, Moscow State University, 1955.

[11] M. L. Bogoyavlenskaya and A. A. Koval'skiĭ, Zhur. Fiz. Khim., 20, 1325 (1946).

[12] F. F. Rust and W. E. Vaughan, J. Org. Chem., 5, 472 (1940).

[13] V. V. Voevodskiĭ and V. A. Poltorak, Dok. Akad. Nauk SSSR, 91, 589 (1953); D. H. R. Barton, Jour. Chem. Soc., 148 (1949).

[14] M. Szwarc and J. W. Taylor, J. Chem. Phys., 23, 2310 (1955).

[15] M. Szwarc, Chem. Rev., 47, 75 (1950).

[16] W. L. Haden and O. K. Rice, J. Chem. Phys., 10, 445 (1942).

[17] J. R. E. Smith and C. N. Hinshelwood, Proc. Roy. Soc., A 180, 237 (1942).

[18] A. Boyer, M. Niclause and M. Letort, J. Chim. Phys., 49, 337 (1952).

[19] C. N. Hinshelwood and P. J. Askey, Proc. Roy. Soc., A 115, 215 (1927).

[20] P. Gray, Fifth Symposium on Combustion, c. 535, 1955.

[21] L. B. Arnold and G. B. Kistiakowsky, J. Chem. Phys., 1, 166 (1933).

[22] E. T. Butler and M. Polanyi, Trans. Far. Soc., 39, 19 (1943).

[23] R. A. Ogg and M. Polanyi, Trans. Far. Soc., 31, 482 (1935).

[24] M. S. Kharash and F. R. Mayo, J. Am. Chem. Soc., 55, 2468, 2521, 2531 (1933).

[25] V. V. Markovnikov, 'On Interactions Between Atoms in Chemical Compounds", Kazan' 1869.

[26] V. V. Markovnikov, Compt. Rend., 81, 668 (1875).

[27] J. B. Conn, G. B. Kistiakowsky and E. A. Smith, J. Am. Chem. Soc., 60, 2764 (1938).

[28] G. B. Kistiakowsky and C. H. Stauffer, J. Am. Chem. Soc., 59, 165 (1937).

[29] A. Maccoll and P. T. Thomas, J. Chem. Phys., 19, 977 (1951).

[30] C. H. Leigh and M. Szwarc, J. Chem. Phys., 20, 403 (1952).

[31] A. T. Blades and G. W. Murphy, J. Am. Chem. Soc., 74, 6219 (1952).

[32] N. N. Semenov, G. B. Sergeev and G. I. Kapralova, Dok. Akad. Nauk SSSR, 105, 301 (1955).

[33] G. B. Sergeev, Thesis, Moscow State University, 1955.

[34] P. Agius and A. Maccoll, J. Chem. Phys., 18, 158 (1950).

[35] A. Maccoll, J. Chem. Soc., 965, 1955.

[36] P. J. Agius and A. Maccoll, J. Chem. Soc., 973, 1955.

[37] A. Maccoll and P. J. Thomas, J. Chem. Soc., 979, 1955.

[38] A. Maccoll and P. J. Thomas, J. Chem. Soc., 2445, 1955.

[39] J. H. S. Green and A. Maccoll, J. Chem. Soc., 2449, 1955.

[40] G. D. Harden and A. Maccoll, J. Chem. Soc., 2454, 1955.

[41] J. H. S. Green, G. D. Harden, A. Maccoll and P. J. Thomas, J. Chem. Phys., 21, 178 (1953).

[42] G. B. Sergeev, Dok. Akad. Nauk SSSR, 106, 299 (1956).

[43] D. H. R. Barton and K. E. Howlett, J. Chem. Soc., 155, (1949).

[44] K. E. Howlett and D. H. R. Barton, Trans. Far. Soc., 45, 735 (1949).

[45] D. H. R. Barton, J. Chem. Soc., 148, (1949).

[46] D. H. R. Barton and P. F. Oryon, J. Am. Chem. Soc., 72, 988 (1950).

[47] R. J. Williams, J. Chem. Soc., 113, (1953).

[48] D. H. R. Barton, A. J. Head and R. J. Williams, J. Chem. Soc., 2033, (1951).

[49] D. H. R. Barton and K. E. Howlett, J. Chem. Soc., 165, (1949).

[50] D. H. R. Barton and A. J. Head, Trans. Far. Soc., 46, 114 (1950).

[51] D. H. R. Barton, A. J. Head and R. J. Williams, J. Chem. Soc., 2039, (1951).

[52] K. E. Howlett, J. Chem. Soc., 4487, (1952).

[53] D. H. R. Barton, Nature 157, 626 (1946).

[54] F. J. Stubbs and C. Hinshelwood, Proc. Roy. Soc., 200, 458 (1949); Disc. Far. Soc., No 10, 129 (1951).

[55] F. O. Rice and K. F. Herzfeld, J. Am. Chem. Soc., 56, 284 (1934).

[56] A. I. Dintses and A. V. Frost, Zhur. Obsh. Khim., 3, 747 (1933).

[57] A. D. Stepukhovich and E. S. Shver, Zhur. Fiz. Khim., 27, 1013 (1953); A. D. Stepukhovich and A. M. Chaikin, Zhur. Fiz. Khim., 27, 1737 (1953); A. D. Stepukhovich and G. P. Vorob'ev, Zhur. Fiz. Khim., 28, 1361 (1954).

[58] V. A. Poltorak, Thesis, Moscow State University, 1952.

[59] V. V. Voevodskii, Thesis, Inst. Chem. Phys., Moscow, 1954.

[60] F. O. Rice and O. L. Polly, J. Chem. Phys., 6, 273 (1938).

[61] V. I. Gol'danskiǐ, Usp. Khim., 15, 63 (1946).

[62] L. A. Wall and W. J. Moore, J. Am. Chem. Soc., 73, 2840 (1951); J. Phys. Coll. Chem., 55, 965 (1951).

[63] F. J. Stubbs, K. U. Ingold, B. C. Spall, C. I. Danby and C. N. Hinshelwood, Proc. Roy. Soc., A 214, 20 (1952).

[64] V. A. Poltorak and V. V. Voevodskiǐ, Dok. Akad. Nauk SSSR, 91, 589 (1953).

[65] F. O. Rice and R. E. Varnerin, J. Am. Chem. Soc., 76, 324 (1954).

[66] C. I. Danby, B. C. Spall, F. I. Stubbs and C. Hinshelwood, Proc. Roy. Soc., A 228, 448 (1955).

[67] H. M. Fray, C. I. Danby and C. Hinshelwood, Proc. Roy. Soc., A 234, 301 (1956).

[68] P. Kebarle and W. A. Bryce, Canad. J. Chem., 35, 576 (1957).

[69] F. O. Rice and M. D. Dooley, J. Am. Chem. Soc., 56, 2747 (1934).

[70] M. Szwarc, J. Chem. Phys., 17, 431 (1949).

[71] M. Szwarc, B. N. Ghosh and A. H. Sehon, J. Chem. Phys., 18, 1142 (1950).

[72] J. Murawski, J. S. Roberts and M. Szwarc, J. Chem. Phys., 19, 698 (1951).

[73] M. Szwarc and A. H. Sehon, J. Chem. Phys., 19, 656 (1951).

[74] M. Szwarc and D. Williams, J. Chem. Phys., 20, 1171 (1952).

[75] M. Ladacki, C. H. Leigh and M. Szwarc, Proc. Roy. Soc., A 214, 273 (1952).

[76] M. Ladacki and M. Szwarc, J. Chem. Phys., 20, 1814 (1952).

[77] M. Szwarc and J. W. Taylor, J. Chem. Phys., 22, 270 (1954).

[78] A. Rembaum and M. Szwarc, J. Am. Chem. Soc., 76, 5975 (1954).

[79] M. Szwarc, Proc. Roy. Soc., A 198, 267 (1949).

[80] M. Szwarc, Proc. Roy. Soc., A 198, 285 (1949).

[81] C. H. Leigh and M. Szwarc, J. Chem. Phys., 20, 407 (1952).

[82] C. H. Leigh and M. Szwarc, J. Chem. Phys., 20, 844 (1952).

[83] A. H. Sehon and M. Szwarc, Proc. Roy. Soc., A 202, 263 (1950).

[84] M. Szwarc and J. Murawski, Trans. Far. Soc., 47, 269 (1951).

[85] A. E. Shilov, Dok. Akad. Nauk SSSR, 98, 601 (1954).

[86] A. E. Shilov, Thesis, Inst. Chem. Phys., Moscow, 1954.

[87] F. P. Lossing, K. U. Ingold and J. H. S. Henderson, J. Chem. Phys., 22, 1489 (1954).

[88] G. P. Sameluk and R. B. Bernstein, J. Am. Chem. Soc., 76, 3793 (1954).

[89] J. B. Farmer, J. H. S. Henderson, F. P. Lossing and D. G. H. Marsden, J. Chem. Phys., 24, 348 (1956).

[90] N. N. Semenov, Acta physicochim. URSS, 20, 291 (1945).

[91] M. S. Nemtsov, M. V. Krauze and E. A. Soskina, Zhur. Obsh. Khim., 5, 343 (1935).

[92] M. V. Krauze, M. S. Nemtsov and E. A. Soskina, Zhur. Obsh. Khim., 5, 356 (1935).

[93] V. I. Vedeneev, Thesis, Moscow, 1957.

[94] W. T. Miller and A. L. Dittman, J.A.C.S., 78, 2793 (1956).

[95] W. T. Miller, S. D. Koch and F. W. McLafferty, J.A.C.S., 78, 4992 (1956).

[96] W. T. Miller and S. D. Koch, J.A.C.S., 79, 3084 (1957).

[97] H. J. Schumacher and W. Thurauf, Zt. phys. Chem., A 189, 183 (1941).

[98] D. D. Stewart and D. M. Smith, J.A.C.S., 52, 2869 (1930); T. D. Stewart and W. Weidenbaum, J.A.C.S., 57, 2035 (1935).

[99] E. A. Shilov and I. V. Smirnov-Zamkov, Izv. Akad. Nauk SSSR, Ot. Khim. Nauk 32 (1951).

[100] B. A. Dolgoplosk, 1948 (Proceedings of V.N.I.I.S.K.); E. I. Tinyakova, B. A. Dolgoplosk and V. N. Reikh, Proceedings of V.N.I.I.S.K. p. 87 (1947-1950), Izv. Akad. Nauk. SSSR, Ot. Khim. Nauk, No. 7 (1957); H. W. Melville, Ind. Rubb. Jour., 67, 1558 (1949).

[101] K. Ziegler, Angewandte Chem., 67, 541 (1955).

[102] G. Natta, Gazz. chim. ital., 87, 528, 549, 570 (1957).

[103] A. E. Shilov and N. I. Bubnov, Izv. Akad. Nauk SSSR, Ot. Khim. Nauk, 1958 (in press).

[104] U. Page, H. Pritchard and A. F. Trotman-Dickenson, J. Chem. Soc., 3878 (1953).

PART IV

BRANCHED CHAIN REACTIONS AND THERMAL EXPLOSIONS

INTRODUCTION

In 1927-28, we introduced the concept of thermal and chain explosions and gave experimental proof of these essentially different two types of explosion. During the following twenty years, these concepts served as a starting point for the development of the theory of combustion and explosions. While the concept of thermal explosion was used only in the field of combustion and explosions, that of branched chain explosions exerted a profound influence on the growth of the science of chemical change. These two types of explosion are also found in nuclear transformations.

Let us first consider the first concept.

The sharp transition between slow reaction and violent explosion is explained as follows. Under certain conditions of temperature and pressure, the reaction rate reaches a critical value for which equality between the heat release due to reaction and heat removal into the surrounding medium becomes impossible. The disruption of the heat balance leads to a progressive self-heating that is observed as an explosion.

In a qualitative way, similar views on explosions had been expressed several times before by scientists of various countries, starting with van't Hoff. But these considerations were superficial, were not presented in a quantitative theory nor confirmed by sure facts. As a result they were quickly forgotten and received no further attention.

Even at the beginning of the century, the ignition temperature was still frequently treated in the literature as an independent quantity characteristic of a given substance. The thermal theory showed that the ignition temperature was not a specific property of a substance but is the result of the heat released by a reaction feeding back on the reaction that generates the heat.

In the studies of Semenov, Todes, Frank-Kemenetskiĭ, Zagulis, Apin and Khariton and also of Rice and co-workers, the thermal theory of explosions received a quantitative formulation. This led to numerous conclusions that could be verified by a series of experiments which in turn could not be interpreted in any other way. The thermal theory of explosion soon held a firm scientific position.

From the laws (mono or bimolecular, antocatalytic) governing the rate of a chemical reaction under non-explosive conditions, it is possible to calculate the temperature at which a substance will explode. This has been done in a variety of cases.

This is how the science of combustion became a chapter of chemical kinetics.

In 1927-1929, the concept of chain explosions was also introduced. We observed at that time that for oxygen pressures below some minimum value, phosphorus vapors are not able to burn in oxygen. Data of Khariton and Val't (1926) and later of Semenov and Shal'nikov (1927) showed that in such a case even traces of a reaction between phosphorus and oxygen cannot be detected. Phosphorus vapors burn only when the oxygen pressure exceeds this limit and the reaction soon dies down since the oxygen pressure because of the oxygen consumption, falls to the limit or below.

It was immediately obvious that here was a case of combustion phenomenon that had nothing in common with a thermal explosion. Indeed, at these low pressures (a few hundredths of a millimeter) at which phosphorus burns, the system cannot warm up appreciably. The isothermal character of this combustion was subsequently proved by special experiments and calculations.

In the thirties, when the significance of this discovery became evident, an earlier French investigation (Joubert 1874) was recalled. Joubert made similar observations on phosphorus vapor. His experiments, however, were so poorly defined and his conclusions so unconvincing that the phenomenon itself fell into complete oblivion. So much so, that after publication of the work of Khariton and Val't, Bodenstein, the foremost authority in chemical kinetics who introduced the concept of unbranched chain reactions, expressed his doubts on the validity of these experiments. He argued that our paradoxical results on the onset of a fast exothermic reaction because of an infinitesimal change in an external parameter (the pressure in this case) and the formulation of a reaction limit were in contradiction with the science of equilibria based on incorrect arguments of Duhem.

After this, we studied the phenomenon in more detail and showed that there were no errors of observation in our previous work as suggested by Bodenstein. We also used different methods, repeated and investigated with great care the amazing fact discovered in 1926 by Khariton and Val't in which dilution of oxygen with an inert gas (argon) was found to decrease the pressure limit of oxygen, thus increasing the reactivity of the system.

Simultaneously, we discovered another strange feature of the reaction — the lowering of the limit by an increase in vessel size.

Attempting to explain all these facts that seemed unique and paradoxical at the time, we formulated the theory of thermal explosion and initiated the concept of branched chain reactions. From the very start, the theory attracted a great deal of attention among physical chemists and a wave of investigations of unbranched and branched chain reactions appeared in all countries. Foremost were the studies of the Soviet and English schools (Hinshelwood) who discovered the upper limit of chain explosion in hydrogen-oxygen mixtures.

These studies showed that chain explosion occurs in many chemical systems. The phenomena of an upper limit of oxygen pressure in combustion of phosphorus and phosphine vapors had remained unexplained the first during 300 years, the second for over a century. They received a simple and natural explanation in the chain theory. Besides, many different cases of upper limit were discovered in many chemical systems. Attempts of Haber to find an alternative explanation of these phenomena proved fruitless.

This is how the concept of a reaction limit was introduced in science: a narrow range of variation of an external parameter inside which transition takes place between a practical inert state of a substance* and a state of violent reaction. Such parameters are pressure, density, temperature, vessel size, dilution by an inert gas, addition of some active substances. These limit phenomena have been duplicated perfectly in nuclear branched chain reactions where they make possible the practical utilization of atomic energy. All these considerations have had great significance in the kinetics of chemical change.

Among the relatively slow chemical processes, one must mention especially the reactions with degenerate branching. Their distinctive feature is that their primary products are intermediate molecular species that enter into the reaction more or less rapidly by forming free radicals. The existence of such processes was first proposed by us in the thirties and soon confirmed experimentally. Hydrocarbon oxidations proved to be reactions with degenerate branching. The theory of 'degenerate explosions' developed between 1931-1934 has been widely used since.

Further studies have revealed that a very large number of reactions proceed via a chain mechanism. At the same time, more and more attention has been devoted to the elucidation of the mechanism of chain reactions.

How can one explain the wide occurrence of chain reactions? The

---

* A thermal explosion must be preceded by a slow but perfectly measurable reaction. A chain explosion takes place under conditions of immeasurably slow reaction. In a thermal explosion, the heat liberated by the reaction is the cause of the explosion. In a chain explosion, the evolution of heat is a consequence of the growth of the chain avalanche.

answer lies in general direct principles of chemistry.

First of all, free radicals are very active species and therefore they react much more easily with valence saturated molecules than the molecules themselves (we have here in mind homolytic reactions).

Then, reaction between a monoradical and a molecule cannot destroy a free valence since at least one product of that reaction must be a free radical. This radical reacts now with another molecule, produces a new free radical. A chain of transformations is thereby propagated.

Three basic cases are possible:

1. A molecule reacts with a radical to produce a monoradical that propagates a straight chain, e.g.:

$$\dot{C}H_3 + CH_3CHO \longrightarrow CH_4 + CH_3\dot{C}O$$

$$CH_3\dot{C}O \longrightarrow \dot{C}H_3 + CO \quad \text{etc. ...}$$

2. The reaction produces three free valences which ultimately yield three monoradicals, each one starting its own chain. This leads to a very fast propagation of a branched chain reaction, e.g.:

$$\dot{H} + O_2 \longrightarrow \dot{O}H + \dot{\underset{\cdot}{O}}$$

$$\dot{\underset{\cdot}{O}} + H_2 \longrightarrow \dot{O}H + \dot{H}$$

$$\dot{O}H + H_2 \longrightarrow H_2O + \dot{H} \quad \text{etc. ...}$$

Overall: $$\dot{H} + O_2 + 3H_2 \longrightarrow 2H_2O + 3\dot{H}$$

3. The main reaction chain is unbranched but a reaction product because of unimolecular decomposition or interaction with any other component of the system, forms free radicals, and therefore new chains. Then, even though the reaction may go slowly, it has many features in common with branched chain processes (autoacceleration, limit phenomena). This is degenerate branching. For instance, alkyl hydroperoxides and aldehydes are formed as follows:

I

$$\dot{R} + O_2 \longrightarrow R\dot{O}_2$$

$$R\dot{O}_2 + RH \longrightarrow ROOH + \dot{R}$$

etc.

II

$$\dot{R} + O_2 \longrightarrow R\dot{O}_2$$

$$R\dot{O}_2 \longrightarrow R'C{\lessgtr}^{O}_{H} + R''\dot{O}$$

$$R''\dot{O} + RH \longrightarrow R''OH + \dot{R}$$

etc.

The hydroperoxide with a weak O - O bond (~ 40 kcal) slowly decomposes into free radicals:

$$ROOH \longrightarrow R\dot{O} + \dot{O}H \quad .$$

The aldehyde oxidizes and generates free radicals:

$$R'C{\lessgtr}^{O}_{H} + O_2 \longrightarrow R'\dot{C}O + H\dot{O}_2$$

and the number of chains slowly increases with time as in autocatalytic processes.

In spite of the many types of chemical mechanisms of chain reactions, they may be all classified among one of these three types. Many examples have been treated in the preceding chapters. More will be handled in this part of the book.

CHAPTER VIII: THERMAL EXPLOSIONS (1)*

If a reaction takes place in the gas phase with a velocity w (number of molecules of product made per second per unit volumte), the quantity of heat released per second in the entire volume v is equal to:

$$q_1 = vQ'w \quad .$$

Here Q' is the heat released per elementary reaction act, equal to $Q' = (Q/N)$ where Q is the heat of reaction per gram-mole of product formed and N is the Avogadro number $(N = 6 \cdot 10^{23})$. At the beginning of reaction when the consumption of initial reactants may be neglected, the rate of reaction, as a function of temperature T and number of molecules of reactants per unit volume a, is given by $w = k_1 a \exp(-E/RT)$ for unimolecular reactions, or $w = k_2 a^2 \exp(-E/RT)$ for bimolecular reactions. Thus:

$$q_1 = \frac{vQka^n \exp(-E/RT)}{N} \tag{1}$$

where $n = 1$ or 2 for unimolecular or bimolecular processes respectively.

The quantity of heat transferred from the reaction space to the reactor walls is:

$$q_2 = \kappa(T - T_0)S \tag{2}$$

where κ is the heat transfer coefficient, T the temperature of the reacting gas, T_0 the temperature of the walls determined from the outside and S is the surface of the vessel.

On Figures 6 and 7, on the following page, q_1 and q_2 are represented as a function of temperature.

Figure 6 shows the case where the wall temperature is maintained at temperature T_0 but the pressure of reacting gases i.e., the value of

* The entire theoretical part is taken from Semenov [2], Todes [3] and Frank-Kamenetskii [4].

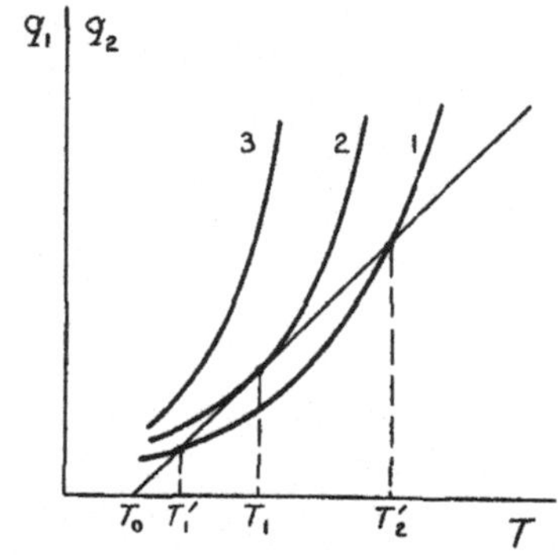

FIGURE 6. Relation between q_1 and q_2, and T.

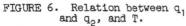

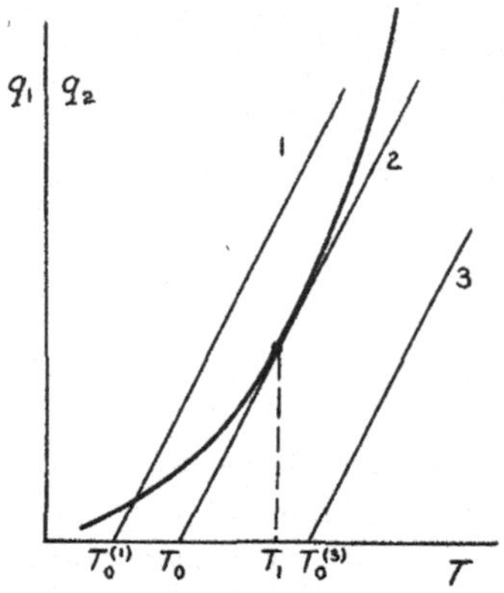

FIGURE 7. Relation between q_1 and q_2, and T.

a in equation (1) is varied. Curve 1 corresponds to the smallest value of a, namely a_1; curve 2 to an intermediate value a_2 and curve 3 to the largest value a_3.

For $a = a_1$, the heat supply q_1 is at the start larger than the heat removal q_2. Therefore, the gas will become warmer than the vessel walls. This continues until the gas temperature reaches some value T_1' (intersection of q_1 and q_2). Then $q_1 = q_2$. The gas will not become warmer since for $T > T_1'$, the heat removal q_2 becomes larger than the heat supply q_1 and if the gas, for some reason, would reach a temperature higher than T_1', it would be brought back down to that temperature.*

Thus, in the case just considered, the reaction does not give auto-ignition and the gas only reaches a temperature T_1' slightly higher than the wall temperature.

If now, with the same wall temperature T_0, the vessel is filled with reactive gas at a sufficiently high pressure such that $a = a_3$, q_1 varies with temperature as shown on curve 3 which does not intersect anywhere the straight line q_2 of heat removal. In this case, the heat supply

* It is only if the gas were heated artificially (say, by adiabatic compression) above the second point of intersection T_2' that q_1 would become larger than q_2. Then the gas would become hotter all by itself and explosion would occur. This second point of intersection does not give a stable temperature level since if T becomes slightly smaller than T_2', the temperature falls back to T_1'. If T becomes slightly larger than T_2', explosion takes place. Therefore, this second point of intersection has no significance in a theory of self-ignition but presents interest only in problems of artificial ignition where the gas is heated for instance by adiabatic compression.

q_1 is larger, at all temperatures, than the heat removal q_2 and thus the gas will become continuously warmer, the reaction faster and faster, and explosion takes place. Thus, for $a = a_3$, thermal auto-ignition occurs.

Curve 2 for $a = a_2$ is tangent to the line of heat removal at one point and therefore defines two domains, one where the reaction proceeds in a steady state and the other where auto-ignition takes place. The quantity $a = a_2$ or the pressure corresponding thereto determines the critical pressure for auto-ignition at a given vessel temperature T_0.

If, on the other hand, the gas pressure is kept constant but the reactor wall temperature is varied $T_0^{(1)} < T_0 < T_0^{(3)}$, the state of the gas is described by Figure 7. Here there is one curve of heat supply q_1 and three lines of heat removal corresponding to three wall temperatures. By a reasoning similar to that above, it can be shown that for $T_0^{(1)} < T_0$ auto-ignition does not occur; for $T_0^{(3)} > T_0$, it takes place. The temperature T_0 at which the curve q_1 is tangent at one point to the line q_2 is the lowest temperature at which auto-ignition takes place at a given pressure p. It is the ignition temperature, for short. To the point at which the curves q_2 and q_1 are tangent to each other corresponds a temperature T_1 and the difference $\Delta T = T_1 - T_0$ will be called the preheat (before ignition). Between the ignition temperature and the pressure (or the quantity a) there exists an analytical relation since at the point of contact $q_1 = q_2$ and

$$\frac{dq_1}{dT} = \frac{dq_2}{dT} :$$

$$\frac{1}{N} vQka^n \exp(- E/RT) = \kappa(T_1 - T_1)S$$

$$\frac{vQka^n E \;\; \exp(- E/RT)}{NRT_1^2} = \kappa S \quad . \qquad (3)$$

From these two relations, the temperature T_1 at the point of contact can first be found as a function of the wall temperature.

Eliminating κS between both equations it is found that:

$$j = \frac{E}{RT_1^2} = (T_1 - T_0) \quad \text{or} \quad \frac{RT_1^2}{E} - T_1 + T_0 = 0$$

$$T_1 = \frac{1 \pm \sqrt{1 - (4RT_0/E)}}{2(R/E)} \quad .$$

But for the majority of cases of interest, (RT_0/E) is a small quantity, usually not exceeding 0.05. The ignition temperature is commonly below 1000°K and the activation energy usually exceeds 20000 cal. Furthermore, to a small activation energy E corresponds a low ignition temperature and vice versa. The solution with the plus sign must be rejected, since it gives T_1 equal to (R/E) which is about 10000° or higher.* On the other hand, the solution with the minus sign corresponds to the point of contact represented on Figures 6 and 7, at much lower temperature T_1. Thus:

$$T_1 = \frac{1 - \sqrt{1 - (4RT_0/E)}}{2(R/E)} = \frac{2(RT_0/E) + 2(RT_0/E)^2 + 4(RT_0/E)^3 + \ldots}{2(R/E)} .$$

With $(RT_0/E) < 0.05$, the higher terms in the expansion may be neglected, starting with $4(RT_0/E)^3$. The error made is then about:

$$2(RT_0/E)^2 \cdot 100\% < 0.0025 \cdot 2 \cdot 100\% = 0.5\%$$

of the measured quantity T_1. Then:

$$T_1 = T_0 + (RT_0^2/E) \quad . \tag{4}$$

The preheat is:

$$\Delta T = T_1 - T_0 = RT_0^2/E \quad . \tag{5}$$

The heating ΔT that takes place when no ignition occurs (when the temperature T_0 is below the ignition temperature) will always be less than $\Delta T_1 = (RT_0^2/E)$.

Then also, when the preheat ΔT is less than RT_0^2/E no thermal explosion is possible and conversely, if $\Delta T > (RT_0^2/E)$, thermal explosion must occur. The preheat preceding the explosion depends on both T_0 and the activation energy E but will not exceed a few tens of degrees in cases of interest here. For instance, if $T_0 = 700°K$ and $E = 30000$ cal, $\Delta T_1 = 33°$. If $T_0 = 700°K$ and $E = 60000$ cal, $\Delta T_1 \approx 16°$. Therefore, the ratio $(\Delta T_1/T_0) \approx (RT_0/E)$ is always small (a few hundreds). In what follows, the following approximation can be used:

$$\frac{1}{T_0 + \Delta T_1} = \frac{1}{T_0}\left(1 - \frac{\Delta T_1}{T_0}\right)$$

---

* Both signs occur because the function exp(- E/RT) has an inflection point at very high temperatures and tends to unity when $T \longrightarrow \infty$. For this reason, the heat removal curve extrapolated to ~ 10000° would intersect the curve q_1 once more (this is not represented on Figure 6).

Substituting the value of T_1 into equation (3), we get the condition for auto-ignition:

$$\frac{Qvka^nE}{NRT_o^2}\left(1 - 2\frac{\Delta T_1}{T_o}\right)\exp\left[-E/RT_o\left(1 - \frac{\Delta T_1}{T_o}\right)\right] = \kappa S$$

or putting

$$\frac{\Delta T_1}{T_o} = \frac{RT_o}{E}$$

and neglecting

$$2\frac{\Delta T_1}{T_o}$$

on the left hand side as compared to unity (this means an error of less than 10% for a, we have also:

$$\frac{Qvka^nEe}{NRT_o^2\kappa S}\exp(-E/RT_o) = 1 \quad . \tag{6}$$

The number of molecules per unit volume is related to the pressure expressed in mm Hg:

$$p = aT \cdot 10^{19} \qquad\qquad a = \frac{p}{T}10^{19} \quad . \tag{7}$$

Substituting into (6), we get a relation between gas pressure and auto-ignition temperature:

$$\frac{Qvkp^nEe \cdot 10^{19}}{NRT_o^{2+n}\kappa S}\exp(-E/RT_o) = 1$$

$$\log\frac{p}{T^{1+\frac{2}{n}}} = \frac{A}{T_o} + B \quad \text{where} \quad A = \frac{0.217E}{n}$$

$$B = \frac{1}{n}\log b \qquad\qquad b = \frac{NR\kappa S}{QvkeE \cdot 10^{19}} \quad . \tag{8}$$

For $n = 1$,

$$\log\frac{p}{T^3} = \frac{A}{T_o} + B \quad \text{where} \quad A = 0.217E \tag{9}$$

For $n = 2$,

$$\log \frac{p}{T^2} = \frac{A}{T_0} + B \quad \text{where} \quad A = 0.11E \quad . \tag{10}$$

When a bimolecular reaction takes place between two components of the mixture, all formulae are still valid provided that the quantity a is multiplied by the product $\gamma(1 - \gamma)$ where γ is the fraction of the first component and $(1 - \gamma)$ that of the second. Moreover, the thermal conductivity will change with mixture composition.

Consider a few examples to verify these equations. Figure 8 shows data of Zagulin [5] on the decomposition of $C\ell_2O$. The quantity

$$\log \frac{p}{T}$$

where p is the pressure at which auto-ignition takes place is plotted versus the inverse of absolute temperature. Zagulin used

$$\log \frac{p}{T} \quad \text{and not} \quad \log \frac{p}{T^2}$$

along the ordinate axis. However it is easy to check that within the temperature range investigated, a plot of $\log (p/T^2)$ versus $(1/T)$ is also a straight line approximately parallel to Zagulin's line. Zagulin found for A a value of 2500. Since the reaction is bimolecular, formula (10) gives $A = 0.11E$ and $E = 22000$ cal. According to Hinshelwood's kinetic data [6], $E = 21000 - 22000$ cal. Theory and experiment are in good agreement.

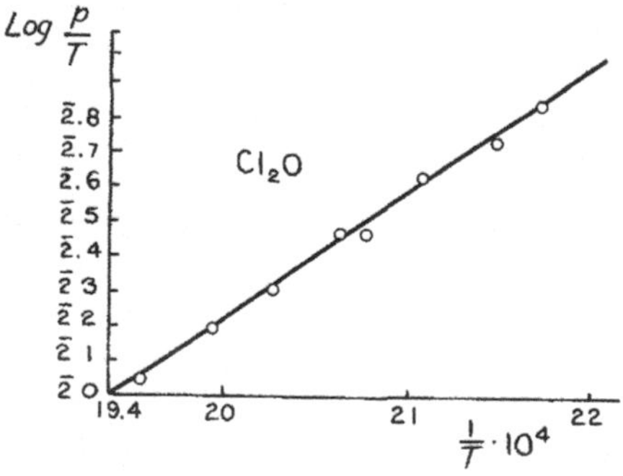

FIGURE 8. Relation between $\log \frac{p}{T}$ and $\frac{1}{T}$ for decomposition of $C\ell_2O$ (5).

Figure 9 shows a plot of

$$\log \frac{p}{T^3}$$

versus $1/T$ for the auto-ignition of azomethane following Rice and Allen [7]. Curve 1 is for pure azomethane, curve 2 for azomethane diluted with helium (76% He in the mixture).

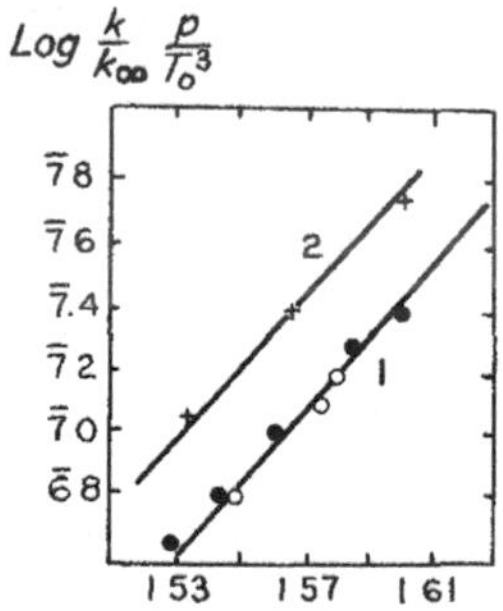

FIGURE 9. Relation between $\log \frac{kp}{k_\infty T_0^{\,3}}$ and $\frac{1}{T}$ for decomposition of azomethane.
Curve 1: azomethane
Curve 2: azomethane + 76% He.

Azomethane decomposes unimolecularly but as pressure increases, the rate constant increases somewhat and tends to a limit k_∞ for sufficiently high pressures. It is therefore necessary to multiply the expression for the critical pressure p by k/k_∞ and to plot not $\log(p/T^3)$ but $\log(kp/k_\infty T^2)$, as was done in Figure 9. The black points on this figure correspond to experiments with pure azomethane, the crosses to a mixture of azomethane and helium (76% helium) and the circles to a mixture with nitrogen (50%). The constant A is 11000. From $E = A/0.217$, the calculated value of E is then 50000 cal, the same figure given by Ramsperger who obtained his value from direct kinetic data.

As shown above, the condition for auto-ignition — equation (6) — contains parameters determined by the reaction rate (the rate constant k, the activation energy E) and heat quantities (the heat of reaction Q, the heat transfer coefficient κ, the vessel size). The condition for auto-ignition may be written in a more simple form if the parameters are properly grouped and related to macroscopic quantities easily determined experimentally. This has been done by Todes [3, 9]. As a parameter characteristic of the heat removal, let us select the thermal relaxation time, i.e., the time t_e during which the excess temperature $\Delta T = T - T_0$ of a heated but non-reacting gas decreases e-fold. This time t_e naturally does not depend on either ΔT or T_0 since κ does not depend on these quantities.

Since

$$\frac{vac}{N} \cdot \frac{dT}{dt} = - \kappa S(T - T_0)$$

where c is the heat capacity, per gram-mole,

$$\ell n \cdot \frac{T - T_0}{T_1 - T_0} = - \frac{\kappa SNt}{cav}$$

where T_1 is the initial temperature of the gas. Thus:

$$T_1 - T_0 = (T_1 - T_0) \exp\left(- \frac{\kappa SNT}{cav}\right)$$

and

$$t_e = \frac{cav}{\kappa SN} \quad . \tag{11}$$

The reaction rate in turn can be characterized by its inverse, namely the reaction time t_r which is by definition the time required for the reactants to be consumed by the reaction if the reaction proceeded at a constant rate corresponding to the initial concentrations:

$$t_r = \frac{a}{ka^n \exp(-E/RT)} \quad . \tag{12}$$

The time t_r may be measured readily, by observing the initial reaction rate at a temperature in the vicinity of the ignition temperature and extrapolating the rate to the temperature of auto-ignition. The time t_r is then obtained by dividing the total number of molecules of reactants per unit volume by the measured initial reaction rate expressed in number of molecules of products per second.

It is easily shown that equation (6) now becomes:

$$\frac{t_e}{t_r} \cdot \frac{QEe}{cRT_o^2} = 1$$

and the condition for auto-ignition can be written as:.

$$\frac{t_r}{t_e} = \frac{QEe}{cRT_o^2} \quad . \tag{13}$$

If

$$\frac{t_r}{t_e} > \frac{QEe}{CRT_o^2} \ ,$$

thermal explosion is impossible and the reaction proceeds steadily. If

$$\frac{t_r}{t_e} < \frac{QEe}{cRT_o^2} \ ,$$

the gas always explodes thermally. However, if explosion takes place when

$$\frac{t_r}{t_e} > \frac{QEe}{cRT_o^2} \ ,$$

this is a sure sign that the explosion is not thermal but is due to a chain mechanism.

Consider now in more detail the problem of heat transfer from the gas to the reactor wall.

If we write $q_2 = \kappa(T - T_o)S$, we assume that heat transfer takes place by convection. In such a case, it is very difficult to calculate κ. However there are many situations where the transfer of heat is practically due to conduction only: when the gas pressure is below atmospheric, the vessel size is small and the preheat before explosion ΔT_1 is not large. Frank-Kamenetskiǐ [4] was the first to analyze the situation and to develop a theory of thermal auto-ignition for the case of conductive — and not convective heat transfer.

Let us note already that, as will be shown below, this does not change the form of the critical conditions. Indeed, the expression $q_2 = \kappa(T - T_0)S$ is still valid but with a different meaning of the coefficient κ which depends on thermal conductivity, the size and shape of the reactor.

The solution of the problem by Frank-Kamenetskiǐ is as follows. Below the explosion limit, the reaction may proceed steadily. Then the distribution of temperatures throughout the reactor space is the solution of the equation for heat conductivity with distributed heat sources. The natural unit of temperature for this problem is the quantity $(RT_0^2/E) = \Delta T$. With the dimensionless temperature:

$$\theta = \frac{E}{RT_0^2}(T - T_0) \quad ,$$

the heat conductivity equation can be written as:

$$\Delta\theta = \delta e^{\theta}$$

where Δ is the Laplace operator and δ a dimensionless parameter:

$$\delta = \frac{Q}{\lambda N}\frac{E}{RT_0^2} r^2 k a^n \exp(-E/RT_0) \tag{14}$$

where λ is the coefficient of heat conductivity, r the radius of the reactor (if the reactor is plane, $2r$ is the distance between its walls).

With the boundary conditions, θ goes to zero at the wall surface. The equation has solutions satisfying the boundary conditions only when δ takes values not exceeding a critical value that depends on the geometrical shape of the reactor. This critical value δ_{cr} also determines the critical conditions for explosion. It has been determined for various simple geometrical configurations, either by analytical [10] or numerical integration.

For a plane reactor

$$\delta_{cr} = 0.88 \tag{15}$$

and consequently the condition for ignition becomes:

$$\frac{QEr^2}{\lambda NRT_0^2} k a^n \exp(-E/RT_0) = 0.88 \quad .$$

This expression coincides in form with equation (6) if one replaces $(\kappa S/v)$ by (λ/r^2). The value of the numerical coefficient takes into account the distribution of temperatures throughout the reactor and its shape. For cylindrical and spherical reactors, Frank-Kamenetskiĭ has found, respectively:

$$\delta_{cr} = 2.00 \quad (16)$$

and

$$\delta_{cr} = 3.32 \quad . \quad (17)$$

Frank-Kamenetskiĭ [4] has calculated the auto-ignition temperature of azomethane at various pressures, starting from the kinetic data, the heat of reaction and the size of the reactors used. Taking $\lambda = 10^{-4} g^{-1} sec^{-1} cm^{-1}$, he compared the calculated value [equation (17)] with the observed values of Rice and Allen [7]. He repeated this for the decomposition of methylnitrate (another unimolecular reaction) studied by Apin and Khariton [11]. The results are shown in Table 42.

As can be seen, the agreement between theory and experiment is not bad at all, indicating the validity of Frank-Kamenetskiĭ's assumption concerning the conductive nature of the heat transfer.

Using data of Volmer [12] on the kinetics of decomposition of nitrous oxide, Frank-Kamenetskiĭ predicted that this gas could exhibit auto-ignition but at very high temperatures.

Zel'dovich and Yakovlev [13] verified this prediction and a comparison between predicted and observed values of the auto-ignition temperature is shown in Table 42. The agreement is good.

TABLE 42

Decomposition of azomethane $(CH_3)_2N_2 \longrightarrow C_2H_6 + N_2$ date of Rice [7]			Decomposition of methylnitrate $2CH_3ONO_2 \longrightarrow CH_3OH + CH_2O + 2NO_2$ data of Apin and Khariton [11]			Decomposition of nitrous oxide N_2O data of Zel'dovich and Yakovlev		
p mm	$T^o_{calcd.}$ °K	$T^o_{observed}$ °K	p mm	$T^o_{calcd.}$ °K	$T^o_{observed}$ °K	p mm	$T^o_{calcd.}$ °K	$T^o_{obs.}$ °K
191	619	614	4.2	590	597	170	1255	1285
102	629	620	8.5	578	567	330	1175	1195
67	635	626	16.5	566	546	590	1110	1100
55	638	630	45.4	551	529			
38	644	636	87	541	522			
31	647	643	107	538	521			
23.5	653	651	163	531	519			
18	656	659						

Consider now the problem of the absolute values of ignition temperatures by means of an example, that of the decomposition of $C\ell_2O$.

This reaction is not simply bimolecular but its rate can be closely approximated by an autocatalytic expression:

$$w = kx(a - x)\exp(-E/RT)$$

where a is the initial number of molecules and x is the number of reacted molecules. The reaction reaches a maximum rate when $x = a/2$. Then:

$$w_{max} = k\frac{a^2}{4}\exp(-E/RT) \quad .$$

Therefore the smallest temperature of auto-ignition is determined not by the initial rate of the process but its maximum value.

As we have seen, the maximum rate changes bimolecularly. This agrees with the results of Hinshelwood [6]. According to the latter, the activation energy E is equal to 22000 cal and the constant k has the value characteristic of a bimolecular process, namely $k = \sqrt{2}\pi\sigma^2 u$.

Following Hinshelwood, $\sigma = 4.8 . 10^{-8}$ and $k \sim 10^{-10}$. This value of k must be substituted in equation (16). Instead of a, we take $a/2$ and $n = 2$. The heat of reaction is 22000 cal and λ may be taken approximately as $5 . 10^{-5}$.

Zagulin, in his work on $C\ell_2O$ ignition used a cylindrical vessel with $r \sim 1$ cm.

Equation (16) for a cylindrical vessel gives:

$$\delta_{cr} = \frac{QEr^2ka^n\exp(-E/RT_0)}{RT_0^2\lambda N} = 2 \quad . \tag{18}$$

In the case of $C\ell_2O$, this becomes:

$$\delta_{cr} = \frac{2.2 \cdot 10^4 \cdot 2.2 \cdot 10^4 \cdot 10^{-10} \cdot a^2 \exp(-22000/2 \cdot T_0)}{4 \cdot T_0^2 \cdot 5 \cdot 10^{-5} \cdot 6 \cdot 10^{23}} = 2 \quad .$$

Since we have already shown that the temperature dependence of the critical ignition pressure of $C\ell_2O$ agrees well with the kinetic data, it is only necessary to show that condition (18) is satisfied for one value of the critical pressure at some arbitrary temperature. Then all absolute values of the experimental data may be calculated from kinetic data. Let us select one of Zagulin's points: $T = 454°$ or $(1/T) = 22 \cdot 10^{-4}$ at a critical auto-ignition pressure of 250 mm Hg. The number of molecules per unit volume is:

$$a = \frac{2.7 \cdot 10^{19} \cdot 250 \cdot 273}{760 \cdot 454} = 5.4 \cdot 10^{18}$$

On the other hand: $\exp(-11000/454) = \exp(-24.2) = 3.2 \cdot 10^{-11}$. Whence $\delta_{cr} \sim 1$ in almost complete agreement with the theoretical value of two. An error in T of 10 to 15° would explain the discrepancy, as would also a small error in E or a factor of two for either the rate constant κ or the heat conductivity λ.

Consider now the reaction $2H_2 + O_2 \longrightarrow 2H_2O$ under conditions such that the ignition mixture is of a thermal nature. The domain of explosion of hydrogen-oxygen mixtures has a complex shape: at a given temperature, there are generally three critical pressures called lower, upper and third limits. The lower and upper limits are due to chain processes and they will be treated in detail in Chapter X. Above the upper limit, an autocatalytic process takes place. It has been studied by Chirkov [14]. The kinetics of the reaction for stoichiometric mixtures is well described by the expression:

$$w = \frac{d\Delta p}{dt} = kp_{H_2}^2 p_{H_2O} \tag{19}$$

where

$$k = 10^{13.9} \exp(-73000/RT) \text{ (mm Hg)}^{-2} \text{ sec}^{-1} \quad . \tag{20}$$

Using equation (16), Frank-Kamenetskiǐ [15] has calculated the temperatures at which the autocatalytic process of Chirkov should lead to thermal explosion. For an autocatalytic reaction, the thermal explosion limit is determined by the maximum value of the rate of reaction. Chirkov's expression tells us that the maximum rate, in a stoichiometric mixture, is reached when 1/3 of the hydrogen is burned. Thus the concentrations of hydrogen and water at the maximum rate are:

$$\left[p_{H_2}\right]_{max} = \frac{2}{3} p \frac{2}{3} = \frac{4}{9} p \quad .$$

$$\left[p_{H_2O}\right]_{max} = \frac{2}{3} p \frac{1}{3} = \frac{2}{9} p$$

where p is the total initial pressure. The maximum rate is then equal to:

$$w_{max} = k \frac{16}{81} \cdot \frac{2}{9} p^3 = \frac{32}{729} kp^3 \quad .$$

This can now be expressed in molar concentrations and substituted in condition (16) where $Q = 57800$ cal/mole and $\lambda_{2H_2} \simeq 5 \cdot 10^{-4}$.* We then obtain:

$$\frac{2.3 \cdot 10^{18}}{T} 10^{-\frac{16000}{T}} \left(\frac{273}{T}\right)^3 p^3 r^2 = 2 \quad . \qquad (21)$$

This is Frank-Kamenetskiĭ's equation for the thermal explosion limits of hydrogen-oxygen mixtures.

After Frank-Kamenetskiĭ derived this relation, Ziskin [17] measured the temperatures of explosion of $H_2 - O_2$ at atmospheric pressure, in cylindrical vessels of various sizes. Figure 10 shows a line corresponding to Frank-Kamenetskiĭ's equation and circles which represent Ziskin's *experimental data. The excellent agreement* between calculated and experimental values leaves no doubt as to the thermal nature of explosions at the third limit of hydrogen-oxygen mixtures.

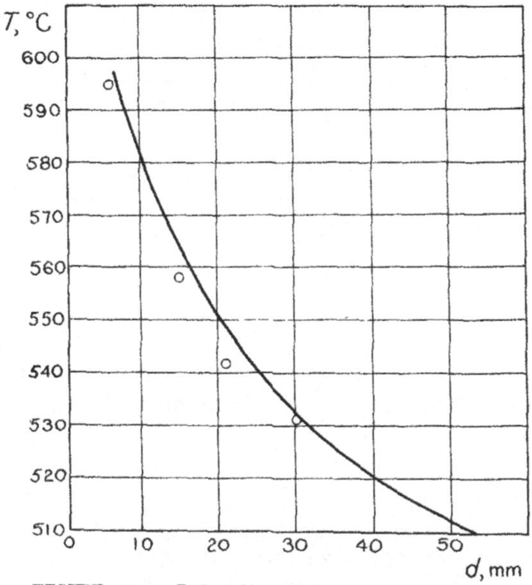

FIGURE 10. Relation between explosion and reactor size. The continuous curve is calculated from Frank-Kamenetskiĭ's formula (15). The circles are the experimental data of Ziskin [17].

In a subsequent study, V. Voevodskiĭ and V. Poltorak [18] used equation (14) to calculate the position of explosion limits for $2H_2 + O_2$ mixtures in a spherical pyrex vessel at various temperatures, from data on reaction kinetics and relation (17) with $\delta_{cr} = 3.32$.

The results of Table 43, on the following page, confirm again the validity of equation (14).

In the same investigation, it was shown that when the reaction rate is decreased by a treatment of the walls with $KC\ell$ solutions, the character of the ignition is changed sharply and thermal explosions are not observed. When

---

* This value is obtained by a calculation based on data by Vasil'eva [16] at room temperature ($\lambda_{2H_2+O_2} \simeq 2.5 \cdot 10^{-4}$) assuming a Sutherland constant

$$G = \frac{2}{3} G_{H_2} + \frac{1}{3} G_{O_2} = 90 \ .$$

TABLE 43

T°C	598	591	597
p mm Hg exptl.	700	660	590
p mm Hg calcd.	685	655	588

pressure is raised somewhat above values calculated from equation (17), ignition fails to occur. Since then the inequality

$$\frac{t_r}{t_e} > \frac{QEe}{cRT_0^2}$$

is satisfied (see [13]), it may be concluded that ignition in reactors treated with KCℓ has not a thermal but a chain origin. The details of the mechanism have been considered by V. Voevodskiĭ [19].

Until now, the analysis has been confined to the critical conditions for auto-ignition and the change with time of the gas temperature or the extent of reaction has not been discussed. Consider now this problem, following Todes [3]. The first simplifying assumption is that the reaction rate does not change with time but retains its initial value. This assumption is of course incorrect since the number of reactant molecules decreases with time. But it will be shown presently that this assumption does not introduce any substantial error in the calculation.

On this basis, the equation describing the change of temperature in the gas with time, is:

$$\frac{cav}{N} \cdot \frac{dT}{dt} = \frac{1}{N} \cdot ka^n \exp(-E/RT)Qv - \kappa S(T - T_0) \quad . \qquad (22)$$

At pressures lower than the critical, this equation leads to the establishment of a stationary temperature:

$$T_1' = T_0 + \Delta T \quad .$$

When the pressure is considerably higher than the critical, the second term on the right hand side of the equation may be neglected and the simplified equation:

$$\frac{dT}{dt} = \frac{1}{c} ka^{n-1}Q \exp(-E/RT) \qquad (22')$$

may be integrated.

Integration gives a particular form of the temperature change with time. Figure 11 represents a solution for a unimolecular reaction with the constants:

$$k = 10^{14}, \qquad \frac{E}{RT} = 40, \qquad \frac{Q}{cT_0} = 25 \quad .$$

As can be seen, during some rather long lapse of time t_i (2 to 3 sec.), the temperature increases very slowly until it reaches the value

$$T_0 + \frac{RT_0^2}{E}$$

(where $\frac{RT_0^2}{E}$ is the preheat before explosion). From this moment onward, the temperature further increases to the explosion temperatur (several thousands of degrees) is practically instantaneous, or more exactly it takes place in a time negligibly small as compared to t_i. Thereafter, since the entire initial charge has burned away, the gas starts to cool down. The diagram of Figure 11 is in two parts, since it is not possible to represent on a single scale a temperature increase of several thousands of degrees and one of a few degrees. The time elapsed from the start to the preheat before explosion is called the induction period t_i. It practically coincides with the time from the beginning of reaction to the explosion. This is what is easily recorded experimentally.

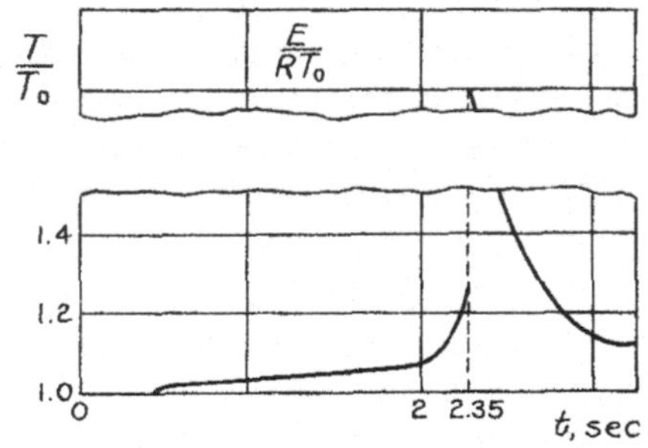

FIGURE 11. Integration of equation (22) for a unimolecular reaction $k = 10^{14}$, $\frac{E}{RT} = 40$, $\frac{Q}{cT_0} = 25$.

Todes has found the following approximate expression for t_i:

$$t_i = \frac{RT_0 cT_0 a}{EQka^n \exp(-E/RT)} = t_r \frac{cRT_0^2}{EQ} \quad .$$

Now, since RT_0/E is a small quantity not exceeding 0.05 in cases of interest here and cT_0/Q is also small, of the order of $5.500/20000 \sim 0.1$ or less, the quantity cRT_0^2/EQ is of the order of 0.01 or 0.001.

Thus $t_i = 0.01 - 0.001 t_r$. This is the order of magnitude of the induction period. But since also

$$t_r \leq \frac{QEe}{cRT_0^2} t_e \quad ,$$

we have $t_i \leq e t_e$ signifying that induction period is of the same order of magnitude as the thermal relaxation time. Near the critical pressure, t_i

becomes several times larger but stays of the same order of magnitude.

This discussion leads to an important corollary: since it takes only a fraction (0.01 to 0.001) of the reaction time from the beginning of reaction to the end of the induction period, at the moment of the sudden increase in temperature or at the moment of explosion, not more than 1% of the original substance has reacted away. This circumstance justifies completely the assumption made concerning the constancy of the reaction rate until the explosion occurs.

This not only justifies the calculation of induction period but also the earlier derivations of the explosion conditions where it was assumed that the quantity of reacting molecules is the same at the beginning and at the point of contact between the curves of heat release and heat removal. Therefore the much more complex problem of predicting explosion conditions taking into account the consumption of reactants need not be considered here.

If on the other hand either E or Q are small, the situation is different. The typical picture of an explosion disappears. If the activation energy is very small, and reaction takes place on every collision, the gas itself cannot be prepared or if two gases are involved, they cannot be mixed. This happens for instance when vapors of sodium are mixed with chlorine. If the activation energy is small but the reaction is nevertheless not too fast (because of a small steric factor), the preheat ΔT becomes very large and there is no quantitative difference left between reactions occurring before or after the explosion limit.

With a small value of E, the absence of a sharp distinction between steady and explosive conditions is due to a large preheat and the reaction rate is very large even in the steady part of the process. When on the other hand Q is small, the difference disappears because the explosion necessitates high temperatures and since Q is small, the substance must react before these temperatures are reached.

Consequently, typical explosive gases are those that possess a large heat of reaction (20000 cal/mole or more) and are also sufficiently stable thermally, being not decomposed upon considerable heating, i.e., they must have a rather high activation energy (20000 cal or more). Typically explosive gases met in practice satisfy these conditions and all the results derived are applicable to them. Let us note besides that a small activation energy E makes the explosion non-typical only when the rates of unimolecular or bimolecular processes also have an abnormally low rate constant k. When this is not the case (for instance when for a bimolecular process $k = \sqrt{2}\pi\sigma^2 u$), a small activation energy does not interfere with the sharpness of the explosion and does not obscure explosion phenomena. The only change is that explosion will take place at very low temperatures. The quantity (E/RT) stays large in spite of the low value of E. Cases of this sort are the explosion of HBr with ozone that occurs very sharply but at a temperature in the vicinity of 100°G or the explosion of hydrogen with fluorine. With respect to small values of Q when E has a normal value, the difference between the steady process and the explosion becomes always indistinct. Todes and Melent'ev [3] have analyzed accurately (by numerical integration) the character of the phenomenon for a monomolecular process (with a rate constant $k = 10^{13}$), the exhaustion of reactants being taken into account.

Two curves will illustrate the consumption of reactants ($\xi = b/a$ where b is the quantity of unreacted molecules and a their initial amount) 1) for the case of a large heat of reaction (typical explosion: Figure 12) and 7) for the case of a small heat of reaction (Figure 13).

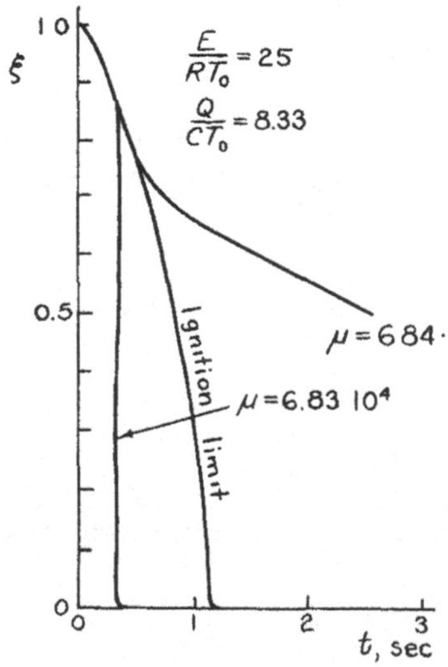

FIGURE 12. Relation between ξ and t- large heat of reaction.

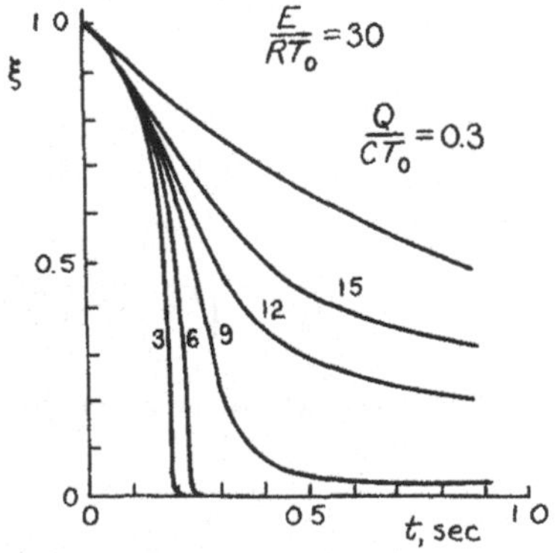

FIGURE 13. Relation between ξ and t- small heat of reaction.

These curves illustrate very well the preceding qualitative discussion. Figure 14, on the following page, also shows the temperature-time curve for the case corresponding to Figure 12 (typical explosion). We see that a 0.2% change in pressure near the auto-ignition limit modifies the trend of the reaction in a qualitative way. Nothing like this is observed for the case of low heats of reaction (Figure 13; the figures on the curves correspond to various values of the dimensionless parameter μ).*

To sum up, a typical explosion occurs at high values of the dimensionless number:

$$\beta = \frac{Q}{c}\frac{E}{RT_0^2} .$$

The higher this number, the less are the ignition limits displaced by combustion of reactants during the induction period. What happens at moderate values of β has been considered by Frank-Kamenetskiĭ [20]. For a reaction of order m, the maximum value of the induction period is equal to:

$$t_{max} = t_e \sqrt[3]{2\pi^2\beta/e^2m} = t_r \sqrt[3]{2\pi^2/e^2m\beta} .$$

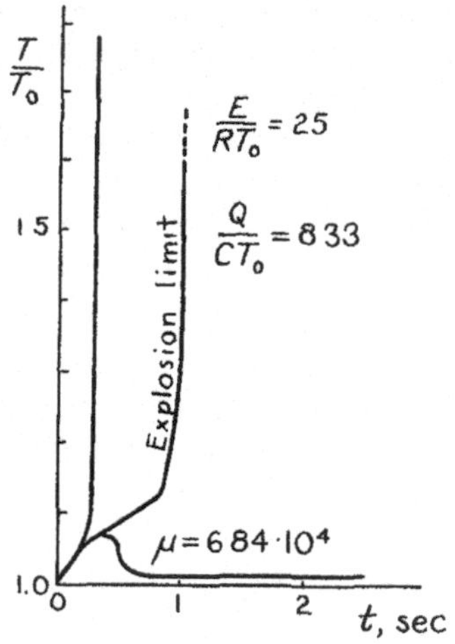

FIGURE 14. Relation between T/T_0 and t.

The displacement of the ignition limit as a result of combustion during the induction period may be expressed by a change in the value of the critical parameter δ (14):

$$\delta = \delta'(1 + w_m)$$

where

$$w_m = \sqrt[3]{2\pi^2m^2/e^2\beta^2} = 1.39\left(\frac{m}{\beta}\right)^{2/3} .$$

The ignition temperature at a given pressure and vessel size changes by the amount:

$$\Delta T = (RT_0^2/E)w_m$$

where T_0 is the average temperature of the experiments.

As an example of an explosive process for which correction due to combustion during the induction period is appreciable, is the explosive

* The parameter $\mu = \kappa/cav$ is inversely proportional to pressure p. It is seen that when Q is large, the combustion changes sharply when μ goes from $6.84 \cdot 10^4$ to $6.83 \cdot 10^4$ (a 0.14% change). When $\mu = 6.83$ the reaction goes slowly but with $\mu = 6.84$, after an induction period, the mixture burns instantaneously (explosion).

decomposition of pure acetylene [21]. Below the limit, the reaction proceeds at a well investigated rate [22]. The kinetics is very well described by a dimerization rate equation of second order. Slow secondary processes take place only after a twofold decrease in pressure. The experimental facts strongly indicate that the primary process in the gas phase is the formation of a dimer of acetylene. Other processes leading to the formation of a complex mixture of aromatic hydrocarbons take place only after cooling and condensation of the dimer, in the liquid phase. The structure of the dimer and its heat of formation may not be determined because it is unstable with respect to further chemical change when condensed.

As the temperature increases, the dimerization process gives way to a thermal explosion. In spite of the complex nature of the phenomenon, the ignition limits are well described by the simple theory. The reaction rate prior to explosion can be well represented by the equation:

$$-\frac{d[C_2H_2]}{dt} = k[C_2H_2]^2$$

where $k = 4.1 \cdot 10^{10} \exp(-29000/RT)$ cm^3/mole. sec.

The fact that the reaction is not in fact bimolecular but a chain process, has an effect only on the initial stages of the induction period where the radical concentration is building up. Subsequently, and during the part of the induction period where there is a steady state concentration of radicals, the reaction is very well described by a bimolecular law.

The ignition experiments were conducted in spherical and cylindrical vessels. To exclude the effect of convection, the experiments in cylindrical reactors were carried out with the reactor in both horizontal and vertical position. The observed limits together with equations (16) and (17) give a means of calculating the only unknown: the heat of dimerization. Without a correction for combustion, the value of 63 to 66 kcal per mole of dimer is found. With the correction for combustion during the induction period which is quite sizeable in this case, this figure becomes 78 kcal per mole of dimer. Even though this value lacks precision, it leads to a satisfactory picture of the structure of the dimer. According to the usual rules of thermochemistry of organic compounds, the heat of formation may be calculated for various possible structures. If the dimer is linear, its heat of formation ought to be much smaller than the value suggested by the explosion limits. Thus for vinylacetylene and divinylacetylene (butatriene), one obtains respectively 35.6 and 54.0 kcal per mole of dimer. For a dimer with a cyclic structure, two single and two double bonds, the value is 73 kcal per mole of dimer, without any correction for bond strain and conjugation energy. The product of dimerization of acetylene in the gas phase is very likely to be cyclobutadiene:

$$
\begin{array}{ccc}
HC & = & CH \\
| & & | \\
HC & = & CH
\end{array}
$$

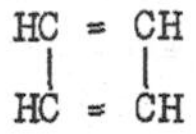

in which apparently the conjugation energy compensates for bond strain.

With this we conclude our treatment of thermal auto-ignition for the simplest reactions. Let us now consider thermal explosions for auto-catalytic processes.[*]

It has been shown that, at pressures exceeding the critical, the induction period is very short and is of the order of 1 sec or less. But in many cases, auto-ignition occurs after a much more considerable delay. Thus for instance, auto-ignition of methane-oxygen mixtures at 730°C and a pressure of 40 mm Hg takes place after a delay of four minutes; similar delays are found with other hydrocarbons and also for the oxidation of carbon disulfide and hydrogen sulfide. Particularly long delays are found with liquid and solid explosives, where they reach tens of minutes and even hours. According to Roginskiĭ [23], for auto-ignition of trotyl and nitro-glycerine in sealed vessels, the delay sometimes reaches 5 to 10 hours.

A study of the kinetics of all these reactions[**] has revealed that they all proceed auto-catalytically i.e., the reaction rate as a function of amount reacted x is represented by the equation $(dx/dt) = \varphi x + n_0$, during the first stages of reaction: n_0 is the number of final or intermediate products produced per second per unit volume. This may be the result of mono- or bimolecular homogeneous processes or of hetero-geneous reactions. In the majority of cases, n_0 is very small and may be neglected for arbitrary values of φ. If however we want to express x or w as a function of time, n_0 must be known since, upon integration the equation becomes:

$$
x = \frac{n_0}{\varphi} (e^{\varphi t} - 1) \qquad \text{and} \qquad \frac{dx}{dt} = n_0 e^{\varphi t} \ . \qquad (23)
$$

The constant φ decreases during the course of the process as a result of the exhaustion of reactants. In particular: $\varphi = \varphi_0(a - x)$. The reaction rate can then be represented as a function of x (Figure 15) and of time (Figure 16) where curves 1, 2, 3, 4 correspond to different pressures $p_1 > p_2 > p_3 > p_4$.

If it is necessary for the occurrence of a thermal explosion that

---

[*] For autocatalytic reactions, there is no correction for combustion since the explosion takes place when the maximum rate is reached.

[**] The theory of all these phenomena has been developed at the Institute of Chemical Physics — see Semenov's 'Chain Reactions' [2].

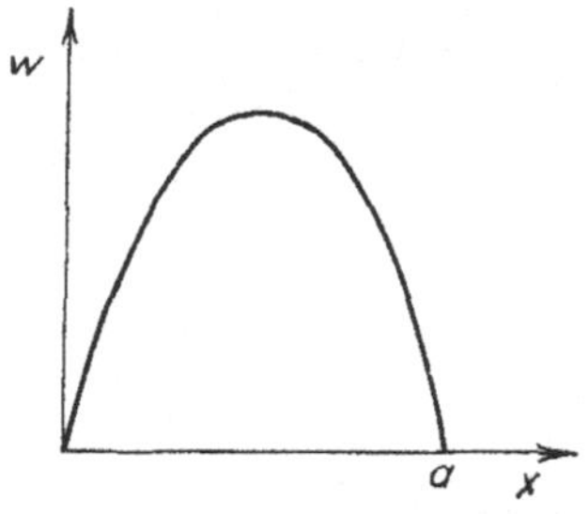

FIGURE 15. Relation between w and x.

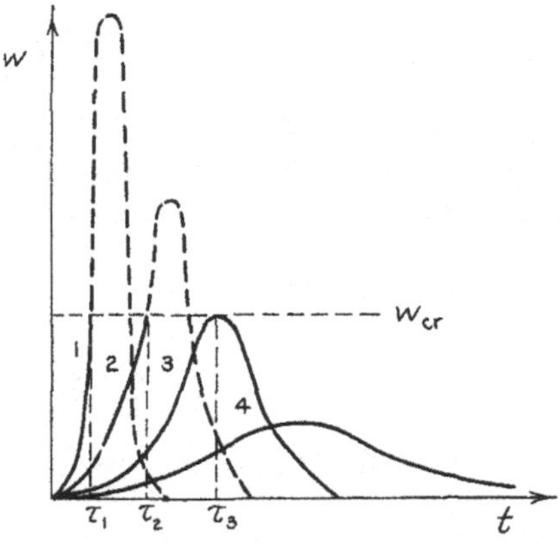

FIGURE 16. Relation between w and t.

the reaction rate reaches a critical value w_{cr}, then at a pressure p_1 explosion occurs after time τ_1, at pressure p_2 after time τ_2; at pressure p_4 no explosion takes place. The smallest pressure at which explosion takes place is p_3 to which corresponds the longest possible delay τ_3. Here, the induction period is not due to the heating of the mixture but to the time required for the combustion reaction to build up isothermally to a rate required for the explosion. Therefore, the induction period may now be quite prolonged.

Let us express mathematically the condition for ignition. Consider only the first stages of the reaction (10 to 20%) during which the constant φ does not vary appreciably and is determined by the amount of reactants and the initial temperature. Usually, the constant φ is proportional to the first or second power of the pressure of the mixture (or also the number of reactant molecules per unit volume). Also, φ increases exponentially with temperature according to the law $\exp(-E/RT)$. Another restriction will be that the thermal relaxation time (of the order of 0.01 to 0.5 sec) be short as compared to the time for the autocatalytic acceleration of the reaction and may be neglected.

With these assumptions, the entire theory of thermal auto-ignition developed earlier remains valid. But instead of $w = ka^n \exp(-E/RT)$, it is necessary to use in the equations:

$$w = fa^n x \exp(-E/RT) + n_0 = \varphi x + n_0$$

Thus for instance, with conductive heat transfer in a cylindrical reactor, the critical condition [equation (16)] for ignition becomes:

$$\delta_{cr} = \frac{Er^2Q(\varphi x+n_0)}{RT_0^2\lambda N} = 2 \quad . \tag{24}$$

From (24) and a knowledge of n_0 and $\varphi = fa^n \exp(-E/RT)$, it is possible to calculate x, i.e., the extent of conversion that must be reached for explosion. Usually a more interesting quantity is the delay that can be readily measured. But

$$x = \frac{n_0}{\varphi}(e^{\varphi t} - 1) \quad .$$

Substituting this expression into (24), we get:

$$n_0(e^{\varphi t} - 1) + n_0 = n_0 e^{\varphi t} = \frac{2RT_0^2}{Er^2Q}\lambda N$$

where τ is the ignition delay. Hence:

$$0.434\tau + \log\frac{n_0}{T_0^2} = \log\frac{2R\lambda N}{Er^2Q} = \text{const.} \tag{25}$$

provided that we neglect the small variation of λ with pressure and temperature.

The quantity

$$\frac{2R\lambda N}{Er^2Q} \approx \frac{10^{-5}\cdot 10^{23}}{10^4\cdot 10^4} \sim 10^{10}$$

(in order of magnitude). Thus the constant in (25) is about 10. When n_0 is small, say smaller than 10^8,

$$\log\frac{n_0}{T_0^2} \sim 2 \quad .$$

For smaller values of n_0 e.g., 10^4,

$$\log\frac{n_0}{T_0^2} = -2 \quad .$$

Thus $0.434\varphi\tau$ changes only by 20% on both sides of $0.434\varphi\tau = 10$ when n_0 is varied by a factor of 10^4. Therefore, in first approximation, we may write

$$\varphi\tau = f\tau a^n \exp(-E/RT) = \text{const.} \tag{26}$$

when temperature and pressure are changed in a rather wide range.

Consequently, at constant pressure, the relation between delay and absolute temperature must be of the form:

$$\log \tau = \frac{A}{T} + B \tag{27}$$

where

$$A = \frac{0.434E}{R} = 0.22E \quad .$$

At constant temperature and variable pressure, the following relation must be obeyed.

$$\log \tau = C - n \log p \quad .$$

This formula has been verified many times in experiments with various explosive systems.

REFERENCES

[1] N. N. Semenov, Usp. Fiz. Nauk, 23, 251 (1940).

[2] N. N. Semenov, 'Chain Reactions', Oxford (1935).

[3] O. M. Todes, Acta physicochim. URSS, 5, 785 (1936).
O. M. Todes and B. M. Melent'ev, Acta physicochim. URSS, 11, 153 (1939); Zhur. Fiz. Khim., 13, 868, 1594 (1939).

[4] D. A. Frank-Kemenetskiĭ, Zhur. Fiz. Khim., 13, 738 (1939).
'Diffusion and Heat Transfer in Chemical Kinetics', Princeton University Press (1953).

[5] A. V. Zagulin, Zeit. phys. Chem., 1, 275 (1928).

[6] C. N. Hinshelwood, J. chem. Soc., 123, 2730 (1923); 125, 1841 (1924).

[7] F. O. Rice and P. E. M. Allen, Jour. Am. Chem. Soc., 57, 310 (1935); 57, 2212 (1935).

[8] H. C. Ramsperger, Jour. Am. Chem. Soc., 49, 912, 1495 (1927).

[9] O. M. Todes, Zhur. Fiz. Khim., 4, 78 (1933).

[10] D. A. Frank-Kamenetskiĭ, Dok. Ak. Nauk SSSR, 18, 413 (1933); Zhur. Fiz. Khim., (in press).

[11] A. Ya. Apin and Yu. B. Khariton, Zhur. Fiz. Khim., 8, 866 (1936).

[12] M. Volmer, Zeit. phys. Chem., 9, 141 (1930).

[13] Ya. B. Zel'dovich and V. I. Yakovlev, Dok. Ak. Nauk SSSR, 19, 699 (1938).

[14] N. M. Chirkov, Acta physicochim., 6, 915 (1937).

[15] D. A. Frank-Kamenetskiĭ, PhD Thesis, Kazan' (1943).

[16] Vasil'eva, Physik. Zeit., 5, 737 (1904).

[17] M. S. Ziskin, Dok. Akad. Nauk SSSR., 34, 279 (1942).

[18] V. V. Voevodskiĭ and V. A. Poltorak, Zhur. Fiz. Khim., 24, 299 (1950).

[19] V. V. Voevodskiĭ, Zhur. Fiz. Khim., 20, 1285 (1946).

[20] D. A. Frank-Kamenetskiĭ, Zhur. Fiz. Khim., 20, 139 (1946).

[21] E. A. Blyumberg and D. A. Frank-Kamenetskiĭ, Zhur. Fiz. Khim., 20, 1301 (1946).

[22] D. A. Frank-Kamenetskiĭ, Zhur. Fiz. Khim., 18, 329 (1944).

[23] S. Z. Roginskiĭ, Phys. Z. Sow., 1, 640 (1932).

CHAPTER IX: CHAIN IGNITION

§1. Generalities on Chain Ignition*

The theory of chain ignition, born in the late twenties and early thirties mainly as a result of the work of Soviet and English scientists, is based on the following postulates that have now been well verified by experimental studies.

1. The rate of chain generation (i.e., that of the process giving active centers — the chain carriers —) is negligibly small. At any rate it is much smaller than the smallest rates accessible to kinetic measurement.

2. The chains are material in nature, i.e., it is propagated by reacting atoms and radicals. It is assumed that chain propagation by means of excited molecules is impossible.

3. The active centers — atoms and radicals — may react in three different ways:

a. Chain termination involving homogeneous or heterogeneous reactions that destroy active centers by transforming them into stable molecules or inactive radicals;

b. chain propagation involving elementary steps in which the number of active centers is preserved;

c. chain branching with processes that increase the quantity of active centers in the system.

The rate of chain generation will be designated by n_0 while α, β, δ are, respectively, the probabilities of chain propagation, termination and branching.

Let us now introduce the concept of kinetic chain length. This is the number of elementary reactions in which an active center, whatever

---

* This chapter was written together with Voevodskiĭ and Nalbandyan. It contains a brief exposé of the general principles of chain theory similar to that given in more detail in the book on 'Chain Reactions'.

its mode of creation in the system, takes part. The rate of formation of these centers is the rate of chain generation n_0. In the absence of branching, an active center may react in only two ways: in chain propagation that produces usually a molecule of final product and a new center or chain termination in which the center disappears. Since a reaction involving an active center is either propagation or termination, we have:

$$\alpha + \beta = 1 \quad .$$

The probability for a chain to have s links is equal to the product of the probability of $s - 1$ propagation steps i.e., α^{s-1} by the probability of termination β:

$$P_s = \alpha^{s-1}\beta = \alpha^{s-1}(1 - \alpha) \quad . \qquad (1)$$

Since

$$\sum_{s=1}^{\infty} P_s = 1 \quad ,$$

the chain length ν is the average value of s and is equal to:

$$\nu = \bar{s} = \sum_1^{\infty} P_s s = \sum_1^{\infty} s\alpha s^{-1}(1 - \alpha) = \frac{1}{1 - \alpha} = \frac{1}{\beta} \quad . \qquad (2)$$

If besides termination and propagation, branching can also occur, the branching factor δ must be taken into account in the expression for ν since it decreases the factor β.

For instance, if one branching takes place together with two termination steps, this is equivalent on the whole to only one termination; two chains have ceased to exist but a new one has appeared. In the expression for ν, one must therefore have not the quantity β but the difference $\beta - \delta$*:

$$\nu_b = \frac{1}{\beta - \delta} = \frac{\nu}{1 - \delta\nu} \qquad (2')$$

---

* Branched chain reactions may be of two types:

1. Long chains with rare branching (Figure 17).

2. Continuously branched chains when branching occurs at every cycle (Figure 18).

These two types were described earlier in our book 'Chain Reactions'. Intermediate types are of course possible.

where ν_b is the chain length with branching taken into account but ν is, as before, the length of a straight unbranched chain.

In all chain processes, the reaction rate is equal to the initial rate of generation of centers multiplied by the chain length:

$$w = \frac{n_0}{\beta - \delta} = \frac{n_0 \nu}{1 - \delta\nu} \tag{3}$$

This equation shows already the possibility of ignition limits. If β and δ depend in a different way on external parameters (pressure, temperature, mixture composition, etc.). It is possible that there exist values of these parameters for which $\delta = \beta$. Then ν_b and also the reaction rate become infinite. The condition

$$\beta - \delta = 0 \tag{4}$$

determines a chain ignition limit.

The discussion might be carried one step further: cases can be

* (cont.)

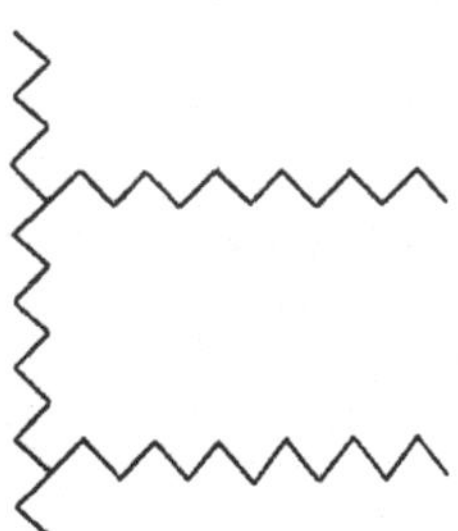

FIGURE 17. Schematic representation of a chain reaction with rare branching.

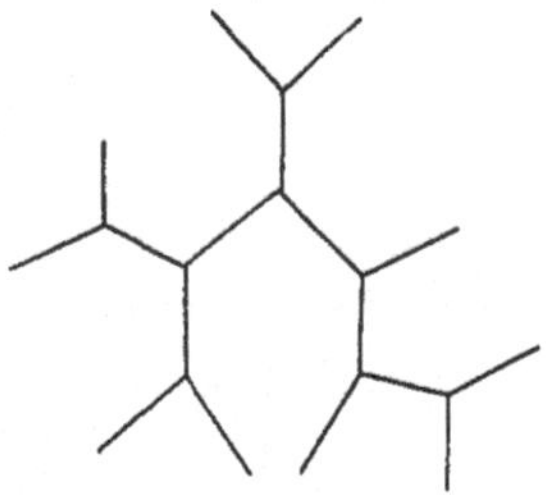

FIGURE 18. Schematic representation of a continuously branched chain reaction.

In reactions of the first type, the branching probability may be taken as constant. It does not depend on β and α.

In reactions of the second type: $\delta = \alpha - 1 - \beta$ and $\delta - \beta = 1 - 2\beta$. Generally: $\delta = \delta^*(1 - \beta)$ where δ^* represents the probability of branching if termination did not take place, while δ is the same probability with chain termination taken into account. In continuously branched chains where chain propagation is always associated with branching $\delta^* = 1$. In the case of very long chains where β is very small, $1 - \beta \sim 1$ and $\delta = \delta^*$. If each branching step gives birth not to two but to three active centers $\delta = 2\delta^*(1 - \beta)$. If four active centers are produced $\delta = 3\delta^*(1 - \beta)$ etc.

imagined where δ would become larger than β. But this leads to an unacceptable result: the chain length ν_b and the reaction rate w would become negative, an impossible situation. The paradox is due to the fact that the probability arguments used in the discussion are valid only when the chain length is finite. Indeed, what is the physical meaning of the probabilities α, β and δ? Each probability is equal to the number of reactions of a given type divided by the total number of reactions taking place in the system. If this ratio is infinitely large, the probability loses its meaning. The concept retains its validity only at the limit, when $\delta \longrightarrow \beta$, for large, yet finite values of the chain length.

Consider now what happens in a system when the rate of the branching process becomes equal to or larger than that of chain termination. If β is larger than δ, this case need not be considered since, for not too long chains, the properties of the system do not change even during long periods of time. For this reason, a steady concentration of active particles is rapidly established together with a constant value of the chain length. If, on the other hand, $\delta > \beta$, the number of active centers increases all the time, the properties of the system change continuously, the concentration of active centers, the reaction rate and the chain length increase continuously. This conclusion may be expressed mathematically. The rate of change of concentration of active particles is:

$$\frac{dn}{dt} = n_0 - (g - f)n \tag{5}$$

where g and f are the kinetic coefficients of the rates of chain termination and branching proportional respectively to β and δ.*

Integrating and remembering that $n = 0$ at $t = 0$, we get:

$$n = \frac{n_0}{g - f}\left(1 - e^{-(g-f)t}\right) .$$

For sufficiently large values of the difference $g - f$, i.e., when branching is less important than termination, the exponential term in equation (6) may be neglected even after long periods of time. Then:

$$n = \frac{n_0}{g - f} . \tag{7}$$

The reaction rate is equal to a kinetic factor a characterizing the propagation step, times the concentration of active centers:

---

* The rate of propagation does not enter in the expression for $\frac{dn}{dt}$ since the quantity of active centers is not modified by propagation (one disappears, another replaces it).

$$w = an = \frac{an_o}{g - f} \quad . \tag{8}$$

Equating (8) and (3), we find:

$$\delta = \frac{f}{a} \qquad \text{and} \qquad \beta = \frac{g}{a} \quad .$$

The reaction rate, in the case $\beta > \delta$, rapidly reaches a constant value, as was expected.

Consider now the case of interest here:

$$f - g \geq 0 \quad .$$

When $f - g = 0$, equations (5) and (7) become respectively:

$$\frac{dn}{dt} = n_o \tag{9}$$

$$n = n_o t \quad . \tag{10}$$

The reaction rate which now becomes:

$$w = an = an_o t \tag{11}$$

increases slowly (for small values of n_o) with time: the process is no more stationary.

When $f - g > 0$, integration of equation (5) gives:

$$n = \frac{n_o}{f - g} \left(e^{(f-g)t} - 1\right) \quad . \tag{12}$$

The reaction rate is now:

$$w = \frac{an_o}{f - g} \left(e^{(f-g)t} - 1\right) \quad . \tag{13}$$

It increases exponentially with time, without limit.

Figure 19 (2) shows how the chain length $\nu = (w/w_o)$ changes with the dimensionless coordinate at, for various values of the ratio $[(f - g)/a] = x$.

The amount of reacted material is:

$$x = \int_0^t w \, dt \tag{14}$$

if $f - g < 0$

$$x = \frac{an_o}{f - g} t \tag{15}$$

$f - g = 0$

$$x = \frac{an_o}{2} t^2 \tag{16}$$

$f - g > 0$

$$x = \frac{an_o}{(f-g)^2} (e^{(f-g)t} - 1) - \frac{an_o t^2}{2} \tag{17}$$

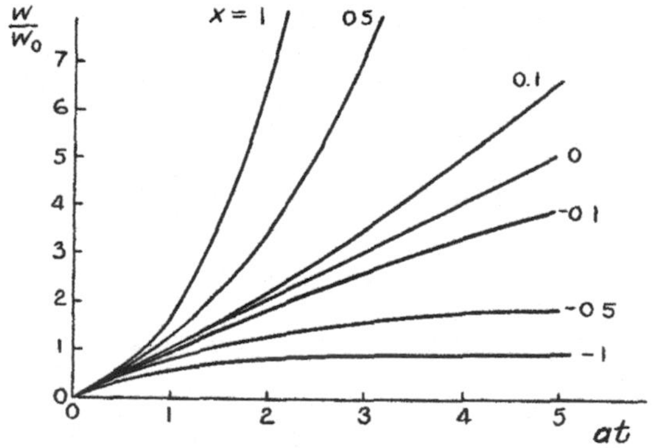

FIGURE 19. Relation between the chain length $\nu = (w/w_o)$ and the dimensionless coordinate at, for various values of the ratio $x = \frac{f - g}{a}$.

It can be seen that even when $f > g$, the reaction rate never becomes negative. The larger is the difference $f - g$, the steeper is the kinetic curve and the sooner is a large rate of reaction reached in the system.

Is it then possible for the reaction rate to become infinite? Formally, this is expected from equation (13). But, when the reaction rate reaches such large values, while the concentration of active centers increases, the fresh reactants from which they come also disappear at a fast rate in the system (if it is closed as is usually the case).* At a given

* In a flow system when fresh reactants are continuously fed to the reactor, the reaction rate does not become infinite either. A fraction of the active centers is continuously carried forward with the combustion products and w may only reach a maximum value.

moment, the concentration of reactants decreases so much that the reaction does not accelerate any more; the rate reaches a maximum, then decreases and goes to zero. Inflammation does not mean at all that the reaction rate is infinitely large. Ignition or combustion is rather a process that transforms some reactants during a finite but very short time, of the order of a fraction of a second.

When the quantity n_o of active centers produced per unit time per cm^3 is very small, the rate of the stationary process, in the case $f - g < 0$ is also very small, even though a is large and $g - f$ small [see equation (8)]. But if $f - g > 0$ and a is sufficiently large, the reaction rate, according to equation (13) reaches a very large value even though only one active particle is generated in the entire reacting volume.

Indeed, if the rate constants of the reactions involving the active centers are very large, the quantity f is also large. In the case $f - g = 0$, let $f = 5 \cdot 10^2$. Since both f and g depend on the conditions of the experiment, e.g., on the pressure of the reacting mixture, a small change in these conditions will modify somewhat the quantity

$$f - g = f(1 - \frac{g}{f}) \quad .$$

Suppose that we start with $(g/f) = 1$ and change its value by 1% on either side. Then $1 - (g/f) = \pm 10^{-2}$ and $f - g = f(1 - g/f) = \pm 5$. Suppose also that $n_o = 10$. Then if $f - g = - 5$ and with the assumption that $f = a$:

$$w = \frac{an_o}{g - f} = \frac{5 \cdot 10^2 \cdot 10}{5} = 10^3 \quad .$$

Thus only 1000 molecules react per second. If it is recalled that even at a pressure of 0.01 mm Hg there are more than 10^{14} molecules per cm^3 in a gas mixture, it is clear that the rate just obtained is practically equal to zero. At such a rate, a thousand years would not be enough to react an appreciable fraction of the molecules. But if $f - g = + 5$, according to equation (13), with $n_o = 10$:

$$w = \frac{an_o}{f - g} e^{(f-g)t} = 10^3 e^{5t} \quad .$$

Thus after $t = 5$ sec., w reaches a large value $\sim 10^{14}$ molecules/second. After $t = 4$ sec., the quantity of reacted substance x is, with $f = a$, according to equation (17) in which the second term is neglected:

$$x = \frac{an_o}{f^2(1 - \frac{g}{f})^2} e^{(f-g)t} = \frac{n_o}{f(1 - \frac{g}{f})^2} e^{(f-g)t} = \frac{10}{5 \cdot 10^2 \cdot 10^{-4}} e^{5t} = 10^{11}$$

Thus only 0.1% of the reactants have been consumed. After $t = 5$ sec., x becomes equal to:

$$x = 2.10^2 \cdot 7.1 \cdot 10^{10} = 1.4 \cdot 10^{13} \quad .$$

Thus between 4 and 5 seconds, 14% of the original substance has burned away. After $t = 5.35$ seconds, $x = 2.10^2 \cdot 4 \cdot 10^{11} = 8.10^{13}$ and 80% of the reactants has reacted. Consequently, the bulk of reaction takes place in less than one second. Naturally such a reaction will be observed as an explosion. It is also seen that a 1% variation of (g/f) on both sides of $(g/f) = 1$ changes the rate from a negligible value of 1000 molecules per second to a large value of 10^{13} molecules per second after 5 sec. (explosion).*

Consequently, one may indeed speak of a chain explosion reached when $f - g \geq 0$ and of the practically complete absence of reaction when $f - g$ is even only slightly negative.

If n_0 is equal, not to 10 but to 10^{10} active centers per second, the result hardly changes. The reaction rate is multiplied by 10^9: it becomes 10^{12} molecules/sec. below the limit and 10^{16} molecules/second above the limit after $t = 1.5$ sec.

It is evident that when n_0 is small and $f - g < 0$, the reaction rate is negligibly small. When $f - g$ is close to zero but slightly above zero, the reaction rate grows rapidly with time. Nevertheless, it stays very small in absolute value and reaches an appreciable value only after some time (4 or 5 seconds in the case treated here). After that it reaches very high values very quickly (Figure 20, following page).

In comparing Figure 20 with Figure 19, it must be kept in mind that in the latter, all curves correspond to very small reaction rates. Indeed, even if n_0 is equal to 10^{10}, the maximum rate shown on this figure is only $7 \cdot 10^{10}$ molecules/sec., a negligibly small figure. Therefore this figure depicts only the initial phase of the transformation when detecting devices will not record any change in the system.

The theory is presented here as if only one active particle participated in chain propagation. In fact, there are usually no less than three such particles (e.g., H and O atoms and the OH radical in the chain ignition of hydrogen-oxygen mixtures). Then, to find the limit, it is necessary to solve a system of three differential equations if the

---

* In chain ignition, at low pressures, f is proportional to the pressure but g is either independent of, or inversely proportional to pressure. Thus f/g is proportional to pressure or the square of it. A 1% variation in (f/g) corresponds then to a 1% change in pressure or the square of it.

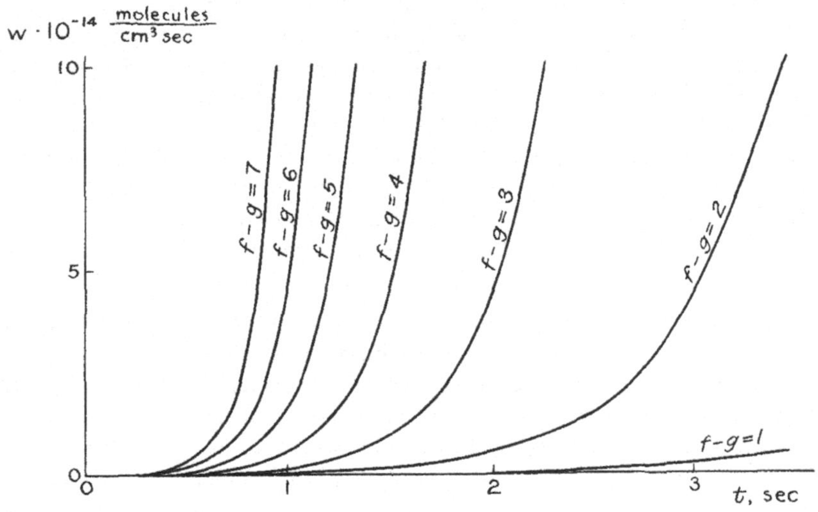

FIGURE 20. Relation between w, calculated by means of equation (13) and the time.

evolution of the reaction with time must be known or of three algebraic equations (for $f - g = 0$) if the chain ignition limit is to be found. In the latter case, the problem presented by three centers offers no difficulty. Nevertheless, the quantities f and g themselves are complex functions of several elementary constants. The theory of chain ignition therefore necessitates a physico-chemical analysis of the reaction mechanism and a mathematical treatment of this analysis in which the quantities f and g are expressed as functions of the elementary constants and the conditions of the problem.

In the subsequent paragraphs and in Chapter X, examples of such an analysis are presented. They treat the cases of the oxidation of phosphorus and of hydrogen. Let us indicate only at this stage that the quantity g depends strongly on the conditions of the experiment. Therefore chain ignition is a much more complex and sensitive problem than thermal ignition. For instance, g is often determined by the capture of atoms and radicals on solid surfaces which are more or less effective in this regard. Consequently, the pressure at which chain ignition takes place depends not only on the size and shape of the reactor but also on minute changes in the state of the surface of the reactor walls. A glass surface is much less effective in capturing active particles than a metal surface. The quantity g is frequently determined by the disappearance of radicals in triple collisions in which an active particle is transformed into a much less active particle, thus terminating the chain (e.g.,

$O + O_2 + M \longrightarrow O_3 + M$). This circumstance often leads to an upper explosion limit below which ignition takes place, above which the system is inert. Similar phenomena do not exist in thermal explosions. The addition of a completely inert gas often leads to chain ignition because it increases the time for diffusion of active particles to the wall and decreases g without having any effect on f. The rapid introduction of an inert gas into a reactive mixture at high pressure (if the reaction is then slow enough and the heating moderate), may in principle stop a developing chain ignition in its early phase, by increasing the number of triple collisions and breaking chains at the upper limit.

It is known that the presence of some foreign active species even in small amounts, strongly inhibits unbranched chain reactions because they capture chain radicals or transform active ones into inactive species. The inhibition monotonically increases with the amount added. In branched chain reactions, this effect leads to much sharper results. The quantity g increases with the concentration of the foreign substance. As soon as the quantity added has reached a value such that the difference $f - g$ becomes negative (f is not affected by the addition), inflammation becomes impossible and the reaction stops altogether.

This is particularly striking when a reaction product is an effective inhibitor. An incipient explosion is rapidly quenched by the build-up of the inhibitor concentration although fuel and oxygen are still in abundant supply. One could multiply at will the examples of peculiar phenomena associated with chain ignition. The strength of chain theory is that it explains, often quantitatively, a variety of reaction effects related to causes that are not easily controlled and that frequently show up after minute changes in reaction conditions. The most striking effects of this kind are observed in branched chain reactions and, in particular, in chain explosions. This is why the studies of chain ignition have been so meaningful historically in the development of chemical kinetics.

More complex than a theory of chain ignition limits, is a theory of the kinetics of branched chain reactions within the domain of autoignition. Equations (13) and (17) have been derived for an idealized case, when the reaction is propagated by means of a single active center. In the case of several centers, it is necessary to solve a system of several differential equations of the type:

$$\frac{dn_i}{dt} = f(n_1, n_2, \ldots, n_i) \quad .$$

This presents some difficulty even to describe the initial part of the kinetic curves (when the concentration of reactants may be taken as constant). A calculation of the course of the process taking the exhaustion of reactants into account is practically impossible. Fortunately in many cases, one of the active centers reacts much more rapidly than the others and the system of equations may be replaced in good approximation by one or two differential equations written for the particles that react more slowly while for the others, one writes

$$\frac{dn_i}{dt} = 0 \quad .$$

Then it is frequently possible to complete the solution of the problem in terms of simple functions.

In the case of rapid combustion reactions, such a kinetic treatment is particularly important in the application of chain theory to the problem of flame propagation (deflagration and detonation). The treatment is also essential for reactions with degenerate branching where the time evolution of the process is quite slow (see Chapter X).

For fast chain reactions and degenerate branching processes, the time evolution of the reaction is described by an S-shaped curve as can be seen on Figure 21. The rate of pressure increase as a function of extent of reaction is shown on Figure 22.

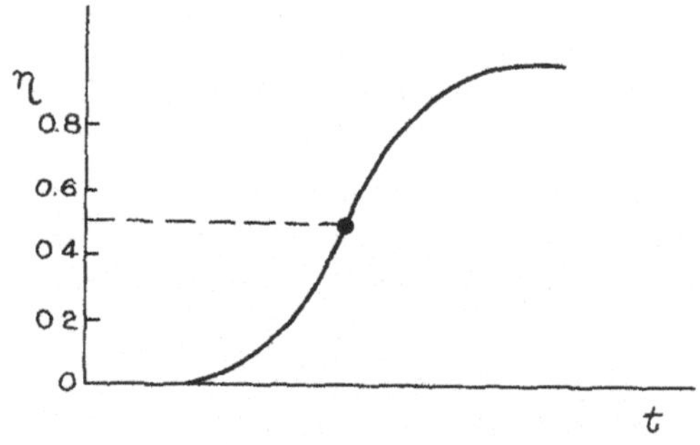

FIGURE 21. Relation between η and t.

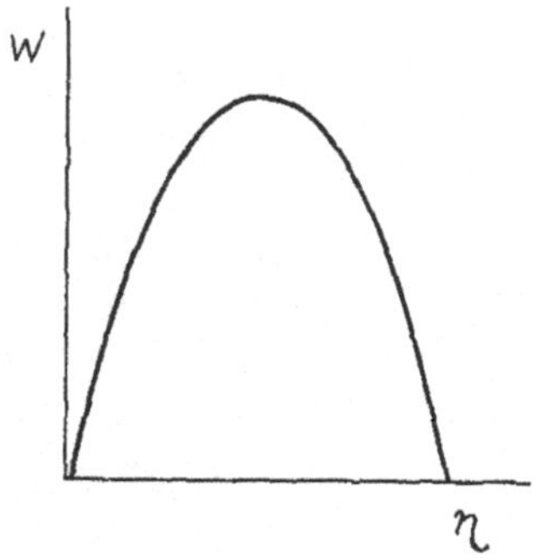

FIGURE 22. Relation between w and η.

§2. Oxidation of Phosphorus

As was already indicated, chain theory grew in two stages. The first stage, starting in 1913 evolved the theory of chains in photochemical reactions; the second stage, beginning in 1927, started its wide application to thermal explosions.

The role played by the $H_2 + C\ell_2$ reaction in the first stage, was played by the oxidation of phosphorus and oxygen in the second. Chronologically, the first studies on phosphorus oxidation were carried out (1926 and 1927) before the subsequent kinetic investigations on $H_2 + O_2$ (1928).

1. Experimental data on Phosphorus Oxidation [1]. The oxidation of phosphorus presents two characteristics. It is a violent process accompanied by flames but it does not go at all if the oxygen pressure exceeds a certain limiting value or is below a certain minimum limit.

The first of these effects was already observed by Boyle three hundred years ago and the second by Goubert, in a poorly defined way, in 1874. The phenomenon of the upper limit was relatively well studied in some detail by Zentnerschwer in 1898 but that of the lower limit was not studied quantitatively at all and even qualitatively it was barely mentioned by Goubert whose experiments did not warrant any conclusion.

The upper limit in phosphorus oxidation corresponds to a pressure of about one atmosphere, the pressure of phosphorus being then equal to 10^{-2} mm Hg. Thus the upper limit occurs at very high dilutions, 99.998% of oxygen and 0.002% of phosphorus.

The lower and upper limits in phosphorus combustion were studied quantitatively and interpreted theoretically in 926-1929 by the Leningrad group of physicists (Khariton and Val't [3], Semenov [4]*, Koval'skiǐ [5] and others).

A series of investigations was carried out to establish firmly the reality of the fact that oxygen practically does not react with phosphorus unless the pressure of oxygen reaches a lower limiting value ($\sim 10^{-2}$ to 10^{-3} mm Hg).

In order to prove that the transition from complete absence of reaction to fast reaction with explosion is a sudden one and corresponds to a strictly determined oxygen pressure, two experimental methods were used.

A. The method of capillary flow. Oxygen is slowly introduced through a narrow capillary from a flask into the vessel containing phosphorus. The oxygen pressure in the vessel is measured by means of a sulfuric acid manometer (the position of the meniscus is observed through a low power microscope). Then curves are obtained as shown on Figure 23, on the following page. The instant at which emission of light occurs is indicated by an arrow. The light then continues while the flow proceeds. We see that, before the flash, the oxygen reaching the vessel is accumulated in it and does not react. Indeed, the curve is then linear with a slope determined by the width and length of the capillary and the oxygen pressure in the flask.

* together, in part, with Shal'nikov.

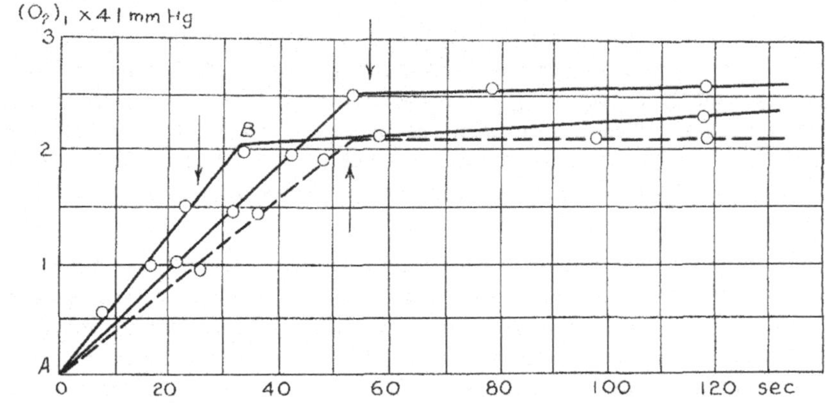

FIGURE 23. Pressure change as oxygen is introduced in a vessel containing phosphorus.

The flash is accompanied by a break in the curve. After that the oxygen pressure in the vessel does not increase any more with time although oxygen is continuously introduced through the capillary (the line is now almost parallel to the horizontal axis). This means that the oxygen introduced burns away, forming P_2O_5 that condenses on the walls. If the oxygen flow is interrupted, the flame immediately disappears and the pressure does not go below $[O_2]_1$. It may therefore be concluded that for reaction to occur and also to subside, the oxygen pressure must reach the same limiting value $[O_2]_1$ (as will be seen later, the quenching pressure is slightly below the ignition pressure).

Special experiments show that the gas remaining after the cessation of the light emission is indeed oxygen and not P_2O_3 or any other volatile compound of phosphorus with oxygen.

It can also be shown that, at a pressure slightly below $[O_2]_1$, oxygen may be kept in contact with phosphorus vapor during days, without any appreciable reaction to take place.

Indeed, after a 24-hour contact between phosphorus and oxygen at a pressure slightly below $[O_2]_1$, it is necessary to introduce through the capillary a negligible quantity of oxygen which is nevertheless sufficient to bring its pressure to $[O_2]_1$ and to cause explosion.

B. The method of compression. The second method is based on the fact that the volume of the vessel containing phosphorus and oxygen at a pressure $[O_2] < [O_2]_1$ is decreased by the introduction of mercury. Then the flash always occurs completely suddenly when, as a result of the compression, the oxygen pressure reaches its critical value $[O_2]_1$. The flash in this case is instantaneous; it disappears since the pressure falls very rapidly below $[O_2]_1$. In these experiments, it is possible to observe a minute difference between the pressure $[O_2]_1$ for ignition and the pressure $[O_2]_1^*$ for quenching. The first is a fraction of a percent larger than the latter and the difference is larger when the flash is more intense.

A study was made of the relation between the critical pressure $[O_2]_1$ and the other factors: temperature, pressure of phosphorus vapors, pressure of inert gases, vessel size. This succeeded in establishing for the first time the amazing facts that proved subsequently to be general characteristics of chain reactions.

Already in 1926, Khariton and Val't observed that addition of argon into a vessel containing phosphorus vapor and oxygen at a pressure $[O_2] < [O_2]_1$ would cause explosion. It appears then that addition of argon lowers the critical pressure of oxygen. A systematic investigation of the phenomenon showed that this was indeed the case.

The relation between $[O_2]_1$ and the argon pressure $[A]$ is of the form:

$$[O_2]_1 \left(1 + \frac{[A]}{[O_2]_1 + [P_4]} \right) = \text{const.}$$

where $[P_4]$ is the pressure of phosphorus vapors. Figure 24 shows the experimental relation between $[O_2]_1$ and $[A]$ while Figure 25 shows the averaged curve representing the data of Figure 24, in the coordinates

$$\frac{1}{[O_2]_1} \quad \text{and} \quad 1 + \frac{[A]}{[O_2]_1 + [P_4]} .$$

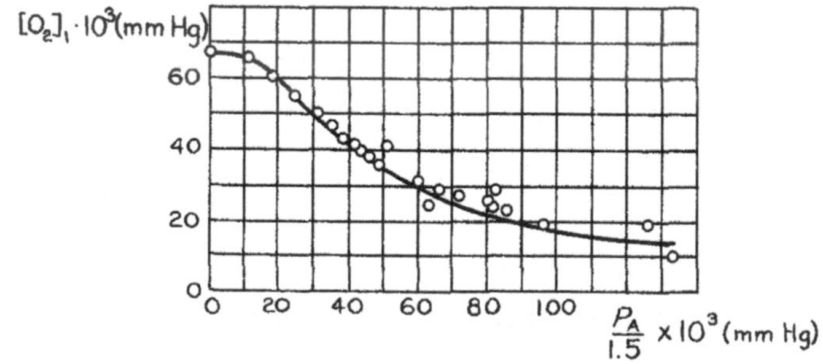

FIGURE 24. Lowering of the lower limit $[O_2]_1$ as a function of argon pressure (data of Semenov).

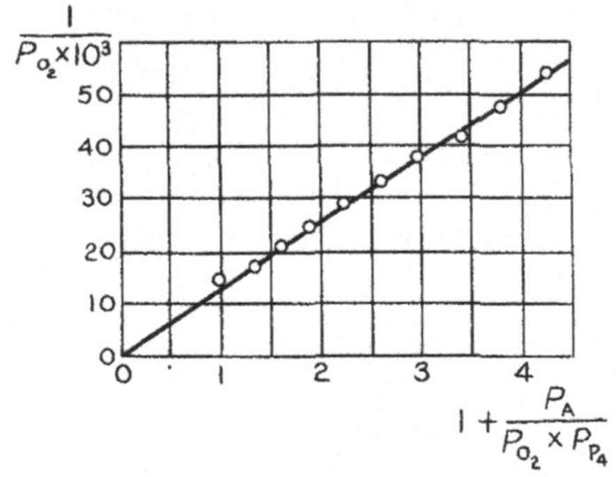

FIGURE 25. Effect of argon on the lower limit $[O_2]_1$ (following data of Figure 19).

Later, Shal'nikov, using a vessel of very large diameter found a value of $[O_2]_1$ much lower than usual.

It was then quickly discovered that the critical pressure is a function of the diameter d of the vessel and decreases as the inverse of d^2.

The table below illustrates this fact.

A similar relation holds for cylindrical vessels. But in cylinders and spheres of identical diameters, the values of $[O_2]_1$ are distinctly different ($[O_2]_1$ is smaller in a cylinder).

diameter of the bulb, cm	$[O_2]_1$ mm Hg	$[O_2]_1 \cdot d^2$
6	$5.27 \cdot 10^{-3}$	$190 \cdot 10^{-3}$
13.4	$1.16 \cdot 10^{-3}$	$209 \cdot 10^{-3}$
18.1	$0.61 \cdot 10^{-3}$	$200 \cdot 10^{-3}$

When the temperature is changed between 17 and 50°C, the phosphorus pressure being kept constant, $[O_2]_1$ stays practically the same. Therefore if there is a dependence of $[O_2]_1$ on T, it must be a very weak one.

When the pressure of phosphorus is increased, $[O_2]_1$ decreases. The relation $[O_2]_1 \sqrt{[P_4]}$ = const. was found at first, in a restricted range of $[P_4]$, but subsequently the more accurate data of Koval'skiĭ were found to fit the relation $[O_2]_1\ [P_4]$ = const.

Summarizing all the experimental evidence, we proposed the purely empirical relation:

$$d^2[O_2]_1[P_4]\left(1+\frac{[A]}{[O_2]_1+[P_4]}\right) = \text{const.} \qquad (18)$$

between the critical pressure of oxygen $[O_2]_1$, the pressure of phosphorus $[P_4]$, the pressure of argon [A] and the reactor diameter d.

In cylindrical reactors, the constant in (18) is approximately equal to 1 to $1.5 \cdot 10^{-3}$ if d is expressed in cm and all pressures in mm Hg.

In 1931, Melville and Ludlam [6] made a careful check of equation (18) as expressing a relation between the critical pressure $[O_2]_1$ and the pressure of an inert gas. Fourteen different gases were tried.

It turned out that equation (18) was quite satisfactory if a correction was made, namely: the concentration of inert gas [A] must be multiplied by some constant a, different with different gases (Figure 26).

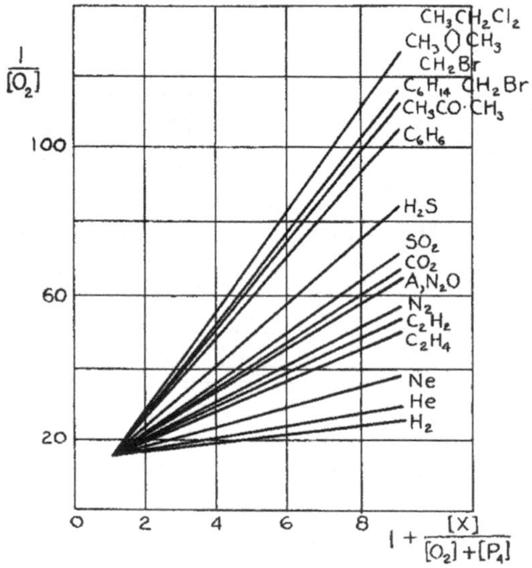

FIGURE 26. Effect of various foreign gases on the lower limit (data of Melville and Ludlam [6]).

Thus instead of equation (18) we have:

$$d^2[O_2]_1[P_4]\left(1 + \frac{a[A]}{[O_2]_1 + [P_4]}\right) = \text{const.} \qquad (18')$$

The constant a increases with molecular weight but does not depend on the chemical structure of the gas. The gases studied were CO_2, N_2, NO, ethylene, neon, hydrogen, argon, acetone, mesytilene, methylene chloride, benzene, hexane, hydrogen sulfide, SO_2 and acetylene. In order of magnitude, a is close to unity. For the different gases, it changes between 0.1 and 1.

The upper limit $[O_2]_2$ of oxygen pressure was studied by Koval'skiǐ in the following way. A vessel containing phosphorus was put in a liquid air bath and pumped down to a high vacuum. Then, without removing the liquid air bath, oxygen was introduced in the vessel to a pressure $p > [O_2]_2$. The liquid air bath was then removed, the vessel heated to some temperature T (depending on what pressure of phosphorus vapor was desirable) and the oxygen was slowly removed through a capillary. At a perfectly determined oxygen pressure $[O_2]_2$, the phosphorus vapor in the volume exploded. This was followed by violent combustion of the solid phosphorus in the vessel. The pressure was recorded manometrically.

By changing T, the relation between $[O_2]_2$ and the pressure of phosphorus could be found. It can be expressed by the equation:

$$\frac{[O_2]_2}{[P_4]} = \text{const.} \tag{19}$$

Thus

$$[O_2]_2 = \frac{1}{C_2}[P_4]$$

and the pressure of oxygen $[O_2]_2$ is directly proportional to the pressure of phosphorus vapor.* The constant C_2 is independent of vessel size and equal to $2.8 \cdot 10^{-5}$.

Since $[O_2]_1 = C_1/[P_4]$ and $[O_2]_2 = (1/C_2)[P_4]$, when the pressure $[P_4]$ is lowered, the lower limit is raised while the upper limit becomes smaller so that the region of oxygen pressures where combustion takes place, disappears altogether.

The work of Koval'skiǐ was the first to relate the phenomena of the lower and upper limits and to show that these limits are the boundaries of the same process of combustion of phosphorus vapors in oxygen.

In a doubly logarithmic plot of oxygen pressure versus phosphorus vapor pressure, the explosion region occupies an area bounded by the curve shown in Figure 27 where the points represent the experimental values at the lower and upper limits.

In this diagram, shown on the following page, a line a - b corresponding to a change in oxygen pressure at constant pressure of phosphorus vapor, intersects twice the limits at points a and b representing respectively the lower and upper explosion limits $[O_2]_1$ and $[O_2]_2$. Over an extended range, the left and right branches of the curve are practically straight lines with slopes of $45°$ and $90° + 40°$. Thus the equation of

---

* In fact, experiment gives a slightly different relation: $[O_2]_2 = C[P_4]^{0.85}$. It is conceivable that experimental errors account for departure of the exponent from unity. But, as will be seen presently, the relation $[O_2]_2 = C[P_4]$ obtains only at low pressures $[O_2]$ and $[P_4]$. When $[P_4]$ becomes larger, it tends to the relation $[O_2]_2 = C[P_4]^{0.5}$.

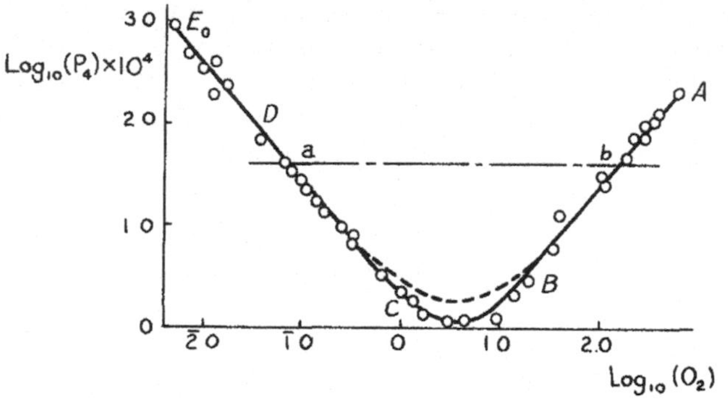

FIGURE 27. Explosion domain for phosphorus [5].

the left branch (lower limit) is $\log [O_2]_1 = -\log [P_4] + \log c_1$ or $[O_2]_1 = C_1/[P_4]$ and that of the right branch is $\log [O_2]_2 = \log [P_4] + \log (1/C_1)$ or $[O_2]_2 = (1/C_2)[P_4]$, in agreement with the experimental relations discussed above. Near the minimum, both the left and the right branches become curved and for a certain value of $[P_4]$, both limits coincide. At pressures below 10^{-4} mm Hg, combustion becomes impossible at any value of the oxygen pressure.

The entire curve is well represented by the quadratic equation:

$$C_1 - [O_2][P_4] + C_2[O_2]^2 = 0 \; ^* \tag{20}$$

the roots of which are:

$$[O_2]_1 = \frac{[P_4] - \sqrt{[P_4]^2 - 4C_1C_2}}{2C_2} \quad \text{and} \quad [O_2]_2 = \frac{[P_4] + \sqrt{[P_4]^2 - 4C_1C_2}}{2C_2}$$

the lower and upper limits respectively.

When $[P_4] < \sqrt{4C_1C_2}$, the roots become imaginary and combustion becomes impossible.

If $[P_4] > \sqrt{4C_1C_2}$:

$$[O_2]_1 = \frac{C_1}{[P_4]} \quad \text{and} \quad [O_2]_2 = \frac{[P_4]}{C_2} \quad .$$

---

*

$$C_1 = \frac{10^{-3}}{d^2} \quad \text{and} \quad C_2 = 2.8 \cdot 10^{-5} \quad .$$

Even in the early work, it was found that the upper limit is very sensitive to the presence of traces of certain substances with a specific action.

Koval'skiĭ discovered that immeasurably small quantities of ozone have a strong effect on the upper limit. They raise the latter and since they enlarge the combustion region, they make the mixture more explosive.

Zentnerschwer [7] and Tausz and Gerlacher [8] showed that a series of substances strongly decrease the upper limit. Among these are molecules such as ethylene that have absolutely no specific action on the lower limit.

Let us call $[O_2]_2^x$ or p_x the value of the upper limit in the presence of x% of an additive. Then, according to Tausz and Gerlacher,

$$p_x = \frac{c}{a + x} \cdot$$

The quantity $\frac{1}{c}$ was called by them the poisoning coefficient.

They give the following table (Table 45) of values of constants for various additives:

TABLE 45

	a	c	$\frac{1}{c}$
Hydrogen Sulfide	23	13520	0.000074
Benzene	21	10805	0.000093
Cyclohexane	20	10107	0.000099
Acetylene	11.3	7017	0.00014
Methylcyclohexane	8.2	4577	0.00022
Ethylene	1.75	1073	0.0093
Propylene	0.35	168	0.006
Cyclohexene	0.11	58	0.017
Isoprene	0.025	20	0.05
Iron Carbonyl	0.0033	1.7	0.59

It is likely that this specific effect of certain substances (including ozone that is always formed during combustion) explains the fact discovered by Koval'skiĭ that definite values of $[O_2]_2$ are obtained only in a fresh vessel.* If two experiments are run in a given vessel, the second value of $[O_2]_2$ is considerably larger than the first.

Before considering the theory, it must be noted that the

* After each experiment, the reaction vessel was removed and replaced, so that the number of points on the right branch of the curve of Figure 22 represents the number of vessels used.

phosphorus flame at pressures slightly in excess of the lower limit, is a cold flame. This follows from the fact that when phosphorus vapors are ignited (e.g., by letting oxygen in through a capillary), practically no pressure increase takes place. Upon quenching, the pressure does not decrease.

The absence of heating is understandable since at these low pressures, the quantity of fuel is very small but the heat capacity of the system is not so small.

It has been emphasized already that the existence of such cold flames with a sharply defined ignition point, is one of the strongest arguments in favor of a pure chain type of explosion. In such cases, a trivial thermal theory fails completely.

2. Theory of the Lower Limit. The analysis of the experimental data obtained in these investigations on the oxidation of phosphorus vapors led to the establishment of the concept of chain branching and to the first formulation of the theory of branched chain reactions. An important role in the development of the theory was the newly introduced idea of chain termination on the walls of the reactor. This idea had been proved directly already by the study of Trifonov [9] (see Chapter VI, Section 1). As was explained already, it was shown there that chlorine atoms are captured by the walls at a rate determined by their diffusion to the wall. The average time τ for diffusion may be found from the Einstein relation $D\tau = x^2$ where x is the distance between the point of generation of the free radical and the wall and D is the diffusivity.

In a vessel with plane parallel walls, with a distance d between walls, the generation of chlorine atoms takes place at an average distance $x = d/2$ and $\tau = d^2/4D$. When radicals generated uniformly throughout the volume, the average time required for transport to the wall is:

$$\bar{\tau} = \frac{\int_{-d/2}^{+d/2} \frac{x^2}{D}\, dx}{d} = \frac{d^2}{12D} \quad . \tag{21}$$

The average rate of chain termination in the entire volume is then $n_0'(1/\bar{\tau}) = 12\, Dn_0'/d^2$ where n_0' is the total number of primary radicals in the reaction vessel. The average rate of chain termination per unit volume is

$$kn_0 = \frac{12D}{d^2}\, n_0$$

where

$$n_0 = \frac{n_0'}{V}$$

(v, reactor volume).

Let us now consider the oxidation of phosphorus vapors. It is assumed that this is a chain reaction similar to that between hydrogen and chlorine in which H and Cℓ atoms are the chain carriers. There are good reasons to believe that oxygen atoms are one of the radicals propagating the chain of oxidation of phosphorus vapors, since a large quantity of ozone is formed during the process. The following reaction scheme had been proposed [4, 10, 11]:

$$\text{Chain}\begin{cases} 1)\ O + P_4 \longrightarrow P_4O & \text{rate: } k_1(P_4)(O) = a_1(O)^* \\ 2)\ P_4O + O_2 \longrightarrow P_4O_2 + O & \text{rate: } k_2(P_4O)(O_2) = a_2(P_4O) \text{ etc.} \end{cases}$$

The new features that we introduced are the following:

1. P_4O_2 is further oxidized to P_2O_5 via a series of intermediate oxides, with regeneration of O atoms: this introduced the concept of a branching chain with overall reaction:

$$3)\ P_4O_2 + O_2 \longrightarrow P_2O_5 + \mu O \ .$$

The rate of process (3) is $k_3(O_2)(P_4O_2) = a_3(P_4O_2)$.

2. The active species O and P_4O are trapped by the wall upon collision with it. This introduced chain termination steps on the walls:**

---

* Curved brackets (e.g., (O_2), (P_4), (O) are used to denote concentrations (number of particles per cm^3). Square brackets (e.g., $[O_2]$, $[P_4]$, $[O]$ are used to denote partial pressures, in mm Hg. There is thus a relation $(x) = N[x]$ where N is the number of particles per unit volume at a pressure of 1 mm Hg and at a temperature T°K.

$$N = \frac{2.7 \cdot 10^{19} 273}{760 \cdot T} = \frac{0.97 \cdot 10^{19}}{T}$$

where T is the absolute temperature. At 0°C or 273°K, $N = 3.55 \cdot 10^{16}$. The pressure of the mixture $P_4 + O_2$ will be represented by p, the fraction of oxygen by γ, that of phosphorus by $(1 - \gamma)$. Then $[O_2] = \gamma p$ and $[P_4] = (1 - \gamma)p$. The total pressure of the mixture diluted with an inert gas is $P = [O_2] + [P_4] + [N_2] = \frac{1}{N}\{(O_2) + (P_4) + (N_2)\}$.

** At the present time, it is believed that the reaction mechanism is somewhat different. Thus in Chapter III, we proposed the scheme:

$$1)\ O + P_4 \longrightarrow PO + P + P_2$$

$$2)\ P + O_2 \longrightarrow PO + O$$

$$3)\ P_O + O_2 \longrightarrow PO_3;\ PO_3 + P_4 \longrightarrow P_2O_3 + P + P_2$$

$$P + O_2 \longrightarrow PO + O_2 \quad \text{(etc., branching)} \ .$$

4) O + wall rate: $K_1 = \frac{12D_O}{d^2}(O)$

5) P_4O + wall rate: $K_2 = \frac{12DP_4O}{d^2}(P_4O)$.

For simplicity, let us put $K_1 = K_2 = K$, i.e., assume that the diffusivities of O and P_4O are equal. This does not introduce a large error since, in the more likely mechanism shown in the footnote, K_2 is related to the diffusion of a P atom. Assume also that some small quantity of O atoms is formed primarily in the vessel, e.g., as a result of spontaneous endothermic process taking place in the volume or on the wall:

$$P_4 + O_2 \longrightarrow P_4O + O \quad .$$

This process releases an oxygen atom in the volume. Such reactions are very slow since oxygen and phosphorus may coexist during days below the limit without the slightest trace of reaction.

Let us designate by n_o the small number of O atoms formed per second per unit volume. Then:

$$\frac{d(O)}{dt} = n_o - (a_1 + K)(O) + a_2(P_4O) + \mu a_3(P_4O_2) \qquad (a)$$

$$\frac{d(P_4O)}{dt} = a_1(O) - (a_2 + K)(P_4O) \qquad (b) \qquad I$$

$$\frac{d(P_4O_2)}{dt} = a_2(P_4O) - a_3(P_4O_2) \quad . \qquad (c)$$

Summing the first two equations gives:

$$\frac{d[(O) + (P_4O)]}{dt} = n_o - K[(O) + (P_4O)] + \mu a_3(P_4O_2) \quad ,$$

* (cont.)

When phosphorus is oxidized at low pressures, appreciable deposits are observed which consist of polymerized lower oxides. The mathematical treatment of this scheme leads to just about the same results as the simpler scheme just recalled that was proposed by us in 1928-1929.

In our book 'Chain Reactions' (1934), we treated this scheme by means of the method of probabilities. This appeared clearer at the time, in view of the novelty of the concept. Now it is simpler to use the kinetic equations. Of course, the same results are obtained.

where μ is the number of O atoms formed per branching step. The quantity n_0 is very small and constant. The second term represents the rate of destruction of both active particles on the wall i.e., the rate of chain termination, the third represents the rate of generation of new chains by branching i.e., the rate of branching. If the rate of termination is larger than that of branching, then some constant concentration of active particles is quickly established. It is a very small amount since n_0 is small. When the third term is larger than the second, i.e., when the rate of branching is larger than that of termination a stationary solution is not possible, the quantity of active particles will grow with time and quite rapidly if a_3 is large, which is precisely the case here. The rapid acceleration of the rate leads to a very rapid reaction of the type of an explosion.

Since K is directly proportional to the diffusivity D, which in turn is inversely proportional to pressure, and a_3 is directly proportional to pressure, as pressure increases, the rate of termination decreases and that of branching increases. Therefore, when the initial pressure of the mixture is raised, we go over from a negligibly slow reaction to a very fast one. This transition, as was shown in the first section, takes place over a very narrow range of pressure, if the quantity a is large. In the vicinity of this interval lies the pressure corresponding to the equality between rates of termination and branching. Thus the remarkable phenomenon of chain ignition receives a simple explanation: a substance is completely inert at a pressure less than a critical value p_1 but reacts violently at a pressure slightly in excess of p_1.

To determine the lower limit p_1, it is necessary to find the conditions for which no stationary solution is possible. To do this, it is not necessary to solve the system of differential equations (although this is also easy). It is sufficient to find the solution for the steady state process, according to the system of equations I:

$$\frac{d(O)}{dt} = \frac{d(P_4O)}{dt} = \frac{d(P_4O_2)}{dt} = 0$$

and to find the conditions under which such a solution is impossible.

Let us solve these three algebraic equations. From equations (b) and (c) of I, one finds:

$$(P_4O) = \frac{a_1}{a_2 + K}(O); \qquad (P_4O_2) = \frac{a_2}{a_3}(P_4O) = \frac{a_1 a_2 (O)}{a_3(a_2+K)}$$

Substituting in (a), we get:

$$n_o - (a_1 + K)(0) + \frac{a_1 a_2}{a_2 + K}(0) + \frac{\mu a_1 a_2}{a_2 + K}(0) = n_o + R(0)$$

where

$$R = - \frac{\{K^2 + K(a_1 + a_2)\} + \mu a_1 a_2}{a_2 + K} \quad . \tag{22}$$

A steady state solution $(0) = n_o/R$ is possible if $R < 0$ and impossible if $R \geq 0$. The oxygen pressure at the lower limit $[O_2]_1$ or the limiting pressure p_1 of the mixture of oxygen and phosphorus, are determined by the condition $R = 0$. The first term of the numerator of equation (22) is proportional to the rate of termination, the second term to the rate of branching.

Let us determine $[O_2]_1$ from the condition $R = 0$ or

$$\mu a_1 a_2 - K^2 - K(a_1 + a_2) = 0 \quad . \tag{23}$$

Before introducing the value of the constants into equation (23), let us note that a close analysis of the diffusion equations for the case of branching, shows that $K = \pi^2 D/d^2$ instead of the value $K = 12D/d^2$ obtained earlier for the case of unbranched reactions. Both values are almost the same. A small decrease of the constant is due to the fact that the concentration of active particles drops from the center of the reactor to the wall. Therefore the rate of branching will also be larger in the center while n_o has the same value everywhere. As a result of the averaging, $\bar{\tau}$ is larger and therefore K smaller.

Thus:

$$K = \frac{\pi^2 D}{d^2} = \frac{\pi^2 D_o}{pd^2} = \frac{K_o}{p} \tag{24}$$

for a plane parallel reactor. Here D_o is the diffusivity of oxygen and phosphorus atoms at a pressure of the mixture $p = 1$ mm Hg. Calculation shows that for a cylindrical reactor:

$$K = \frac{23D}{d^2} = \frac{23D_o}{pd^2} = \frac{K_o}{p} \tag{25}$$

and for a spherical reactor:

$$K = \frac{4\pi^2 D}{d^2} = \frac{4\pi^2 D_o}{pd^2} = \frac{K_o}{p} \quad . \tag{26}$$

In what follows, we will use the expression for a cylindrical reactor. The constants are:

$$a_1 = k_1(P_4) = k_1N[P_4] = k_1N(1 - \gamma)p$$
$$a_2 = k_2(O_2) = k_2N[O_2] = k_2N\gamma p \quad .$$

Solution of the equation $R = 0$ gives the critical value $[O_2]_1 = p_1\gamma$:

$$\mu a_1a_2 - K(a_1 + a_2) - K^2 = 0$$
$$\mu k_1k_2N^2\gamma(1 - \gamma)p^2 - K_0N\{k_1(1 - \gamma) + k_2\gamma\} - \frac{K_0^2}{p^2} = 0 \quad . \tag{27}$$

Solving, we get:

$$p_1^2 = c(1 + \sqrt{1 + B}) \tag{28}$$

where

$$C = \frac{K_0\{k_1(1-\gamma) + k_2\gamma\}}{2\mu k_1k_2\gamma(1-\gamma)N}$$

and

$$B = \frac{4\mu k_1k_2\gamma(1-\gamma)}{[k_1(1-\gamma) + k_2\gamma]^2} \quad .$$

In the oxidation of phosphorus, k_1 is close to k_2. Indeed, the observed values of $[O_2]_1$ do not depend on temperature, within experimental error, in the range $- 40$ to $+ 40°$. This means that both reactions of chain propagation proceed practically without activation energy at a rate equal to that of collision between O and P_4O on the one hand and O_2 and P_4 on the other hand. Since the collision cross-sections are close to each other, k_1 and k_2 are also practically equal. If thus $k_1 = k_2$, we get

$$C = \frac{K_0}{2\mu k\gamma(1-\gamma)N}$$

and

$$B = 4\mu\gamma(1 - \gamma) \quad .$$

It is not easy to estimate μ. It seems that it is close to unity since P_4O_2 may disappear via alternative paths (e.g., polymerization) that do not yield additional O and P atoms.

For an equimolar mixture $\gamma(1-\gamma) = 0.25$ and thus $B \sim 1$. For non-stoichiometric mixtures $B < 1$. Even when $B = 1$, we get a value $1 + \sqrt{1+B} = 1 + \sqrt{2} = 2.4$ only 20% in excess of that corresponding to $B \sim 0$ when $1 + \sqrt{1+B} = 2$. Therefore, within the degree of approximation of interest to us here, we may neglect B and write an approximate expression for p_1:

$$p_1^2 = \frac{23D_o}{d^2 k\gamma(1-\gamma)N} \tag{29}$$

or

$$d^2[P_4][O_2] = \frac{23D_o}{kN} = \text{const.} \tag{30}$$

This corresponds to the experimental relation (compare (18) with $[A] = \text{const.}$). Experimentally, at room temperature and a phosphorus pressure $2.5 \cdot 10^{-2}$, $\text{const.} = 1 - 2 \cdot 10^{-3}$.

Let us try to calculate from theory an absolute value of the critical oxygen pressure $[O_2]_1$; at 20°, the vapor pressure of phosphorus is $[P_4] = 2.5 \cdot 10^{-2}$; for the diffusivity of O atoms in the mixture at 760 mm, we take ~ 0.3. Then $D_o = 0.3 \cdot 760 = 230$, $N = 3.3 \cdot 10^{16}$, $d \sim 2$ cm. From the relation

$$\frac{23D_o}{kN} = 1.25 \cdot 10^{-3},$$

we find $k = 10^{-10}$. Thus the reactions 1) $O + P_4$ and 2) $P_4O + O_2$ occur on each collision as we have assumed on the basis of the absence of a temperature coefficient and of the fact that P_4 and O_2 enter symmetrically in equation (30). Substituting into (30) the values of D_o, k, d and $[P_4]$, we obtain $[O_2]_1 = 1.5 \cdot 10^{-2}$. This is also approximately the experimental result at room temperature (1). This agreement demonstrates rather well the correctness of the theory.

Consider now the theoretical effect of dilution by an inert gas on the limiting pressure. The constants a_1 and a_2 depend only on partial pressures $[P_4]$ and $[O_2]$. As to the diffusivity D, it depends on the total pressure P of the mixture ($K = K_o/P$). Thus, in the last term of (27), K_o^2/p^2 must be replaced by K_o^2/P^2 and in the second term the factor p/P must be introduced. As a result, we get a quadratic equation in Pp. Thus, instead of (28), we get:

$$Pp = c(1 + \sqrt{1+B}) \tag{31}$$

with B and C as before.

Neglecting B, we get:

$$Pp = \frac{23D_0}{d^2k\gamma(1-\gamma)N} \tag{32}$$

and since $P = p + [A] = [P_4] + [O_2] + [A]$:

$$\frac{P}{p}\, p^2\gamma(1 - \gamma)d^2 = d^2[P_4][O_2]\left\{1 + \frac{[A]}{[P_4] + [O_2]}\right\} = \text{const.}$$

$$= 23D_0/kN = 1.5 \cdot 10^{-3} \quad .$$

But this is precisely the experimental relation (compare (18')).

Therefore, we have obtained theoretically the empirical formula (18) which summarizes all experimental facts related to the study of the lower ignition limit of phosphorus vapors.

3. Theory of the Upper Limit. The upper limit is caused by homogeneous chain termination. Then the rate of termination must depend on oxygen pressure more strongly than branching.* At some pressure $p_2 = [O_2]_2$, the rate of termination becomes equal to that of branching. At higher pressures, termination exceeds branching and reaction stops. This is why ignition of phosphorus vapors exhibits an upper limit. At the high pressures involved, it is possible of course to neglect wall termination, just as homogeneous termination may be neglected at low pressures.

Let us add to the system I of equations, a term $F(O)$ describing homogeneous chain termination due to destruction of oxygen atom. The terms corresponding to wall termination may now be omitted in equations (a) and (b) by putting $K = 0$. The limit can now be obtained from the relations:

$$\frac{d(O)}{dt} = \frac{d(P_4O)}{dt} = \frac{d(P_4O_2)}{dt} = 0 \quad .$$

Solution puts the explosion condition in the form:

$$\mu a_1 - F = 0 \quad . \tag{33}$$

Koval'skiĭ assumed [12] that homogeneous chain termination is due to destruction of O atoms as a result of their collisions with small

---

* Since the pressure of phosphorus vapors near room temperature is negligibly small as compared to the oxygen pressure at the upper limit, the limiting pressure is practically equal to that of oxygen.

traces of active impurities contained in the oxygen. In this case, $F = k_y y(O_2)$ where y is the fraction of impurities in oxygen and k_y the rate constant for the reaction between oxygen atoms and the molecules of impurities. Then:

$$\mu a_1 - F = \mu k_1(P_4) - k_y y(O_2) = 0 \quad . \tag{34}$$

The upper limit $(O_2)_2$ is:

$$(O_2)_2 = \frac{\mu k_1}{y k_y}(P_4) \tag{35}$$

and

$$[O_2]_2 = \frac{\mu k_1}{y k_y}[P_4] = c_1[P_4] \quad . \tag{36}$$

This is also the relation found experimentally by Koval'skiĭ for very small values of $[P_4]$ and not too high values of $[O_2]_2$. This can be seen from Figure 27 where the right branch of the curve is, at first, a straight line with slope equal to unity. But at higher pressures, the line curves over to a straight line of slope equal to two. This can also be seen in Figure 28, representing the data of Tausz and Gerlacher [8]. This means that $[O_2]_2 = c_1'\sqrt{[P_4]}$. In his analysis of upper limit data in the oxidation of phosphine, Dalton [13] proposed that the termination step was the formation of ozone in triple collisions: $O + O_2 + M \longrightarrow O_3 + M$. It is quite conceivable that the same termination occurs in the oxidation of phosphorus. Any third body may serve as M. Since the pressure of phosphorus is small, in undiluted mixtures $(M) = (O_2)$. The rate of termination is then $x(O_2)^2(O)$ where x is the rate constant for triple collisions. The condition for the limit becomes:

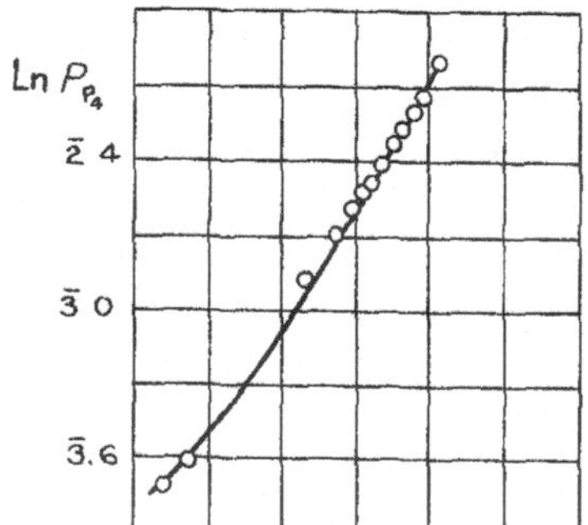

FIGURE 28. Upper limits in the oxidation of phosphorus [8].

$$\mu a_1 - x(O_2)^2 = \mu k_1(P_4) - x(O_2)^2 = 0 \tag{37}$$

Thus:

$$(O_2)_2 = \sqrt{\frac{\mu k_1}{x}}\sqrt{(P_4)} \tag{38}$$

$$[O_2]_2 = \sqrt{\frac{\mu k_1}{x}}\,\sqrt{[P_4]} \tag{39}$$

It is important to decide which slope, one or two, is the right one, since in the first case only the ratio

$$[O_2]_2/[P_4] = (\mu k_1/\gamma k_y) = \frac{\gamma}{1-\gamma}$$

has a limiting value. Thus reaction stops above a certain oxygen content in the mixture and there is no upper limit in the $P_4 + O_2$ mixture with respect to total pressure. In the second case:

$$p_2 = \frac{1}{\gamma}\sqrt{\mu k_1/xN}\,\sqrt{p_2(1-\gamma)} \tag{40}$$

has a limiting value.

For the oxidation of phosphorus vapors, it is difficult to decide this question directly from the experiment, because the pressure of phosphorus is considerably lower than that of O_2. But, by analogy with phosphine, silane and other molecules, an upper limit with respect to pressure ought to exist. This then is an argument in favor of the second type of homogeneous chain termination.

The most difficult point is to estimate the value of x. Experiments on third body recombination of H and $C\ell$ atoms, give $x = 10^{-32}$ for $H + H + M$. For normal cross-sections, this figure is one hundred times higher than theoretical. This must be due to some peculiar feature of the recombination of simple particles. In our case where an atom reacts with a molecule, it is quite likely that the constant x will be 10 to 50 times smaller (especially since ozone may perhaps not be formed on every triple collision).

The limit $[O_2]_2$ is modified if oxygen is diluted by an inert gas, e.g., nitrogen. Nitrogen molecules are as effective third bodies as oxygen molecules. Therefore equation (37) becomes:

$$\mu k_1(P_4) - x(O_2)[(O_2 + N_2)] = \mu k_1[P_4] - x[O_2]\{[O_2] + [N_2]\}N = 0 \tag{41}$$

or

$$[O_2]_2 P = \frac{\mu k_1[P_4]}{Nx} \tag{42}$$

i.e., at a given (P_4), the product $P \cdot p_{(O_2)_2}$ is a constant, where P is the total pressure and $P_{(O_2)_2}$ is the partial pressure of oxygen at the limit.

TABLE 46

$\% N_2$	$p_{(O_2)_2}$	P	$P \cdot p_{(O_2)_2}$
0	600	600	360000
37	500	800	400000
60	400	1000	400000
75	300	1200	360000
90	200	2000	400000

In columns 1, 2 and 3 of Table 46, the data of Tausz and Gerlacher [8] are given. Column 4 shows the product of total pressure and partial pressure of oxygen at the upper limit. It is seen that, in agreement with formula (41), this product remains constant when the nitrogen content of the mixture changes from 0 to 90%. Theory is therefore upheld by the experiment.

From the experimental value of

$$P \cdot p_{(O_2)_2} = \text{const.} = 4.10^5 \quad ,$$

the quantity x may be estimated. From formula (42):

$$x = \frac{\mu k_1 [P_4]}{N \cdot \text{const.}} \quad .$$

At room temperature $[P_4] = 2.5 \cdot 10^{-2}$, $k_1 = 10^{-10}$, $\mu = 1$, $N = 3.35 \cdot 10^{-16}$. Hence:

$$x = \frac{10^{-10} \cdot 2.5 \cdot 10^{-2}}{4.10^5 \cdot 3.25 \cdot 10^{16}} = 1.8 \cdot 10^{-34} \quad .$$

Such a value of x is very acceptable for the process

$$O + O_2 + M \longrightarrow O_3 + M \quad .$$

We have outlined the theory of the upper limit, neglecting chain termination on vessel walls. This is perfectly legitimate when the oxygen pressure at the upper limit is considerably larger than at the lower limit. However, when the pressure of phosphorus becomes small, the value of the lower limit $(O_2)_1$ which is inversely proportional to the concentration (P_4) of phosphorus vapors, increases while the upper limit, directly proportional to (P_4) or to $\sqrt{(P_4)}$ decreases. Thus both limits converge. In this region, it is necessary to consider both types of termination, homogeneous and heterogeneous. It is not difficult to write the corresponding equation which will be quadratic in the case considered by Koval'skiĭ and cubic in that of Dalton. As we have seen, it is a quadratic equation of the type predicted by theory, that describes the experiments of Koval'skiĭ (see Figure 27) where both limits were measured simultaneously down to very low values of phosphorus pressure ($\sim 10^{-4}$ mm Hg). As we have also seen, at higher pressures, chain termination starts to be determined by triple collisions. If both types of homogeneous termination are taken into account, a cubic equation is obtained that describes all the facts. Naturally, at the lower pressures the mechanism of Koval'skiĭ (proportional to (O_2)) will predominate while at the higher pressures, it will be that of Dalton (proportional to $(O_2)^2$). It is apparently necessary to take into account the impurities present in the oxygen. If it were possible to purify oxygen completely, the equation of the limit would be of degree higher than two, the minimum on the experimental curve of Koval'skiĭ would be more extended and the branch of the upper limit would start right away with a slope equal to two.

Consider now from the theoretical viewpoint the lowering of the upper limit by any additive that captures O atoms. If z is the fraction of additive in the oxygen, the concentration of additive is $z(O_2)$ and the rate at which it reacts with O atoms is $k_z z(O_2)(O)$. Therefore a term $k_z z(O_2)$ must be added to the equation (37) of the upper limit. This equation then becomes: $\mu k_1(P_4) - x(O_2)^2 - k_z z(O_2) = 0$, or in terms of partial pressures $[P_4]$ and $[O_2]$:

$$xN^2[O_2]^2 + k_z zN[O_2] - \mu k_1 N[P_4] = 0 \quad .$$

Hence:

$$[O_2]_2 = \frac{-z + \sqrt{z^2 + b}}{c} \quad . \qquad (43)$$

In Figure 29, on the following page, are shown the experimental data of Tausz and Gerlacher [8] relative to the dependence of the upper limit on the % of various impurities in the oxygen [methylcyclohexane (a), ethylene (b), propylene (c), isoprene (d)]. The authors propose an empirical formula

$$p = \frac{k}{a + 100z} \quad .$$

It is easy to show that the theoretical formula just derived does not give a worse fit of the experimental results, with a proper choice of the constants b and c. However, the values of b and c that fit the data are several orders of magnitude larger than theory allows. It appears that the oxygen used in the work of Tausz and Gerlacher was not clean enough. Then, using the formula of Koval'skiĭ,* we get:

$$(O_2)_2 = \frac{\mu k_1(P_4)}{k_y y + k_z z} = \frac{\dfrac{10^2 \mu k_1(P_4)}{k_y}}{\dfrac{10^2 k_y}{k_z} + 100z}$$

$$[O_2]_2 = \frac{(O_2)}{N} = \frac{k}{a + 100y}$$

We thus obtain the empirical formula of Tausz and Gerlacher.

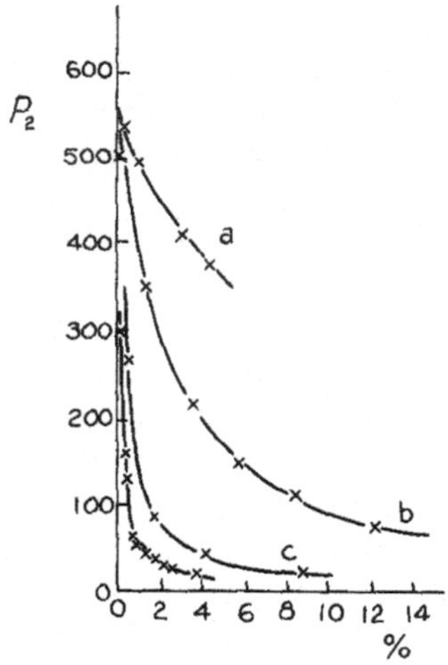

FIGURE 29. Effect of additives on the upper limit in phosphorus oxidation [8].

§3. Oxidation of Sulfur

Sulfur vapors [14] behave in their oxidation with oxygen very much like phosphorus vapors. Also observed in this case are lower and upper limits of oxygen pressure and the reaction between limits proceeds with a weak emission of light (weak because of the very small vapor pressure of sulfur, $p \sim 10^{-4}$ mm Hg at 50° C). Outside these limits, practically no reaction takes place and even at atmospheric pressure, oxidation proceeds at a measurable rate only at about 300°C. But reaction between limits is readily observed at as low a temperature as 50°C. The upper limit with respect to oxygen is distinctly proportional to the square root of the pressure of sulfur vapor.

During the study of this reaction, two facts of fundamental interest were discovered. The first experimental fact is that ignition proceeds only very rarely when oxygen is introduced into a vessel containing sulfur. Usually, it does not take place even though the oxygen pressure lies between both limits, within the domain of explosion. It must be noted that the rate of generation of primary radicals n_0 does not play any important role theoretically since the rate of growth of the chain avalanche

---

* In the presence of an inhibitor, the pressure of oxygen at the upper limit becomes lower and the system operates in the region where the disappearance of oxygen atoms in triple collisions becomes much less effective than their removal by collision with the molecules of inhibitor p initially present in the oxygen.

is so rapid that even a single O atom suffices to start an avalanche. Nevertheless, if this single O atom fails to appear, the avalanche naturally cannot be formed. The generation of an oxygen atom by homogeneous initiation of an oxygen molecule is completely excluded since one thousand years would not be enough time for a single such step to occur at a temperature of 50°C. But this initiation can take place catalytically on the walls, with an O atom being thrown back into the gas phase. In sulfur oxidation such a process is apparently also extremely slow. In this fashion we meet a case where, during a long time, not a single active center is formed in the system.

In order to initiate ignition of sulfur, one might think of producing active centers artifically. For that purpose, a weak electrodeless discharge from a Tesla coil was passed through a mixture of sulfur and oxygen during a few fractions of a second. This was always enough to provoke explosion and combustion of the sulfur then took place as long as there was oxygen left in the vessel which contained solid sulfur. In order to decide whether sulfur or oxygen was responsible for explosion, the discharge was passed during one second through the vessel containing sulfur vapor or through a side tube containing oxygen. It turned out that only a discharge through oxygen is capable of starting ignition when the oxygen is introduced into the vessel containing sulfur. These experiments support the view that oxygen atoms are the active particles of the branched chain reaction.

The second interesting fact is that the main reaction product SO_2 strongly inhibits the reaction and if it is not frozen in a special side-tube, the process following the initial flash is soon arrested completely. Thus SO_2 is an active partner that inhibits strongly the branched chain process and even in small amounts prevents the development of the chain avalanche. Since O atoms are the active particles in chain propagation, this result can be understood if O atoms disappear in the process: $SO_2 + O \longrightarrow SO_3$. Indeed, SO_3 is always observed as a reaction product (up to 20% of the SO_2 produced). This is another fact supporting the view that O atoms participate in the propagation of the chain reaction.

§4. Limits in the Oxidation of Phosphine, Silane, Carbon Disulfide, Carbon Monoxide and in the Decomposition of NCl_3

The typical phenomena of lower and upper limit are widespread among oxidation reactions such as: $PH_3 + O_2$ [15]; $SiH_4 + O_2$ [16]; $CS_2 + O_2$ [17]; $H_2 + O_2$ [12]; $CO + O_2$ [12] etc.. In the cases just quoted, both components are gaseous and for this reason limit phenomena are studied not with respect to oxygen pressure but with respect to total pressure at various oxygen to fuel ratios. In all these cases, there are both an upper and a lower limit with respect to total pressure. The upper limit increases with temperature while the lower limit decreases slightly. At some temperature T_M, both limits coincide to form the typical inflammation peninsula. Below temperature T_M, chain ignition does not take place. As a rule, in these cases, the lower limit is much larger than in the case of phosphorus oxidation. This circumstance, as well as the temperature dependence of the limits, indicates an appreciable activation energy for the propagation and branching steps of the chain. This is in contrast with the situation prevailing in phosphorus oxidation.

Let us mention a few results of studies of ignition regions for a variety of fuels, at various fuel to oxygen ratios (Figures 30 to 34).

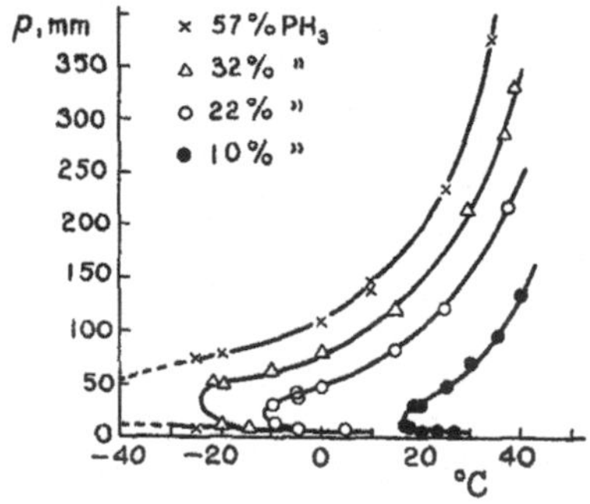

FIGURE 30. Explosion domain for phosphine [15].

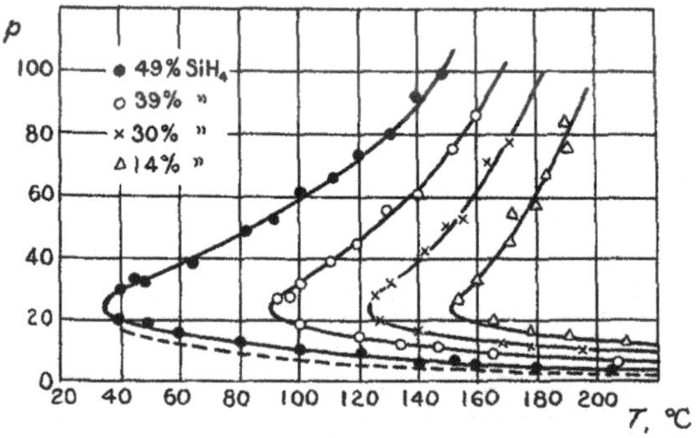

FIGURE 31. Explosion domain for silane [16].

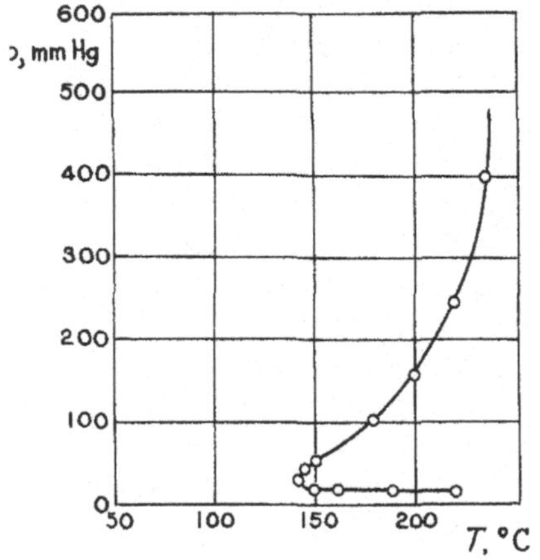

FIGURE 32. Explosion domain for carbon disulfide [17].

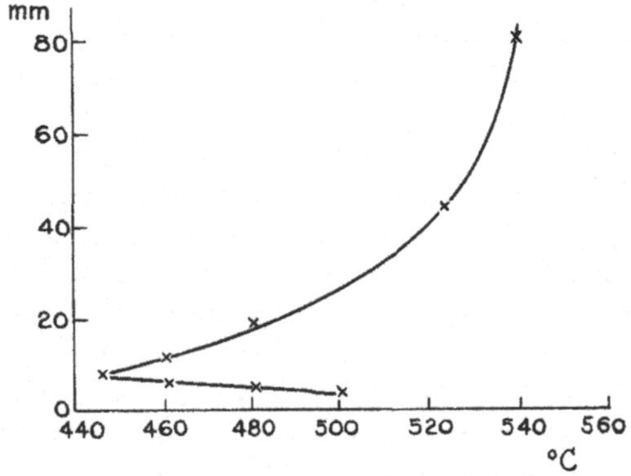

FIGURE 33. Explosion domain for hydrogen [12].

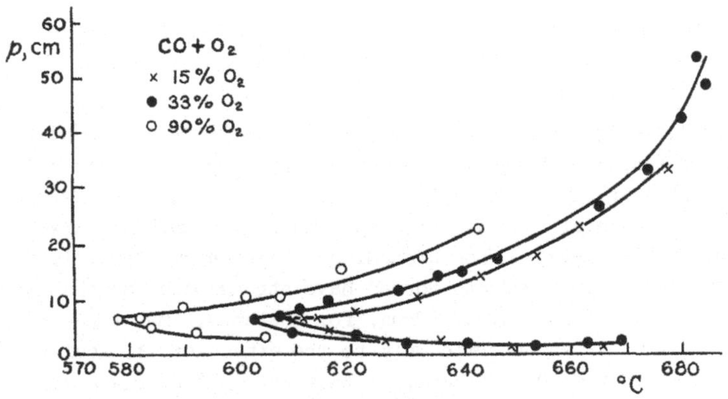

FIGURE 34. Explosion domain for carbon monoxide [12].

In all cases, above the upper limit and below the lower limit, reaction is practically totally absent. This shows directly that ignition is of the chain type and not of the thermal type. Only in the cases of hydrogen and carbon monoxide oxidations, is a slow, apparently heterogeneous, reaction taking place outside the limits at sufficiently high temperatures.

Note that in PH_3 and SiH_4 oxidation, enrichment in fuel broadens the limits. Reaction is therefore made easier. On the contrary, in H_2, CO and CS_2 oxidation, mixtures richer in oxygen correspond to wider mixtures. Only in very lean mixtures (~ 0.03% CS_2) is the opposite trend observed. This all means that in the first group (PH_3, SiH_4), the rate determining step is the reaction between an active particle and a molecule of fuel, while in the second group it is the reaction between an active particle and oxygen. In the case of phosphorus, as was seen earlier, both reactions proceed at equal speed. The qualitative trend of the limits p_1 and p_2 with temperature (Figures 30 to 34) is well described in terms of chain termination on the walls, and within the volume as a result of triple collisions, if the elementary rate constants k_1 and k_2 have an appreciable activation energy and follow the law $\exp(-E/RT)$. This is easy to verify by applying the formulae of the preceding section to an analysis of the branches of upper and lower limits. In order to describe the tip of the peninsula where p_1 and p_2 are equal, it is necessary to take into account simultaneously homogeneous and heterogeneous termination. Then one obtains higher degree equations that are not easily soluble but that give qualitatively the position of the tip of the peninsula. In the oxidation of hydrogen, an equation of second or third degree is obtained:

$$p^3 - \frac{2k_2}{k_6} p^2 + \frac{D_0}{k_6 d^2 \gamma} = 0$$

where k_2 is the rate constant of $H + O_2 \longrightarrow OH + O$, k_6 the rate constant of $H + O_2 + M \longrightarrow HO_2 + M$. This fully describes ignition of hydrogen-oxygen mixtures in vessels for which $\gamma \sim 1$.

All examples considered thus far concern oxidation reactions. In other types of reaction, branching is usually not observed. There is little doubt, however, that similar phenomena exist in other classes of reactions. The formation of several radicals from one radical (branching) always requires considerable energy. When branching takes place by reaction of a radical with oxygen, the products formed have strong bonds so that the energy required is not too large. Thus, for instance, the step $H + O_2 \longrightarrow OH + O$ is endothermic only to the extent of 15 kcal thanks to the formation of the O - H bond, so that it can take place at an appreciable rate at the temperatures of hydrogen oxidation.

If the step leads to the formation of a nitrogen molecule from a nitrogen containing compound, branching may be expected thanks to the strength of the $N \equiv N$ bond (225 kcal).

Apin [18] has found such a case. It is the decomposition of $NC\ell_3$ vapors that explode readily. Apin noticed that there is here a distinct upper limit above which $NC\ell_3$ does not decompose, while below it auto-ignition takes place, (Figure 35, on the following page). The lower limit,

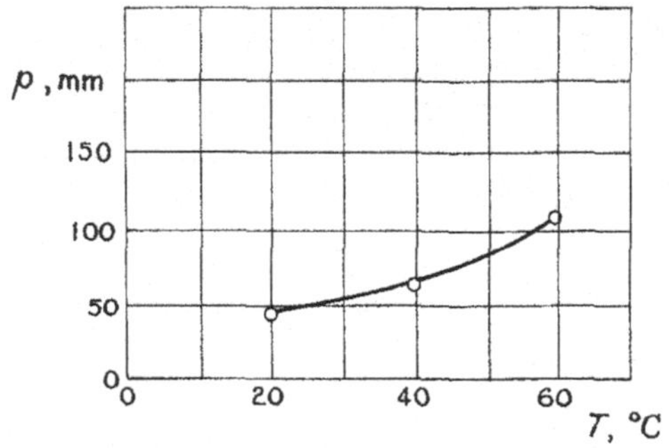

FIGURE 35. Upper limit for decomposition of NCl_3 [18].

if there is one, must correspond to very low pressures.

The mechanism of NCl_3 decomposition may be represented in the following way:

1) $Cl + NCl_3 \longrightarrow NCl_2 + Cl_2$
2) $NCl_2 + NCl_3 \longrightarrow N_2 + Cl_2 + 3Cl$
3) $Cl + NCl_3 + M \longrightarrow NCl_4 + M$.

The radical NCl_4 appears to be stable, does not react further homogeneously and dies on the walls.

Branching is due to step 2) which is apparently strongly exothermic. Therefore this reaction is quite fast so that NCl_3 ignition takes place in the vicinity of room temperature. It is step 3) that is responsible for the upper limit.

REFERENCES

[1] N. N. Semenov, 'Chain Reactions', Oxford (1935).
[2] A. B. Nalbandyan and V. V. Voevodskiĭ, 'Mechanism of oxidation and combustion of hydrogen', Moscow (1949).
[3] Yu. B. Khariton and E. F. Val't, Z. Phys., 19, 517 (1926).
[4] N. N. Semenov, Z. Phys., 46, 109 (1927).
[5] A. A. Koval'skiĭ, Z. physik. Chem., B 4, 180 (1929).
[6] H. W. Melville and E. B. Ludlam, Proc. Roy. Soc., A 132, 108 (1931).
[7] Zentnerschwer, Z. physik. Chem., 26, 9 (1898).
[8] J. Tausz and H. Gerlacher, Z. anorg. Chem., 190, 95 (1930).
[9] A. Trifonov, Z. phys. Chem., B 3, 195 (1929).
[10] N. N. Semenov, Z. phys. Chem., B 2, 161 (1929).
[11] N. N. Semenov, Zhur. Rusk. Fiz. Khim. O., Ch. Fiz., 60, 271 (1928).

[12] A. V. Zagulin, A. A. Koval'skiĭ, I. D. Kopp and N. N Semenov, Zhur. Fiz. Khim., 1, 263 (1930), Z. phys. Chem., B 6, 307 (1930).

[13] R. H. Dalton, Proc. Roy. Soc., A 128, 263 (1930).

[14] N. N. Semenov and Yu. N. Ryabinin, Z. physik. Chem., B 1, 192 (1928).

[15] P. S. Shantarovich, Acta physicochim. URSS, 2, 633 (1935).

[16] P. S. Shantarovich, Acta physicochim. URSS, 6, 65 (1937).

[17] V. G. Voronkov and N. N. Semenov, Zhur. Fiz. Khim., 13, 1695 (1939).

[18] A. Ya. Apin, Zhur. Fiz. Khim., 14, 494 (1940).

CHAPTER X: CHAIN IGNITION IN HYDROGEN-OXYGEN MIXTURES

In order to verify the quantitative aspects of the theory of branched chain reactions, scientists in several countries have studied in detail the combustion processes of hydrogen-oxygen mixtures. These studies have established that, at sufficiently high temperatures, the chain propagation and branching steps are the following:

$$\text{(I)} \quad \begin{aligned} &1)\ OH + H_2 \longrightarrow H_2O + H + 14 \text{ kcal} \\ &2)\ O_2 + H \longrightarrow OH + O - 15 \text{ kcal} \\ &3)\ O + H_2 \longrightarrow OH + H + \sim 0 \text{ kcal.}^* \end{aligned}$$

Alternation of steps 1) and 2) gives a simple chain in which OH radicals and H atoms succeed each other. In step 2), an O atom is also formed, that brings about branching by means of step 3) in which additional OH and O are formed. This is a continuously branching chain.

The rate of elementary steps 1), 2) and 3) is expressed by the relations: $w_1 = a_1(OH) = k_1(H_2)(OH)$; $w_2 = a_2(H) = k_2(O_2)(H)$; $w_3 = a_3(O) = k_3(H_2)(O)$.

The rate constant k_1 has been determined by Avramenko [1]. From line absorption spectra, he measured the change in concentration of OH radicals from a discharge after mixing them with hydrogen. Avramenko proposes the following value for k_1:

$$k_1 = 7 \cdot 10^{-12} \sqrt{T} \exp(-10000/RT)$$

or approximately:

$$k_1 = 2 \cdot 10^{-10} \exp(-10000/RT) \qquad (600 < T°K < 800)$$

$$k_1 = 4 \cdot 10^{-13} \text{ cm}^3/\text{molecule} \cdot \text{sec.} \qquad \text{at } T = 800°K \text{ .}$$

* The heats of reaction are calculated with the help of the heats of dissociation: $H_2 \longrightarrow 2H - 103$ kcal; $O_2 \longrightarrow 2O - 118$ kcal; $OH \longrightarrow O + H - 103$ kcal.

The rate constant k_2 has been determined by Semenov [2] from experiments by Koval'skiĭ who first measured the kinetics of hydrogen oxidation at low pressures.

At $T = 793°K$, $k_2 = 1.12 \cdot 10^{-14}$. Recently [4], Koval'skiĭ's work has been repeated in a quartz reactor with walls treated with HF and potassium tetraborate. As will be seen presently, under these conditions, the lower limit is one order of magnitude smaller than in Koval'skiĭ's experiments. The reaction propagates more slowly, permitting a more accurate study of its kinetics in a wider temperature interval (733 - 873°K). For the constant k_2, the following expression was found:

$$k_2 = 0.94 \cdot 10^{-10} \exp(-15100/RT) \text{ cm}^3/\text{molecule} \cdot \text{sec.}$$

This gives $k2 = 6.7 \cdot 10^{-15}$ at 797°K, a value 1.5 times smaller than the figure of Koval'skiĭ. Later, in approximate calculations, we shall use the value $k_2 = 10^{-14}$ at $T = 800°K$.

The quantity k_3 has not been determined directly. Harteck and Kopsch [5] have studied the reactions of oxygen atoms from a discharge and found an activation energy of 6000 cal/mole for reaction 3). Assuming that the pre-exponential factor is the same as for OH radicals, we obtain: $k_3 = 2.10^{-10} \exp(-6000/RT)$ for the range $T = 700$ to 800°, or $k_3 = 2.10^{-11.6}$ at 800°K, i.e., $k_3 = 6.10^{-12}$ cm^3/molecule · sec.*

For the ratio of the constants $a_1 : a_2 : a_3$, we have: $k_1(H_2) : k_2(O_2) : k_3(H_2)$

$$= \frac{k_1(H_2)}{k_2(O_2)} : 1 : \frac{k_3(H_2)}{k_2(O_2)} = 2 \exp(5100/RT) \frac{(H_2)}{(O_2)} : 1 :$$

$$2 \exp 8900/RT \frac{(H_2)}{(O_2)} .$$

At $T = 800°K$:

$$a_1 : a_2 : a_3 = 50 \frac{(H_2)}{(O_2)} : 1 : 500 \frac{(H_2)}{(O_2)} .$$

For a stoichiometric mixture:

$$a_1 : a_2 : a_3 = 100 : 1 : 1000$$

---

* There are now some reasons to believe that the pre-exponential factor of this reaction has a considerably smaller value.

For a mixture containing 10% H_2:

$$a_1 : a_2 : a_3 = 10 : 1 : 100 \quad .$$

Therefore, down to hydrogen contents as small as 10%, it may be assumed that $a_2 \ll a_1 \ll a_3$. The quantities

$$\frac{1}{a_1}, \quad \frac{1}{a_2} \quad \text{and} \quad \frac{1}{a_3}$$

are close to the respective life-times of OH radicals, H and O atoms (life-time between their generation in an elementary step and their disappearance in another one). The concentration of these particles during the course of reaction is characteristic of their life-time. Therefore, the concentration of H atoms is considerably larger than that of OH radicals and much larger than that of O atoms. It appears that, following the slower step 2), the O atoms immediately react in step 3) and so do the two OH radicals formed in 2) and 3). It follows that step 2) determines the kinetics of the process as a whole. For one cycle, the steps can be summarized by an overall reaction:

$$H + 3H_2 + O_2 \longrightarrow H + 2H + 2H_2O + 13 \text{ kcal.}$$

In one cycle, two water molecules are formed together with two H atoms and branching at every cycle. The dissociation of hydrogen in two atoms is accomplished with the energy released by the formation of two water molecules. In fact, the overall process still releases 13 kcal. The H atoms sooner or later recombine to form H_2 (at low pressures, this happens on the walls) and 103 kcal are thereby liberated.

Chain initiation is due to very rare elementary reactions between H_2 and O_2 molecules, homogeneously or heterogeneously. This gives primary atoms or radicals:

$$0)\quad H_2 + O_2 \longrightarrow HO_2 + H - 55 \text{ kcal.}$$

or

$$H_2 + O_2 \longrightarrow 2OH - 15 \text{ kcal.}$$

If one of these processes is homogeneous, $n_0 = k_0(H_2)(O_2)$ and from the data of Koval'skiĭ, it is possible to estimate the rate constant k_0. The latter is 10^{-22}, in order of magnitude at $T = 800°K$.

If then hydrogen oxidation proceeded solely by means of reaction 0), at a pressure of about 1 mm Hg, the reaction time would be several days while in reality, at pressures of the mixture of about 1 mm Hg (above the lower limit), the reaction is completed in a few tens of seconds.

Thus the very slow chain initiation step has only the significance of a trigger mechanism which starts the rapid autoaccelerating process of branching via steps 1), 2) and 3) of scheme I.

Besides chain initiation and propagation, one must also take into account chain termination on the walls where OH radicals, H and O atoms are adsorbed. These adsorbed active particles soon recombine with other H, O or OH particles coming from the gas phase. The molecules of H_2, O_2 and H_2O formed in this way are loosely held on the wall, are rapidly desorbed to make room for further adsorption of atoms and radicals.

The rate at which active particles disappear at the wall is determined by their rate of diffusion to the wall as well as by the probability ϵ of capture by the wall on each collision, i.e., the kinetics of adsorption. If ϵ is small, the rate of adsorption is determined almost solely by its kinetics; if it is large, diffusion is largely determining. Let us consider these two extreme cases.

A. Kinetic approximation. In this case, the probabilities ϵ_{OH}, ϵ_H and ϵ_O are small as compared to unity. The rate of chain termination per unit surface is given by the collision frequency times the probability of capture. The time required for diffusion of the particles to the wall is much less than the time necessary for adsorption. The first one may be neglected. Since for small values of ϵ the active particles are reflected many times from the wall before their capture, the concentration of active particles throughout the entire volume is practically constant. The number of OH radicals, H and O atoms colliding with the wall, per unit surface and per unit time is:

$$\frac{u_{OH}(OH)}{4}; \qquad \frac{u_H(H)}{4}; \qquad \frac{u_O(O)}{4}.$$

The adsorption rate is:

$$w_1 = \frac{u_{OH}(OH)}{4}\epsilon_{OH}; \qquad w_2 = \frac{u_H(H)}{4}\epsilon_H; \qquad w_3 = \frac{u_O(O)}{4}\epsilon_O ,$$

where u is the velocity of thermal motion of each particle.

Before we write down the kinetic equations, we must keep in mind that the propagation steps and apparently the initiation steps are homogeneous while the termination steps take place on the surface, so that for the entire reaction vessel, the rate of the first four processes is proportional to its volume v while that of the three termination steps is proportional to its surface s. In order to express the rate per unit volume, we must multiply w_1, w_2 and w_3 by the ratio (s/v). For cylindrical vessels $(s/v) = (2/r) = (4/d)$ where d is the diameter of the reactor. The corresponding rates will then be:

$$w_1' = a_1'(OH); \qquad w_2' = a_2'(H); \qquad w_3' = a_3'(O)$$

with the constants:

$$a_1' = \frac{\epsilon_{OH} u_{OH}}{d}; \qquad a_2' = \frac{\epsilon_H u_H}{d}; \qquad a_3' = \frac{\epsilon_O u_O}{d} \quad .$$

The complete reaction scheme (II) is:

0) $H_2 + O_2 \longrightarrow H + HO_2 - 55$ kcal. — chain initiation

1) $OH + H_2 \longrightarrow H_2O + H$
2) $H + O_2 \longrightarrow OH + O$
3) $O + H_2 \longrightarrow OH + H$
} propagation of the branched chain

1') OH + wall
2') H + wall
3') O + wall
} chain termination.

The kinetic equations (III) for these elementary steps are:

$$\frac{d(OH)}{dt} = -(a_1 + a_1')(OH) + a_2(H) + a_3(O) \tag{1}$$

$$\frac{d(H)}{dt} = n_0 + a_1(OH) - (a_2 + a_2')(H) + a_3(O) \tag{2}$$

$$\frac{d(O)}{dt} = a_2(H) - (a_3 + a_3')(O) \quad . \tag{3}$$

The overall rate, based on oxygen consumption is:

$$\frac{d\Delta(O_2)}{dt} = n_0 + a_2(H) \tag{4}$$

where $\Delta(O_2) = (O_2)_0 - (O_2)_t$: the amount of oxygen reacted at time t.

As was seen above, $a_2 \ll a_1$ or a_3. It may be shown that, if such an inequality is fulfilled, the system III may be radically simplified by putting

$$\frac{d(OH)}{dt} = \frac{d(O)}{dt} = 0 \quad .$$

These two algebraic equations give readily the concentrations (OH) and (O) in terms of the concentration of the most slowly reacting active centers — H atoms —:

$$(O) = \frac{a_2(H)}{a_3 + a_3'}; \qquad (OH) = \frac{a_2(H)}{a_1 + a_1'}\left(1 + \frac{a_3}{a_3 + a_3'}\right).$$

Substituting in equation (1), we get:

$$\frac{d(H)}{dt} = n_0 + a_1(OH) - (a_2 + a_2')(H) + a_3(O) = n_0 + \varphi(H) \tag{5}$$

where:

$$\varphi(H) = a_2 \frac{\left\{2 - \frac{a_2'}{a_2}\left(1 + \frac{a_1'}{a_1} + \frac{a_3'}{a_3} + \frac{a_1'a_3'}{a_1a_3}\right)\right\} - \frac{a_1'a_3'}{a_1a_3}}{1 + \frac{a_1'}{a_1} + \frac{a_3'}{a_3} + \frac{a_1'a_3'}{a_1a_3}} = 2a_2\psi \quad . \tag{6}$$

All quantities a' do not depend on pressure and the quantities a are proportional to pressure, so that φ and also ψ increase monotonically with pressure. Moreover, at small pressures φ and ψ are negative. At some pressure $p = p_1$, ψ and φ are equal to zero. At $p > p_1$, ψ and φ are positive. If the ratios

$$\frac{a_1'}{a_1} \quad \text{and} \quad \frac{a_3'}{a_3}$$

are small enough quantities so that their product may be neglected as compared to unity, then if φ and ψ are positive, they may be expressed, to a good approximation by means of the simple formulae:

$$\varphi = \left(1 - \frac{p_1}{p}\right)2a_2 \qquad \text{or} \qquad \psi = \left(1 - \frac{p_1}{p}\right) .^* \tag{7}$$

---

* It is easy to see that if

$$\frac{a_1'}{a_1} \quad \text{and} \quad \frac{a_3'}{a_3}$$

are sufficiently smaller than unity and

$$\frac{a_1'a_3'}{a_1a_3} \ll 1 \quad ,$$

$$\varphi = 2a_2\left\{1 - \frac{a_2'}{a_2} - \frac{a_1'}{a_1} - \frac{a_3'}{a_3}\right\} \quad .$$

Since $a_2 = k_2(O_2) = c_2\gamma p$, $a_1 = k_1(H_2) = c_1(1 - \gamma)p$ and $a_3 = k_3(H_2) = c_3(1 - \gamma)p$, where p is the pressure of the mixture, γ the fraction of

Integration of equation (5) gives an expression for the increase of the concentration of H atoms with time:

$$\varphi < 0, \quad p < p_1, \quad (H) = \frac{n_o}{|\varphi|}\left(1 - e^{-|\varphi|t}\right) \xrightarrow[\text{increases}]{\text{as } t} \frac{n_o}{|\varphi|} \qquad (8)$$

$$\varphi = 0, \quad p = p_1 \quad (H) = n_o t \qquad (9)$$

$$\varphi > 0, \quad p > p_1 \quad (H) = \frac{n_o}{\varphi}(e^{\varphi t} - 1) \quad . \qquad (10)$$

Formulae (8), (9) and (10) are correct for small changes in the variables (H_2) and (O_2) when the quantities a may be considered as constant (small extents of reaction). The reaction rate based on oxygen consumption is equal to:

$$w = -\frac{d(O_2)}{dt} = \frac{d\Delta(O_2)}{dt} = n_o + k_2(O_2)(H) = n_o + a_2(H) \quad .$$

Hence, the quantity of oxygen consumed at time t is:

$$\Delta(O_2) = \int_0^t a_2(H) \cdot dt \quad ,$$

or:

$$\begin{cases} \varphi < 0 \\ p < p_1 \end{cases} \quad \Delta(O_2) = n_o t + \frac{n_o a_2 t}{|\varphi|} = n_o t + \frac{n_o}{2|\psi|} t \qquad (11)$$

$$\begin{cases} \varphi = 0 \\ p = p_1 \end{cases} \quad \Delta(O_2) = n_o t + \frac{n_o a_2 t^2}{|\varphi|} \qquad (12)$$

$$\begin{cases} \varphi > 0 \\ p > p_1 \end{cases} \quad \Delta(O_2) = n_o t - \frac{n_o t}{\psi} - \frac{n_o t}{\varphi a_2 \psi^2} + \frac{n_o}{\varphi a_2 \psi^2} e^{2a_2\psi t}$$

* (cont.)
oxygen and $1 - \gamma$ the fraction of hydrogen, we get:

$$\varphi = 2a_2\left\{1 - \frac{A_2}{p\gamma} - \frac{A_{1+3}}{p(1-\gamma)}\right\} = 2a_2\left(1 - \frac{\beta}{p}\right) \quad .$$

At $p = p_1$ (lower limit) $\varphi = 0$ and $\beta = p_1$. Consequently,

$$\varphi = \left(1 - \frac{p_1}{p}\right) 2a_2 \quad \text{and} \quad \psi = \left(1 - \frac{p_1}{p}\right) \quad .$$

or when t is large enough:

$$\Delta(O_2) = \frac{n_o}{\varphi a_2 \psi^2} \exp(2a_2 \psi t) \quad . \tag{13}$$

These formulae are valid only for limited combustion ($\Delta(O_2)/(O_2)$ less than 10%), as long as the quantities a may be treated as approximately constant.

§1. Lower Ignition Limit

The lower chain ignition limit p_1 is characterized by the fact that at a pressure slightly below p_1 the time of reaction is very large (the average rate very slow) while at a pressure slightly above p_1, it is very small (the average rate is very fast). The larger the ratio of these times, the more distinct is the ignition limit.

The limit, as we have just seen, is determined by the condition $\varphi = 2a_2\psi = 0$ or alternatively $\psi = 0$ where

$$\psi = 1 - \frac{p_1}{p} = \frac{p - p_1}{p} \quad .$$

Let us examine the reaction times in a stoichiometric mixture when ψ varies between -0.01 and $+0.01$, corresponding to a pressure change of -1 and $+1\%$ of p_1. Let $p_1 = 7.6$ mm Hg. Let us determine the reaction time at $p = p_1 + \Delta p = 7.6 + 0.076 = 7.676$ and $p = p_1 - \Delta p = 7.6 - 0.076 = 7.524$ mm.

Let us designate by $t_{10\%}$ the time required for 10% reaction. According to (11), we get for $\psi = -0.01$:

$$\Delta(O_2)_{10\%} = \frac{(O_2)}{10} = n_o t \left(1 + \frac{1}{2\psi}\right) = \frac{n_o t_{10\%}}{2.10^{-2}} = 50\ n_o t_{10\%}$$

$$t_{10\%} = \frac{(O_2)}{500\ n_o} \quad .$$

If $\psi = +0.01$, formula (13) gives:

$$\Delta(O_2)_{10\%} = \frac{(O_2)}{10} = \frac{n_o}{4a_2^2\psi^2}\, e^{2a_2\psi t_{10\%}}$$

$$= \frac{n_o}{4k_2(O_2)\cdot 10^{-4}}\, e^{2.10^{-2}k_2(O_2)t_{10\%}} \quad .$$

If the mixture is stoichiometric, at 800°K, and $n_0 = 10^{-22}(H_2)(O_2)$, we find that, for $\psi = -0.01$, $t_{10\%} = 330$ sec. For $\psi = +0.01$,

$$t_{10\%} = \frac{2.3 \cdot 4.31}{6.2} \sim 1.6 \text{ sec.}^*$$

It is seen that a 2% pressure change decreases the reaction time by a factor of one hundred. A 20% pressure change i.e., a variation of ψ between -0.1 and $+0.1$, would change $t_{10\%}$ by a factor of 10000.

The value of the pressure p_1 at the lower limit is very sensitive to a treatment of the walls. Special pretreatment may decrease p_1

---

* According to (11), for $\psi = -0.01$,

$$\Delta(O_2)_{10\%} = \frac{(O_2)}{10} = \frac{n_0 t_{10\%}}{2.10^{-2}} = 50\, n_0 t_{10\%}$$

or, with $n_0 = k_0(H_2)(O_2)$,

$$t_{10\%} = \frac{1}{500\, k_0(H_2)} \quad .$$

For $\psi = +0.01$, according to (13):

$$\frac{(O_2)}{10} = \frac{n_0}{4a_2\psi^2}\, e^{2a_2\psi t_{10\%}} \quad ;$$

$$t_{10\%} = \frac{2.3 \log\left[\dfrac{(O_2) \cdot 4a_2\psi^2}{10 \cdot n_0}\right]}{2a_2\psi} = \frac{2.3 \log\left[\dfrac{(O_2) \cdot 4k_2(O_2)\psi^2}{10\, k_0(H_2)(O_2)}\right]}{2k_2(O_2)\psi} \quad .$$

At 800°K and $p_1 = 7.6$ mm, in a stoichiometric mixture, $(O_2) = 0.33 p_1 N$ where N is the number of molecules per cm^3 at $p = 1$ mm:

$$N = \frac{2.7 \cdot 10^{19} \cdot 273}{760T} = \frac{0.97 \cdot 10^{19}}{T} = 1.22 \cdot 10^{16} \text{ at } 800°\text{K} \quad .$$

$$k_0 = 10^{-22}; \quad k_2 = 10^{-4}; \quad \frac{(O_2)}{(H_2)} = \frac{1}{2}; \quad k_2(O_2) = 3.1 \cdot 10^2 \quad .$$

With these values, the result quoted in the text is readily obtained.

by a factor of 100, say to $p_1 = 0.076$ mm. Then, for $\psi = -0.1$, $t_{10\%} = 2.75 \cdot 10^5$ sec and for $\psi = +0.1$, $t_{10\%} = 30$ sec. The ratio of reaction times is still 10^4, giving a relatively sharp limit.

Note that for $\psi = +0.1$, the reaction proceeds much faster but only after a long induction period. Thus, $t_{2\%}$ is only 4 sec. less than $t_{10\%}$.

At such low pressures (~ 0.1 mm), p_1 cannot be measured by the ignition flash which cannot be seen very well. It must be measured by pressure changes recorded by means of a membrane manometer with a limit of sensitivity of at least 10^{-3} mm Hg. The vessel is filled to the required pressure. After a waiting period of 30 to 100 seconds, reaction takes place during a few seconds as observed by the pressure change. The limit p_1 is determined with a precision of $10^{-3} - 10^{-4}$ mm. If the pressure at the limit is of the order of 1 mm or higher, the limit can be easily determined by the flash that is clearly visible in the dark.

As was already pointed out, the value of the lower limit strongly depends on the nature and state of the wall surface. This is natural since ϵ_{OH}, ϵ_H and ϵ_O are determined by the character of the surface. Without any special pretreatment of the wall, it is often impossible to make reproducible measurements. To stabilize the surface and obtain reproducible data on the lower limit, it is necessary to pretreat the wall by melting the glass, or carrying out a large number of successive experiments, or to treat the surface with atomic hydrogen. What is especially effective is a chemical treatment (with e.g., hydrofluoric acid, tetraborates etc.) or a wash with certain salt solutions (e.g., $KC\ell$ solutions).

The lower limit p_1 is determined by the condition $\psi = 0$, or according to formula (16):

$$2 - \left[\frac{a_2'}{a_2} \left(1 + \frac{a_1'}{a_1} + \frac{a_3'}{a_3} + \frac{a_1'a_3'}{a_1a_3} \right) - \frac{a_1'a_3'}{a_1a_3} \right] = 0 \ .$$

Since $a_2 = k_2(O_2)$; $a_1 = k_1(H_2)$; $a_3 = k_3(H_2)$, $(O_2) = N\gamma p$; $(H_2) = N(1-\gamma)p$ and with the notations: $A_1 = (a_1'/k_1)$; $A_2 = (a_2'/k_2)$, $A_3 = (a_3'/k_3)$, we get:

$$2 - \left[\frac{A_2}{Np_1\gamma} \left(1 + \frac{A_1}{Np_1(1-\gamma)} + \frac{A_3}{Np_1(1-\gamma)} + \frac{A_1A_3}{N^2p_1^2(1-\gamma)^2} \right) - \frac{A_1A_3}{N^2p_1^2(1-\gamma)^2} \right] = 0 \ .$$

Note that the quantities A do not depend on pressure.

Solution of this cubic equation gives p_1 as a complex function of mixture composition and the parameters A. But if the ratios a_1'/a_1

and a_3'/a_3 are much less than unity at the limit, the equation takes the simple form:

$$2 - \frac{a_2'}{a_2} = 2 - \frac{a_2'}{2k_2(O_2)} = 2 - \frac{a_2'}{k_2} \cdot \frac{1}{Np\gamma} = c \quad . \tag{14}$$

Thus:

$$\gamma p_1 = \frac{a_2'}{2k_2N} \quad . \tag{14'}$$

The product γp_1 should be a constant at a given temperature, irrespective of mixture composition. This means that the oxygen pressure at the limit is a constant and does not depend on the composition of the mixture. This result is based on the assumption that a_1'/a_1 and a_3'/a_3 are small.

If now the rate of capture by the wall of H, O and OH are of the same order and magnitude and thus the parameters ϵ_H, ϵ_O and ϵ_{OH} have about the same value, as long as $a_2 \ll a_1$ and $a_2 \ll a_3$, the ratios a_1'/a_1 and a_3'/a_3 will indeed be small as compared to a_2'/a_2 and unity.

As we have seen, as long as the hydrogen content of the mixture is above 10% or $\gamma \leq 0.9$, $a_2 \ll a_1$ and $a_2 \ll a_3$ and therefore, formula (14') may be used in the range $0 < \gamma \leq 0.9$. In fact, ϵ_{OH} is generally higher than ϵ_H so that experiment only can decide the maximum oxygen content for which relation (14') stays valid.

In 1937, Nalbandyan and Biron [6] have studied this question, using a pyrex vessel with fused walls. These early data showed that the product γp_1 changes little with γ, in the range $0.1 < \gamma < 0.5$: it changes only from 0.06 to 0.07, provided that $\gamma < 0.5$.

A more thorough study was reported recently by Nalbandyan and Ivanov [7] using a quartz vessel (5.8 cm in diameter) pretreated with hydrofluoric acid and tetraborate. The pretreatment depresses considerably the lower limit so that the reaction proceeds clearly in the kinetic range.

Results at 430, 470 and 520°C are shown in Table 47, on the following page, where γ is the oxygen content of the mixture.

The data indicate that $p_1\gamma$ is indeed constant, within a few percents when γ is changed tenfold between 0.05 and 0.5. At higher values of γ, the product $p_1\gamma$ begins to increase rapidly with γ.

Formula (14') is therefore quite adequate as long as $\gamma < 0.5$. This means that under these conditions, the ratios a_1'/a_1 and a_3'/a_3 may be neglected in front of unity and a fortiori of a_2'/a_2 which is equal to

TABLE 47

T = 430°(703°K)			T = 470°(743°K)			T = 520°(793°K)		
γ	p_1	γp_1	γ	p_1	γp_1	γ	p_1	γp_1
0.05	1.2	0.060	0.05	0.085	0.042	0.05	0.58	0.029
0.1	0.60	0.060	0.1	0.42	0.042	0.1	0.29	0.029
0.15	0.39	0.0585	0.15	0.28	0.042	0.15	0.20	0.030
0.20	0.29	0.058	0.2	0.21	0.042	0.20	0.15	0.030
0.25	0.24	0.060	0.29	0.14	0.040	0.25	0.12	0.030
0.29	0.20	0.058	0.33	0.13	0.043	0.29	0.10	0.029
0.33	0.18	0.059	0.4	0.11	0.044	0.33	0.09	0.029
0.40	0.15	0.060	0.5	0.09	0.045	0.40	0.075	0.030
0.45	0.135	0.061	0.6	0.09	0.057	0.45	0.065	0.029
0.50	0.13	0.065	0.7	0.1	0.07	0.50	0.065	0.032
0.60	0.13	0.078				0.60	0.065	0.039
0.75	0.14	0.105				0.75	0.08	0.060
0.85	0.16	0.135				0.85	0.09	0.076
0.95	0.18	0.171				0.95	0.11	0.104

2 at the limit. Let us recall that, in a stoichiometric mixture $a_1 : a_2 : a_3 = 100 : 1 : 1000$. Evidently a_2' may not be more than five times smaller than a_1' since, according to the data a_2'/a_2 for mixtures with $\gamma < 0.5$ must be 10 to 20 times larger than a_1'/a_1 if the simple formula is well obeyed. Consequently

$$(a_2'/a_2) : (a_1'/a_1) = a_2'a_1/a_2a_1' = \frac{100\ a_2'}{a_1'} \geq 20$$

and a_2'/a_1' must be $\geq 1/5$.

Since on the other hand

$$a_2' = \frac{\epsilon_H u_H}{d}, \qquad a_1' = \frac{\epsilon_{OH} u_{OH}}{d}$$

and $u_H = 4\ u_{OH}$:

$$\frac{a_2'}{a_2} = \frac{4\epsilon_H}{\epsilon_{OH}} \geq \frac{1}{5} \qquad \text{or} \qquad \frac{\epsilon_H}{\epsilon_{OH}} \geq \frac{1}{20} \quad .$$

In the vessels used by Nalbandyan, ϵ_{OH} may not be more than twenty times larger than ϵ_H.

It is easy to determine a_2' and thus ϵ_H from lower limit data. According to formula (14'):

$$a_2' = \frac{\epsilon_H u_H}{d} = 2k_2 w \gamma p_1 \quad .$$

At $T = 800°K$, for a stoichiometric mixture, in a vessel with a diameter equal to 5.8 cm, washed with hydrofluoric acid, $p_1 = 0.1$ mm Hg, $N = 1.22 \cdot 10^{16}$, $k_2 = 10^{-14}$. After substitutions, one gets $a_2' = 8$. With $u_H = 4.3 \cdot 10^5$ at 800°K, we get $\epsilon_H = 10^{-4}$. With the inequality just derived, $\epsilon_{OH} \leq 2 \cdot 10^{-3}$. Formula (14') shows that, for a given value of γ, the product of p_1 and the vessel diameter d should be constant.

The theoretical relation $p_1 d = \text{const.}$ is very well verified by the experiment, as follows from Semenov's data [8] shown in Table 48* in which d is given in centimeters and p_1 in arbitrary units.

TABLE 48

	585°C		482°C		440°C	
d	p_1	$p_1 d$	p_1	$p_1 d$	p_1	$p_1 d$
5.8	16	93	28	160	42	244
10.2	10	102	18	190	28	285
15.7	8	125	14	220	20	314
20.0	6	120	10	200	15	300
30.0	4	120	7	210	10	300

The most decisive test of the question whether the reaction proceeds in the kinetic or in the diffusion region, is an experiment where the combustible mixture is diluted with an inert gas. In the kinetic region, addition of an inert gas should not affect the partial limiting pressures of the $H_2 + O_2$ mixture. On the contrary, in the diffusion range, addition of an inert gas lowers the partial pressure of the combustible mixture at the limit.

Nalbandyan and Biron [6] have shown that in a pretreated pyrex vessel where the pressure of the combustible mixture at the limit was of the order of 1 mm, dilution by an inert gas did not have any effect on the pressure limit. On the contrary, in a quartz vessel where the pressure limit in the pure mixture was 11 mm, an inert gas lowers the partial pressure of the combustible mixture at the limit. Obviously, in the first case chain termination is controlled by the kinetics of sorption while in the second case, it is controlled by diffusion.

Many experiments have shown that the lower limit decreases with

* When the values of p_1 are not too small (this happens at not too high temperatures), one must introduce a correction for homogeneous chain termination and then the condition $p_1 d = \text{const.}$ is not strictly obeyed.

temperature according to the law $p_1 \sim \exp(E/RT)$ where E varies between 6 and 11 kcal depending on the vessel. Formula (14') indicates that theoretically $p_1 \sim (a_2'/k_2) \sim (\epsilon_2/k_2)$ where $k_2 = A\exp(-E_2/RT) = A\exp(-15100/RT)$.

Direct experiments on recombination of atomic hydrogen show that ϵ also obeys an Arrhenius expression $\exp(-E'/RT)$: $\epsilon = \epsilon_0 \exp(-E'/RT)$. In glass vessels between 307 and 602°C, Smith [9] found E' = 8000 kcal. Therefore, theoretically, $p_1 \sim \epsilon/k_2 \sim \exp(E_2 - E')/RT$.

Comparing theory with experiment, we see that, in various vessels $E' = E_2 - E$ varies between 4 and 9 kcal. The data of Smith [9] who made direct experiments with atomic hydrogen fell within the same range. The direct values of Smith are in good agreement with ours as determined from lower limit data. Thus, in a vessel treated with KCℓ, at T = 427°, Smith finds $\epsilon_H = 2 \cdot 10^{-4}$.

§2. The Induction Period

When the reaction is studied at $p > p_1$, the decrease in pressure Δp equal to $\Delta M/N$ can be observed directly. Here ΔM is the decrease in the total number of molecules per cm^3 at a given time. It is easy to show* that

$$\Delta M = \Delta(O_2) - \frac{(H)}{2} .$$

Formulae (10) and (13) give the following expression for the variation of pressure Δp with time:

$$\Delta p = \frac{\Delta M}{N} = \frac{1}{N}\left[\Delta(O_2) - \frac{(H)}{2}\right]$$

$$= \frac{n_0}{N4a_2\psi^2}\exp(2a_2\psi t) - \frac{n_0}{N4a_2\psi}\exp(2a_2\psi t)$$

$$= \frac{n_0}{N \cdot 4a_2\psi^2}(1-\psi)\exp(2a_2\psi t) .$$

Since $a_2 = k_2(O_2) = k_2N\gamma p$ and $\psi = 1 - \frac{p_1}{p}$, we get:

---

* Indeed:

$$\Delta M = \Delta(O_2) + \Delta(H_2) - (H_2O) - (H) ,$$

$$\Delta(O_2) = 2(H_2O); \quad \Delta(H_2) = (H_2O) + \frac{(H)}{2} .$$

Hence:

$$\Delta M = \Delta(O_2) + (H_2O) + \frac{(H)}{2} - (H_2O) - (H)$$

$$= \Delta(O_2) - \frac{(H)}{2} .$$

$$\Delta p = \frac{n_0 p_1}{N^2 4k_2 \gamma p^2 \left(1 - \frac{p_1}{p}\right)^2} \exp\left[2k_2 N\gamma p\left(1 - \frac{p_1}{p}\right)t\right]$$

$$= C \exp[2k_2 N\gamma(p - p_1)t] = C \exp \varphi t \quad . \tag{15}$$

The quantity n_0, and thus C, is small. Therefore Δp will be small during a certain period of time. Only after t has reached some value $t = \tau$ at which Δp reaches the limit of sensitivity $\Delta p'$ of the measuring device, does it become possible to measure Δp. The time $t = \tau$ during which Δp cannot be detected by the measuring instrument is called the induction period. A good membrane gauge records changes of $\Delta p' = 10^{-4}$ mm. The induction period is then:

$$\varphi\tau = 2k_2 N\gamma(p - p_1)\tau = 2.3 \log \frac{\Delta p'}{C} \quad . \tag{16}$$

The induction period can be measured by means of a sensitive glass membrane manometer with a mirror attached to the membrane. The spot from the mirror falls on a film mounted on a fast rotating drum.

The determination of k_2 will be explained later. At $T = 795°$K it is close to 10^{-14}, so that for a stoichiometric mixture ($\gamma = 0.33$) the quantity $2k_2 N\gamma$ at $T = 800°$ is equal to 80. The experimental value of the product $(p - p_1)\tau$ at 795° is 0.128. Hence

$$\varphi\tau = 80 \cdot 0.128 = 10.4 = 2.3 \log \frac{\Delta p'}{C} \; .$$

Since

$$\frac{\Delta P'}{C} \sim \frac{10^{-4}}{C}$$

is under the logarithmic sign, even considerable changes in the quantity C and thus in n_0 [see [15] and [16]] have little effect on the product $\varphi\tau$. Indeed, if C changes tenfold, $\varphi\tau$ changes by 20%. In the measured range, the quantity $p - p_1$ changes only twofold and therefore $\log(\Delta p'/C)$ changes only by about 7%.

Therefore, within the limits of experimental error, it can be said that the quantity

$$2.3 \log \frac{\Delta p'}{C}$$

is practically independent of a two- or threefold change in pressure and retains a constant value.

Karmilova, Nalbandyan and Semenov [4] have determined the

induction period at various initial pressures p for a stoichiometric mixture. The eight Tables 49 to 56 on the following page, correspond to various temperatures between 733 and 873°. The second column of each table shows the values of $p - p_1$ while the fourth one shows the products $(p - p_1)\tau$ where p_1 is the lower limit. No data are shown in the tables, corresponding to pressures larger than 1 mm or values of τ shorter than 0.1 second since the results are then unsatisfactory, in the first case because of consumption of reactants, in the second because of systematic errors in the induction period. All quantities are in mm Hg and seconds.

As seen in the tables, the products $(p - p_1)\tau$ stay constant within 10% when the pressure changes by a factor of two or three. Thus experiment confirms theory and supports the concept behind the induction period. To a small extent, within the 10% limit, a systematic decrease of the product $(p - p_1)\tau$ can be observed as pressure increases. This may be due to a small dependence of C on pressure. For instance

$$C = \frac{A}{p} .$$

This is not completely certain since a small systematic error may be present in the experiments. As the temperature goes up, the product $(p - p_1)\tau$ goes down.

According to equation (16):

$$(p - p_1)\tau = \frac{1}{2k_2N\gamma} \cdot 2.3 \log \frac{\Delta p'}{C}$$

where

$$k_2 = A \exp(- E_2/RT) .$$

Assuming that

$$2.3 \log \frac{\Delta p'}{C}$$

does not change with temperature, we conclude that:

$$\tau(p - p_1) \sim \exp(E_2/RT) = 10^{E_2/4.6T} .$$

The tables contain the average values of $(p - p_1)\tau$ at various temperatures between $T_1 = 733$ and $T_2 = 873$°K. A plot of $\log[(p - p_1)\tau]$ versus $(1/T)$ gives a very good straight line (Figure 36) the slope of which gives $E_2 = 15400$ cal.

Precise measurements of k_2 as a function of temperature, give $E_2 = 15100$ cal as reported in the next paragraph. It follows that the quantity $\varphi\tau = 2k_2N\gamma(p - p_1)\tau$ remains practically constant within the

TABLES 49 to 56

$T = 460°C = 733°K \quad p_1 = 0.16$

p	$p - p_1$	τ	$(p - p_1)\tau$
0.61	0.45	0.72	0.32
0.81	0.65	0.43	0.27
1	0.84	0.35	0.295
1.2	1.04	0.26	0.27
		average	0.29

$T = 480°C = 753°K \quad p_1 = 0.14$

p	$p - p_1$	τ	$(p - p_1)\tau$
0.55	0.41	0.56	0.23
0.73	0.59	0.37	0.22
0.88	0.74	0.29	0.215
1	0.86	0.22	0.19
1.2	1.06	0.18	0.19
		average	0.21

$T = 502°C = 775°K \quad p_1 = 0.12$

p	$p - p_1$	τ	$(p - p_1)\tau$
0.45	0.33	0.5	0.165
0.57	0.45	0.38	0.17
0.73	0.61	0.26	0.16
0.85	0.73	0.18	0.13
		average	0.16

$T = 522°C = 795°K \quad p_1 = 0.11$

p	$p - p_1$	τ	$(p - p_1)\tau$
0.38	0.27	0.51	0.138
0.51	0.4	0.35	0.14
0.63	0.52	0.23	0.12
0.75	0.18	0.18	0.116
0.87	0.16	0.16	0.125
		average	0.128

$T = 540°C = 813°K \quad p_1 = 0.1$

p	$p - p_1$	τ	$(p - p_1)\tau$
0.38	0.28	0.35	0.099
0.5	0.4	0.26	0.104
0.61	0.51	0.19	0.097
0.73	0.63	0.16	0.1
0.84	0.74	0.13	0.097
		average	0.1

$T = 560°C = 833°K \quad p_1 = 0.09$

p	$p - p_1$	τ	$(p - p_1)\tau$
0.33	0.24	0.4	0.096
0.45	0.36	0.23	0.083
0.55	0.46	0.18	0.083
0.69	0.6	0.14	0.085
0.81	0.72	0.11	0.08
		average	0.085

$T = 580°C = 853°K \quad p_1 = 0.08$

p	$p - p_1$	τ	$(p - p_1)\tau$
0.42	0.34	0.21	0.072
0.58	0.5	0.14	0.07
0.7	0.62	0.1	0.062
		average	0.068

$T = 600°C = 873°K \quad p_1 = 0.07$

p	$p - p_1$	τ	$(p - p_1)\tau$
0.39	0.32	0.18	0.05
0.52	0.45	0.1	0.045
0.61	0.54	0.08	0.044
		average	0.046

interval studied.

§3. Kinetics of the Reaction Above the Lower Limit

According to equation (15), at small degrees of conversion $\Delta p = C\ \mathrm{ecp}(\varphi t)$ where $\varphi = 2a_2\psi = 2k_2(O_2)\psi = 2k_2p\gamma\psi$ and

$$\psi = \left(1 - \frac{p_1}{p}\right).$$

The initial pressure of the mixture is p. If Δp is measured as a function of t, a linear relation $\ell n\Delta p = \varphi t + \text{const.}$ must be obtained. For small conversions this is indeed the case and φ can be easily evaluated.

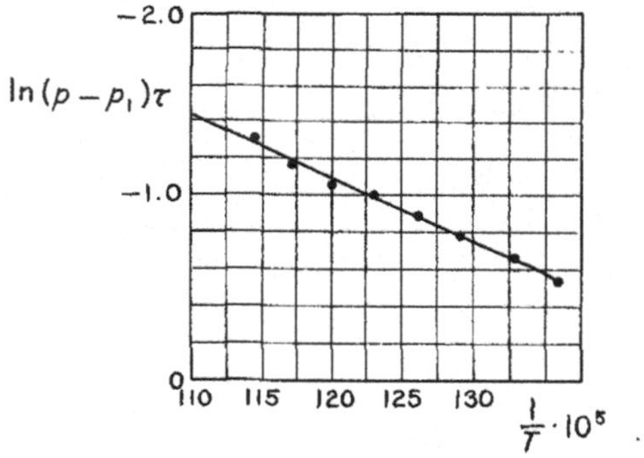

FIGURE 36. Relation between $\log(p - p_1)\tau$ and $1/T \cdot 10^5$ (4).

Equation (15) is approximate since during the course of reaction oxygen is consumed and φ decreases. In fact, at higher conversions, i.e., longer times, the data deviate from the straight line relationship. When φ is large i.e., when the initial pressure is much above the lower limit, the maximum rate is reached sooner and the relation between Δp and t ceases to be obeyed at smaller values of conversion. In the experiments of Karmilova, Nalbandyan and Semenov [4], the lower limit was quite low and for this reason, the measurements were taken at pressures two- or three-fold in excess of the limit. The effect of conversion is then strong, the linear segment of the $\ell n\Delta p$ vs. t curve quite short and deviations are already observed at only a few percents of conversion. The determination of φ by this method is therefore inadequate. Another method, which will be explained presently, must be used.

In the work of Koval'skiĭ [3], the experiments were carried out in a vessel where the value of the lower limit was about ten times larger and the values of φ were small (the initial pressure was close to the lower limit). Under these conditions, curvature of the $\ell n\Delta p$ vs. t line starts at larger relative and absolute values of conversion. Koval'skiĭ was thus able to determine φ without any trouble from the slope of the line $\ell n\Delta p$ - t. Determining the induction period or the time $t_{x\%}$ required to reach a small constant value of conversion of $x\%$, he calculated the products φt from the data and showed they did not depend on pressure. For instance, with a twofold variation in pressure, he obtained the following set of values of φt: 13; 10.6; 11; 11; 10.4. Knowing φ at a given pressure and temperature, he was able to calculate the rate constant k_2 and found $k_2 = 1.1 \cdot 10^{-14}$ at $T = 793°$. He succeeded also in getting an approximate value for $C = \Delta p/\exp(\varphi t)$ and from it calculated the rate constant $k_0 = 1.4 \cdot 10^{-22}$. This figure is not very accurate since a small error in φ gives a very large error in C.

Until now, the kinetic equations have been used in their simplest form, with a constant value of a_2. This is valid for the initial stages of the reaction. It would be interesting to consider now the kinetics of the reaction during its entire course by taking into account the disappearance of reactants. This was done in 1944 by Semenov [1C]. To do this, it is necessary to solve simultaneously equations (4) and (5). In equation (5), one may put $\varphi = 2a_2 - a_2'$ if $\gamma < 0.5$:

$$\frac{d(H)}{dt} = n_0 + (2a_2 - a_2')(H) = n_0 + \left[2k_2(O_2) - a_2' \right](H)$$

$$\frac{d\Delta(O_2)}{dt} = n_0 + k_2(O_2)(H) \quad . \tag{17}$$

A stoichiometric mixture will be considered. The following notations will be used:

$(O_2)_0$: initial concentration of oxygen

(O_2) : concentration of oxygen at any time

$(O_2)_1$: concentration of oxygen at the limit

p_0 and p : pressures, initially and at time t

$$\xi = \frac{(H)}{(O_2)_0}$$

$\eta = \dfrac{\Delta(O_2)}{(O_2)_0} = \dfrac{(O_2)_0 - (O_2)}{(O_2)_0}$: fraction of oxygen reacted at time t.

Then since

$$(O_2)_1 = \frac{1}{3} p_1 N, \qquad a_2 = 2k_2(O_2)_1$$

and

$$\frac{(O_2)_1}{(O_2)_0} = \frac{p_1}{p_0} \quad ,$$

we get:

$$\frac{d(H)}{dt} = n_0 + \left[2k_2(O_2) - 2\frac{p_1}{p_0} k_2(O_2)_0 \right](H)$$

$$\frac{d\Delta(O_2)}{dt} = n_0 + k_2(O_2)(H)$$

or:

$$\frac{d\xi}{dt} = \frac{n_o}{(O_2)_o} + \left[2k_2(O_2)(1 - \eta) - 2k_2(O_2)\frac{p_1}{p_o} \right]\xi \tag{18}$$

$$\frac{d\eta}{dt} = \frac{n_o}{(O_2)_o} + k_2(O_2)_o(1 - \eta)\xi \tag{19}$$

Dividing (18) by (19) and neglecting the small quantity

$$\frac{n_o}{(O_2)_o} ,$$

we get:

$$\frac{d\xi}{d\eta} = 2 - 2\frac{p_1}{p_o}\frac{1}{1 - \eta} .$$

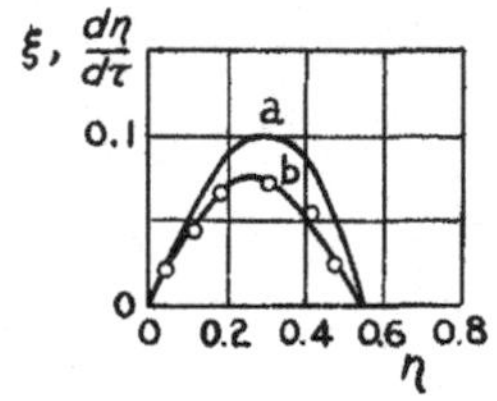

FIGURE 37. Relation between ξ and η (curve a), $\frac{d\eta}{d\tau}$ and η (curve b) at $\frac{2p_1}{p_o} = 1.4$.

Integration gives:

$$\xi = 2\eta + 2\frac{p_1}{p_o}\ln(1 - \eta) . \tag{20}$$

This relation between ξ and η is represented on Figures 37 to 39 (curve a). From the relation between ξ and η, it is easy to calculate the relation between $\frac{d\eta}{dt}$ and η (curve b).

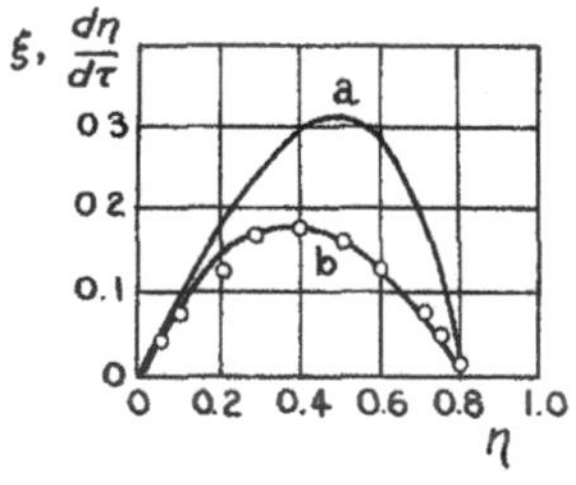

FIGURE 38. Relation between ξ and η (curve a), $\frac{d\eta}{d\tau}$ and η (curve b) at $\frac{2p_1}{p_o} = 1$.

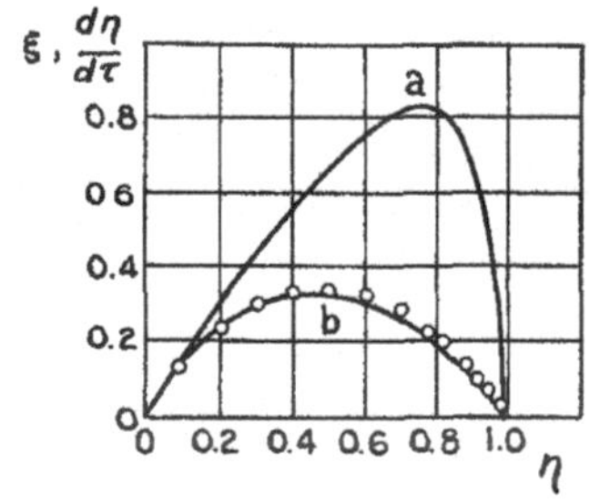

FIGURE 39. Relation between ξ and η (curve a), $\frac{d\eta}{d\tau}$ and η (curve b) at $\frac{2p_1}{p_o} = 0.5$.

It is seen that the relative concentration of H atoms first increases with η and reaches a maximum at

$$(1 - \eta) = \frac{p_1}{p_o}$$

(when the oxygen pressure, because of reaction, falls below the lower limit). The reaction does not stop at this stage but continues owing to the presence of hydrogen atoms. It finally stops when ξ becomes equal to zero at some value $(1 - \eta)$ corresponding to unreacted mixture. According to (20) the residual oxygen concentration is given by:

$$\eta_1 + \frac{p_1}{p_o} \ln(1 - \eta_1) = 0 \quad . \tag{21}$$

Experiments confirm the existence of this residual pressure and its decrease with increasing

$$\frac{p_0}{p_1} \quad .$$

Kinetically, what is measured is not ξ or η but the pressure change:

$$\Delta p = \frac{\Delta M}{N} = \frac{(O_2) - \frac{(H)}{2}}{N} \quad .$$

Let us introduce the dimensionless variable

$$\pi = \frac{\Delta M}{(O_2)_0} = \frac{\Delta p}{(p_0/3)} \quad .$$

This quantity expresses the pressure drop Δp as a fraction of the final pressure drop if the stoichiometric mixture reacted to completion. Note that

$$\frac{p_0}{3} = \frac{(O_2)_0}{N}$$

in the case of a stoichiometric mixture. It is easily seen that:

$$\pi = \eta - \frac{\xi}{2} = \frac{p_1}{p_o} \ln(1 - \eta) \quad . \tag{22}$$

From (17) and (18) we get:

$$\frac{d\pi}{dt} = \frac{d\eta}{dt} - \frac{1}{2}\frac{d\xi}{dt} = k_2(O_2)_0 \frac{p_1}{p_0} \xi$$

or, with the dimensionless time $\tau = k_2(O_2)_0 t$:

$$\frac{d\pi}{d\tau} = \frac{p_1}{p_0} \xi = \frac{p_1}{p_0} \left[2\eta + 2\frac{p_1}{p_o} \ln(1 - \eta) \right] . \tag{23}$$

But from (22):

$$2\,\frac{p_1}{p_0}\,\ell n(1-\eta) = -2\pi$$

or $1-\eta = \exp(-\pi p_0/p_1)$ and $\eta = 1-\exp(-\pi p_0/p_1)$.

Hence, according to (22) and (23):

$$\frac{d\pi}{d\tau} = \frac{2p_1}{p_0}\left[1-\exp(-\pi p_0/p_1)\cdot\pi\right]. \qquad (24)$$

Experimentally, Δp is measured as a function of t. (Figure 40).* The experimental curves are easily transformed into relations between

$$\frac{d\pi}{d\tau}$$

and

$$\pi = \frac{\Delta p}{0.33p_0}.$$

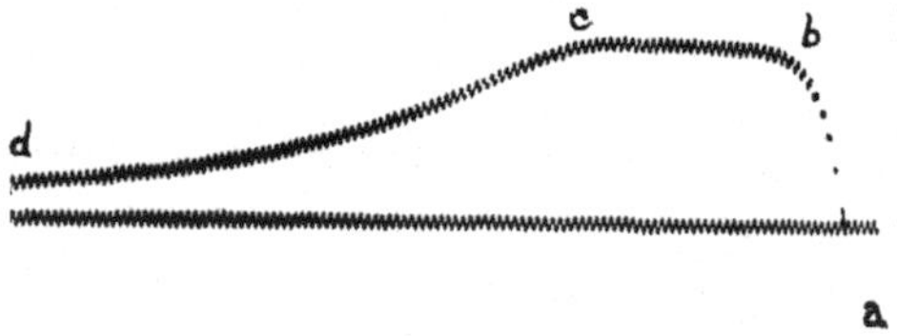

FIGURE 40. Relation between Δp and t. Segment ab: introduction of the mixture; bc: induction period; cd: reaction period.

Since the experiments were carried out (4) at a value of p_0 approximately three times p_1, the combustion was practically complete. A comparison between theory and experiment and a determination of the rate constant k_2 may be made in the following way. First construct, for various values of

$$\frac{p_1}{p_0}$$

the curves of

$$\frac{d\pi}{d\tau}$$

versus π according to equation (24). Then the experimental curves

$$\frac{d\pi}{dt}$$

* The experimental variation of Δp with t is shown on Figure 40. The segment ab corresponds to the introduction of the mixture in the reactor. The segment bc is the induction period; cd corresponds to reaction. This was obtained by means of a glass membrane gauge. A light beam fell on the mirror of the membrane gauge. This beam came from another mirror oscillating with a constant period of 1/300 sec. This was done to increase the precision of the time measurement, an important factor since the experimental curves have to be differentiated.

as a function of π are drawn. Now, a scale factor $(1/\alpha)$ for the experimental curves is determined in such a way that the maximum of the experimental curve coincides with the maximum of the theoretical curve. It is possible then to 1) see how well the experimental curves superpose on the theoretical curves

$$\frac{d\pi}{d\tau} \text{ vs. } \pi$$

and 2) calculate k_2 from the scale factor $(1/\alpha)$.

Indeed:

$$\frac{d\pi}{d\tau} = \frac{1}{\alpha}\left(\frac{d\pi}{dt}\right)_{\text{exp.}} .$$

Since

$$\tau = k_2(O_2)_0 t = \frac{k_2 N p_0}{3}, \qquad \frac{d\pi}{d\tau} = \frac{1}{\frac{k_2 N p_0}{3}} \cdot \frac{d\pi}{dt} .$$

Hence,

$$\alpha = \frac{k_2 N p_0}{3} .$$

Knowing α, N and p_0, we may compute k_2.

Figure 41a shows a family of theoretical curves:

$$\frac{d\pi}{d\tau} = \frac{2p_1}{p_0}\,[1 - \exp(-\,\pi p_0/p_1) - \pi] \tag{25}$$

at $T = 795°K$ and for various values of

$$\psi = \left(1 - \frac{p_1}{p_0}\right) .$$

The points are the data of Karmilova, Nalbandyan and Semenov [4] transformed according to the procedure just explained. There is excellent agreement between theory and experiment for the entire range. From the scale factor $(1/\alpha)$, k_2 can be calculated.

Table 57 shows k_2 thus calculated at $T = 795°K$, $p_1 = 0.106$ and various values of p_0.

Out of the seven experimental curves, six give constant values within experimental error and only the last one yields a slightly high value due apparently to some heating caused by the high reaction rate at a pressure in excess of 1 mm. The average value of k_2 at 795° is $6.6 \cdot 10^{-15}$ which differs by a factor of 1.5 from the value $k_2 = 1.1 \cdot 10^{-14}$ obtained by Koval'skiǐ.

TABLE 57

p_0 mm	p_1/p_0	$k_2 \cdot 10^{15}$
0.38	0.28	6.15
0.51	0.21	6.55
0.63	0.17	6.9
0.75	0.14	6.4
0.87	0.12	7
1	0.105	7
1.2	0.9	8.2

without 8.2, the average is $6.6 \cdot 10^{-15}$

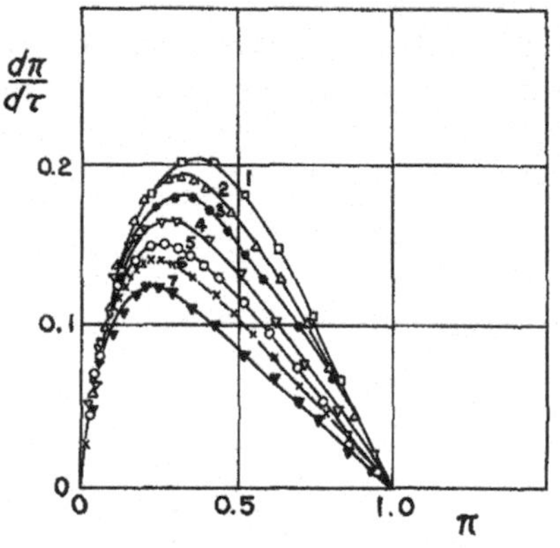

FIGURE 41a. Relation between $\frac{d\pi}{d\tau}$ and π at $2p_1/p_0$ = 0.56, 0.42, 0.34, 0.28, 0.24, 0.21 and 0.18 respectively. Calculations from equation (24). The points are experimental values (4).

Figure 41b shows a family of theoretical curves calculated by means of equation (25) together with the family of experimental curves drawn with the scale factor $(1/\alpha)$ at seven different temperatures. The values of k_2 corresponding to these temperatures and calculated with the appropriate scale factors are given in Table 58 on the following page. From these data, a curve of $\log k_2$ versus $1/T$ can be drawn: as shown on Figure 42, all seven points lie on a straight line the slope of which gives the activation energy E_2 = 15100 cal.

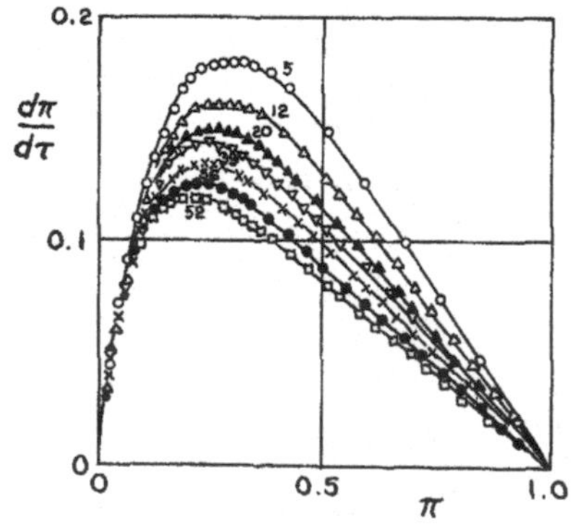

FIGURE 41b. Relation between $\frac{d\pi}{d\tau}$ and π at various temperatures and $2p_1/p_0$ ratios:

	T°C	$2p_1/p_0$
5	460	0.323
12	480	0.28
39	540	0.2
20	552	0.2
56	580	0.17
32	560	0.167
60	600	0.22

Curves are calculated from equation (24). Points are experimental (4).

From E_2 and the absolute value of the rate constant, one obtains the pre-exponential factor A_2 = $0.94 \cdot 10^{-10}$:

$$k_2 = A_2 \exp(-E_2/RT) = 0.94 \cdot 10^{-10} \exp(-15100/RT) . \quad (26)$$

TABLE 58

T°C	T°K	$k_2 \cdot 10^{15}$
460	733	3.12
480	753	4.4
502	775	5.22
522	795	6.6
540	813	8.8
560	833	10
580	853	13.2
600	873	16.2

The pre-exponential factor is about ten times smaller than the number of collisions of H atoms. Therefore, the reaction $H_2 + O_2 \longrightarrow OH + O$ has a steric factor ~ 0.1.

The theoretical formula gives readily the value π_{max} at which the dimensionless rate

$$\frac{d\pi}{d\tau}$$

reaches its maximum:

$$\pi_{max} = \frac{p_1}{p_0} \ln \frac{p_0}{p_1} \quad .$$

It is also easy to find π_{max} from the data. Figure 43 shows such a curve for various values of

$$\frac{2p_1}{p_0} \quad .$$

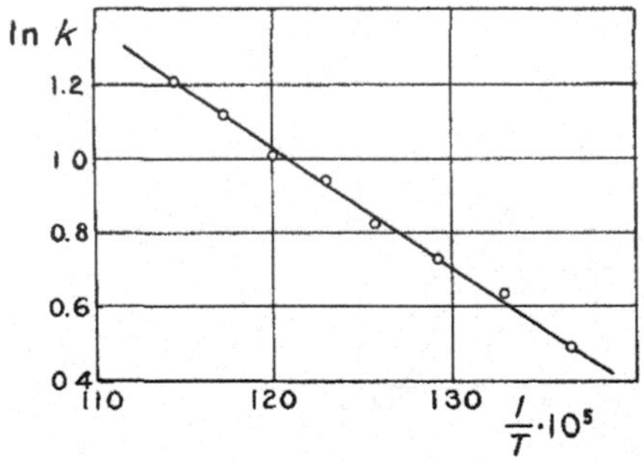

FIGURE 42. Relation between $\log k_2$ and $1/T$ (4).

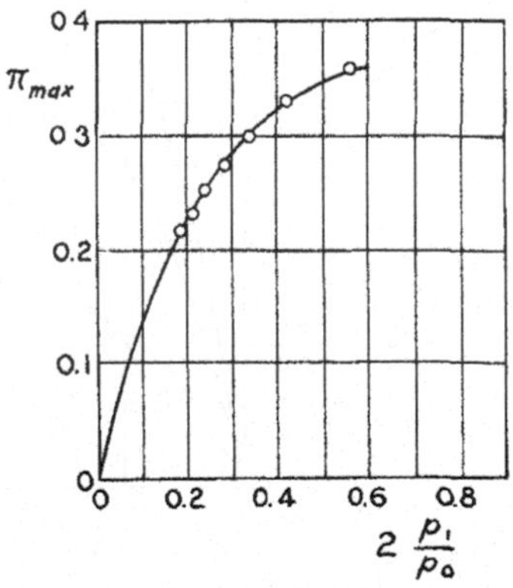

FIGURE 43. Relation between π_{max} and $2p_1/p_0$ (4).

The continuous curve represents the theoretical dependence of π_{max} on $2p_1/p_0$ and the points are experimental data. Agreement between theory and experiment is excellent. To sum up, there remains little doubt concerning the correctness of the assumed reaction mechanism for $\gamma < 0.5$.

Figure 44 represents the variation of η, ξ, π with τ at

$$\frac{p_1}{p_0} = 0.5 \quad .$$

The theoretical formula (20) gives easily the variation of the relative concentration

$$\xi = \frac{(H)}{(O_2)_0}$$

with extent of reaction η. Figure 45 shows the theoretical relation between ξ and π.

It is seen that the maximum concentration of hydrogen atoms when

$$\frac{p_1}{p_0} = 0.5 \quad \text{or} \quad 0.25$$

is comparable to the concentration $(O_2)_0$ and reaches $0.8(O_2)_0$ at $(p_1/p_0) = 0.25$ [see equations (20) and (22)].

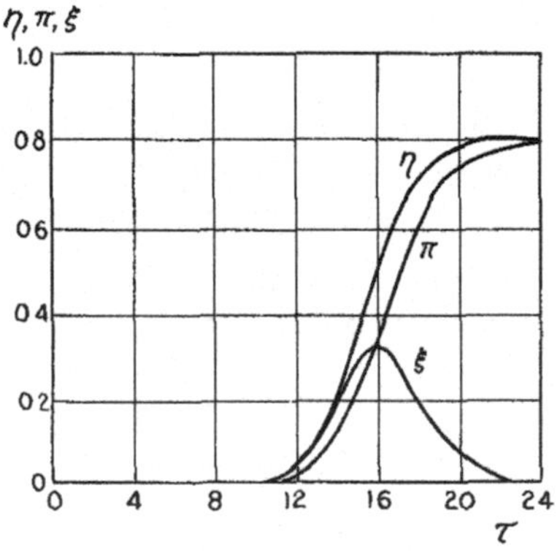

FIGURE 44. Relation between η, π, ξ and τ (10).

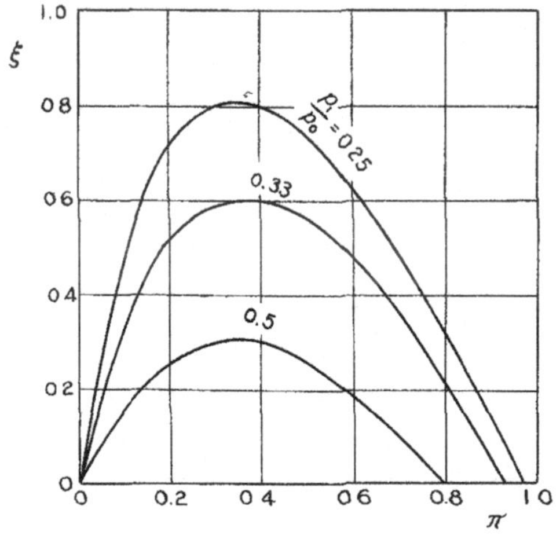

FIGURE 45. Relation between ξ and π.

Kondrat'ev and Kondrat'eva [11] as well as Kondrat'ev and Avramenko [12] have measured directly the concentrations (H) and (OH) in stationary flames at very low pressures where the heating is weak. At $p_1 = 1.5$ mm and $p = 4$ mm in the gas flow, a value of $\xi = 0.5$ was found, in close agreement with the theoretical value. The concentration of OH is about 100 times smaller than the H concentration, i.e., $(2k_1/k_2) = 100$, in agreement with the calculated figures used in the first section. The concentration (H) was determined from the heating of a thermocouple covered with $ZnO \cdot Cr_2O_3$ $(\epsilon_H = 1)$ and the concentration (OH) by means of absorption spectra.

B. The Case of Pure Diffusion. Until now the purely kinetic approximation has been used, when the time for diffusion of the active centers is much shorter than the time required for their adsorption (ϵ small). The diffusion time was therefore neglected.

Consider now the other extreme situation when the time for adsorption is very short $(\epsilon \sim 1)$ as compared to the time required for diffusion of the active centers to the wall. Therefore, the first of these times may now be neglected. Since a particle reaching the wall will collide with it many times before it returns to the interior of the reactor, the purely diffusional approximation is satisfactory, within 10% when ϵ varies between 1 and $2 \cdot 10^{-2}$.

Instead of the quantities $a_1'(OH)$, $a_2'(H)$, $a_3'(O)$ of the previous kinetic equations, we now need terms describing the diffusion of active centers from a volume element within the reactor to the walls under their concentration gradient. These terms were previously omitted since in the purely kinetic region, the concentrations (H), (OH) and (O) are practically constant throughout the reactor.

Since each active particle disappears as soon as it reaches the wall if $\epsilon \sim 1$, $(H)_w = (OH)_w = (O)_w = 0$ at the wall. Then there is a concentration gradient of active particles within the reactor, directed from the center to the wall. For a plane vessel, the kinetic equations are:

$$1)\quad \frac{\partial(OH)}{\partial t} = -a_1(OH) + a_2(H) + a_3(O) + D_{OH}\frac{\partial^2(OH)}{\partial x^2}$$

$$2)\quad \frac{\partial(H)}{\partial t} = n_0 + a_1(OH) - a_2(H) + a_3(O) + D_H\frac{\partial^2(H)}{\partial x^2} \qquad (27)$$

$$3)\quad \frac{\partial(O)}{\partial t} = a_2(H) - a_3(O) + D_O\frac{\partial^2(O)}{\partial x^2} \quad .$$

As before $a_2 \ll a_1$, $a_2 \ll a_3$ and $(H) \gg OH$ $(H) \gg (O)$. The diffusion coefficients D_{OH}, D_H, D_O are of the same order of magnitude

(D_{OH}) ≈ D_O, $D_H \approx 4\ D_{OH}$). Therefore with even more justification than in the purely kinetic case, we may write:

$$\frac{\partial(OH)}{\partial t} = \frac{\partial(O)}{\partial t} = 0.$$

Moreover

$$\frac{\partial^2(OH)}{\partial x^2}\,, \quad \frac{\partial^2(O)}{\partial x^2} \ll \frac{\partial^2(H)}{\partial x^2}$$

and the first two may be neglected. Consequently, as before, we may replace three diffusion equations by a single one and find the relation between (OH), (O) and (H) from algebraic equations:

$$-\ a_1(OH) + a_2(H) + a_3(O) = 0$$

$$a_2(H) - a_3(O) = 0\ .$$

Hence, instead of the differential equation (17) we get:

$$\frac{\partial(H)}{\partial t} = n_0 + 2a_2(H) + D_H \frac{\partial^2(H)}{\partial x^2}\ . \tag{28}$$

A solution to this equation was found by Bursian and Sorokin [13]. It can be found in our book 'Chain Reactions' [14]. If the distance between the plane walls is d and the coordinate of one plane is $x = 0$ and if the other $x = d$, equation (23) must be integrated with the following initial and boundary conditions:

$$t = 0 \quad (H) = 0 \ \text{everywhere}$$

$$t > 0 \quad (H)_0 = (H)_d = 0\ .$$

Consider first the steady state problem with

$$\frac{\partial(H)}{\partial t} = 0\ :$$

$$D\frac{\partial^2(H)}{\partial x^2} + 2a_2(H) + n_0 = 0$$

$$(H)_0 = (H)_d = 0\ . \tag{29}$$

The solution of equation (29) is:

$$(H)_x = \frac{n_o}{2a_2}\left\{\frac{\cos\sqrt{F}\,(2x-d)}{\cos\sqrt{F}\,d} - 1\right\}$$

where

$$F = \frac{2a_2}{D}.$$

As can be expected, $(H)_x$ is maximum at $x = \frac{d}{2}$ and decreases towards the walls where it becomes zero in accordance with the boundary condition.

It is easy to calculate the average $(\bar{H})$ throughout:

$$(\bar{H}) = \frac{\int_0^d (H)_x \cdot dx}{d} = \frac{n_o}{2a_2}\left(\frac{\tan d\sqrt{F}}{d\sqrt{F}} - 1\right) = \frac{n_o}{2a_2}\left(\frac{\tan y}{y} - 1\right) \qquad (30)$$

The quantity [see equation (14)]

$$F = \frac{2a_2}{4D} = \frac{2k_2N\gamma p^2}{4D_o}$$

since

$$D = \frac{D_o}{p}$$

where D_o is the value of D at $p = 1$ mm. Thus y increases with p. When it reaches the value $\pi/2$, $\tan y$ and thus also (H) go to infinity. This means that if

$$y \geq \frac{\pi}{2},$$

a steady state solution is impossible. In other words, the condition

$$y = dp\sqrt{\frac{2k_2N\gamma}{4D_o}} = \frac{\pi}{2}$$

determines the lower limit p_1. Consequently, in the purely diffusional region, the product p_1d has the value:

$$dp_1 = \sqrt{\frac{\pi^2 D_o}{2k_2N\gamma}}. \qquad (31)$$

Another more convenient expression for (H) may be written with the help of the rapidly converging series:

$$\frac{\tan y}{y} - 1 = \frac{8}{\pi^2}\left(\frac{2y}{\pi}\right)^2 \left\{ \frac{1}{1 - \left(\frac{2y}{\pi}\right)^2} + \frac{1}{9\left[9 - \left(\frac{2y}{\pi}\right)^2\right]} + \frac{1}{25\left[25 - \left(\frac{2y}{\pi}\right)^2\right]} + \cdots \right.$$

It is easy to see that when $y < \frac{\pi}{2}$ the series converges so rapidly that, to a good approximation, one may retain only the first term of the expansion. Hence:

$$(\bar{H}) = \frac{0.82\ n_0}{k - 2a_2} \tag{32}$$

where

$$k = \frac{D\pi^2}{d^2} \ .$$

But this is the solution of the simple equation:

$$0.82\ n_0 + 2a_2(\bar{H}) - k(\bar{H}) = 0 \ .$$

Therefore, the diffusion equation leads to an equation of the kinetic type where the constant

$$a_2' = \frac{\epsilon_H u_H}{d}$$

is replaced by the constant

$$k = \frac{D\pi^2}{d^2} \ .$$

The only difference is that n_0 is replaced by $0.82\ n_0$ and that the equation refers to the average value $(\bar{H})$ throughout the reactor. In this way, the term

$$D_H \frac{\partial^2 (H)}{\partial x^2}$$

in equation (29) may be replaced by the term $k(\bar{H})$. For the average reaction rate, we have:

$$w = \frac{d\Delta(O_2)}{dt} = n_0 + a_2(H) = n_0 + a_2 \frac{0.82\ n_0}{k - 2a_2} \ . \tag{33}$$

Again, this is practically the same as in the kinetic case. Only, instead of a_2', we have

$$k = \frac{D\pi^2}{d^2} \ .$$

It can be seen readily that the solution of the non steady problem [integration of equation (28)] may be replaced by the solution of the simple differential equation

$$\frac{d(H)}{dt} = 0.82\, n_o + 2a_2(H) - k(H)$$

$$k > 2a_2 \ : \ (\bar{H}) = \frac{0.82\, n_o}{k - 2a_2}\,[1 - \exp - (k - 2a_2)t] \qquad (34)$$

$$k < 2a_2 \qquad (\bar{H}) = \frac{0.82\, n_o}{k - 2a_2}\,[\exp - (k - 2a_2)t - 1] \ .$$

This is similar to the solution of the kinetic problem, with the only difference that the constant

$$a_2' = \frac{\epsilon_H u_H}{d}$$

is replaced by the constant

$$k = \frac{D\pi^2}{d^2} \ .^*$$

For a cylindrical reactor with radius $R = \frac{d}{2}$, the main equation becomes:

$$\frac{\partial(H)}{\partial t} = n_o + 2a_2(H) + D\,\frac{\partial^2(H)}{\partial r^2} + \frac{D}{r}\,\frac{\partial(H)}{\partial r} \qquad (35)$$

with the conditions:

$$\text{at} \quad t = 0, \qquad (H) = 0 \quad \text{everywhere} \ ;$$

$$\text{at} \quad t > 0, \qquad (H)_{r=R} = 0 \quad \text{and} \quad \left(\frac{\partial(H)}{\partial r}\right)_{r=0} = 0$$

(since (H) must have a maximum value along the axis).

---

* In fact, the exact solution of equation (28) is expressed by the series:

$$(\bar{H}) = \frac{(\pi^2/12)n_o}{D\pi^2/d^2} \sum_{s=0}^{\infty} \frac{1 - e^{-[(2s+1)-\gamma]\theta}}{(2s+1)^2\{(2s+1)^2-\gamma\}}$$

with

$$\frac{\pi^2}{12} = 0.82; \qquad \theta = \frac{\pi^2 D}{d^2}\, t = kt; \qquad \gamma = \frac{2a_2 d^2}{\pi^2 D} = \frac{2a_2}{k} \ .$$

The series converges very rapidly and with a very good approximation, only the first term with $s = 0$ need be retained. Then, the solution is identical with that given in equation (34).

A similar reasoning reveals that, for a cylindrical reactor of diameter d:

$$k = \frac{23D}{d^2} = \frac{23D_o}{d^2 p} . \tag{36}$$

For a spherical reactor:

$$k = \frac{4\pi^2 D}{d^2} = \frac{4\pi^2 D_o}{d^2 p} . \tag{37}$$

In a cylindrical reactor, the lower limit is determined by the condition:

$$\frac{23D}{d^2} = 2a_2$$

and since

$$D = \frac{D_o}{p}, \quad a_2 = 2k_2 N\gamma \ :$$

$$p_1 d = \left(\frac{23D_o}{2k_2 N\gamma} \right)^{1/2} = 4.8 \left(\frac{D_o}{2k_2 N\gamma} \right)^{1/2} = 4.8 \sqrt{B} . \tag{38}$$

In the purely kinetic region, we had a different expression:

$$p_1 d = \frac{\epsilon_H u_H}{2k_2 N\gamma} .$$

At $T = 800°K$, for a stoichiometric mixture $k_2 = 10^{-14}$, $N = 1.22 \cdot 10^{16}$, $\gamma = 0.33$ and $D_o = 7.10^3$. Then $B = 87.$*

---

* D_o is calculated as follows. Tables give the diffusion coefficient of H_2 in O_2, equal to 0.677 under normal conditions. It can be assumed that the diffusivity of H in O_2 will be $\sqrt{2}$ times larger, i.e., $D_{H \to O_2} = 0.95$. In a stoichiometric mixture with 1/3 O_2 and 2/3 H_2, $D_{H \to H_2}$ is about twice $D_{H \to O_2}$ (by analogy with the relative values of $D_{H_2 \to O_2}$ and $D_{H_2 \to D_2}$). Therefore, the diffusivity of H in the stoichiometric mixture is:

$$\frac{1}{D} = \frac{0.33}{0.95} + \frac{0.66}{1.9} = 0.35 + 0.35 = 0.7.$$

Hence

$$D = \frac{1}{0.7} = 1.43 .$$

Consequently, $p_1 d = 4.8\sqrt{B} = 4.8\sqrt{87} = 4.8 \cdot 9.3 = 44$. In a reactor with $d = 2$ cm, $p_1 = 22$ mm; with $d = 3$ mm, $p_1 = 15$ mm etc. This is a relatively high figure for the lower limit, if it is recalled that with a good pretreatment of the walls, at 800°K in the kinetic region, p_1 is of the order of 0.1 - 0.2 mm or below.

The highest observed values of p_1 (e.g., in quartz [6] or iron reactors [15] at T = 800°K) are 10 to 15 mm in agreement with theory.

The lower limit can therefore be calculated in the purely kinetic and diffusional regions. All the intermediate cases, corresponding to various values of ϵ_H may also be treated theoretically. This has been done by Semenov [16]. The results of such a calculation with B = 100 i.e., for a case similar to the one just treated are shown on Figure 46.

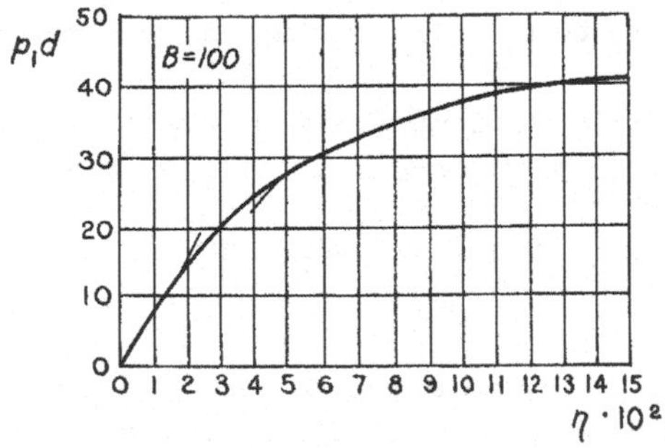

FIGURE 46. Relation between $p_1 d$ and η (B = 100) (16).

The quantity $p_1 d$ is plotted in ordinates and $\eta \cdot 10^2$ in abscissae (so that the unit of the abscissae is $\eta = 10^{-2}$, 10 means $\eta = 10^{-1}$ etc.). The quantity η is related to ϵ by means of the expression:

$$\eta = \frac{3\epsilon}{8\lambda_0} .$$

* (cont.)
At T = 800°K, and p = 1 mm,

$$D_0 = D\ 760 \left(\frac{800}{273}\right)^{1.75} = 1.43 \cdot 760 \cdot 6.35 \sim 7000 .$$

Finally

$$B = \frac{D_0}{2k_2 N\gamma} = \frac{7000}{2.44 \cdot 10^2 \cdot 0.33} = \frac{70}{0.8} = 87 .$$

The quantity B depends on mixture composition $(B \sim \frac{1}{\gamma})$. At 800°K, if $\gamma = 0.5$, B = 58.

At 800°K, in a stoichiometric mixture, $\lambda_o = 4.10^{-2}$. Thus $\epsilon \sim 0.1\eta$. The unit of the abscissae corresponds to $\epsilon = 10^{-3}$, ten corresponds to $\epsilon = 10^{-2}$ etc.*

The Figure shows that up to $\eta = 2.10^{-2}$ i.e., $\epsilon = 2.10^{-3}$, the purely kinetic region is in control and equation (39) may be used. It can be proved that for $\eta \geq 5.10^{-2}$ i.e., $\epsilon \geq 5.10^{-3}$, the following formula is adequate:

$$p_1 d = 4.8\sqrt{B} - \frac{1}{\eta} \quad . \tag{40}$$

This expression reduces to the value found above $p_1 d = 4.8\sqrt{B}$ when $\epsilon = 1$, within a few percents.**

If B is known, the product $p_1 d$ immediately indicates in which region the measurements are being made.

---

* It is difficult to calculate λ_o exactly and the coefficient 0.1 may be a little high.

** It can be shown that in the larger part of the intermediate region between the purely kinetic and diffusional cases, at $\eta > 5.10^{-2}$, the following formula is applicable:

$$k = \frac{23D}{d^2\left(1 + \frac{8\lambda_o}{3\epsilon d}\right)^2} \tag{41}$$

and the limit may be found by equating the rate of branching $2a_2(H)$ and the rate of termination $k(H)$. It is easy to see that, for $\epsilon \sim 1$,

$$\frac{8\lambda_o}{3\epsilon d} \sim \frac{8}{3}\frac{\lambda_o}{d}$$

is much smaller than unit so that (41) reduces to the formula we just used:

$$k = \frac{23D}{d^2} \quad .$$

Putting

$$D = \frac{D_o}{p}$$

and

$$\lambda = \frac{\lambda_o}{p}$$

together with

$$\frac{3}{8\lambda_o} = \eta \quad ,$$

we get:

$$k = \frac{D_o}{pd^2\left(1 + \frac{1}{pd\eta}\right)^2} \quad . \tag{42}$$

For instance, if $B = 100$, a value of p_1d up to 15 corresponds to the purely kinetic region. The limiting formula $p_1d = 4.8\sqrt{B}$ is valid, within 10% for $0.2 < \eta < 1$ or $2.10^{-2} < \epsilon < 0.1$.

In the diffusional region, p_1 depends on T by the law

$$p_1 \sim \sqrt{\frac{1}{k_2}} \sim \exp(7500/RT) \quad .$$

Early measurements of the lower limit in the diffusional range have given a much higher temperature coefficient but this was due to the fact that the data were obtained in a temperature range where homogeneous chain termination exerts a substantial effect on p_1.

The more detailed work of Nalbandyan and Shubina [17] gave for E' the value 9000 cal, close to the theoretical value.

In the kinetic region where the diffusion time of H atoms plays no important role, the limit is determined by the partial pressure of the $H_2 + O_2$ mixture, whether it is diluted by an inert gas or not. In the diffusion region, dilution by an inert gas lowers the lower limit of the partial pressure of the combustible mixture since addition of an inert gas increases the diffusion time to the wall. Indeed, in the formula

$$p_1d = 4.8\sqrt{\frac{D}{2a_2}} \quad ,$$

the diffusion coefficient

$$D = \frac{D_o}{P}$$

where P is the total pressure: that of the $H_2 + O_2$ mixture plus that of the inert gas. But the quantity $a_2 = k_2(O_2)$ depends only on the oxygen content of the $H_2 + O_2$ mixture so that $a_2 \sim p$ where p is the partial pressure of the combustible mixture. Therefore, in the purely diffusional region, the condition for the limit is

$$\frac{23D_o}{Pd^2} = 2a_2$$

** (cont.

Hence the condition at the limit $2a_2 = k$ becomes:

$$2k_2(O_2) = 2k_2N\gamma p = \frac{23\ D_o}{pd^2\left(1 + \frac{1}{pd\eta}\right)^2} \quad \text{and} \quad (pd)^2 = \frac{\frac{23D}{2k_2N\gamma}}{\left(1 + \frac{1}{pd\eta}\right)^2} = \frac{23\ B}{\left(1 + \frac{1}{pd\eta}\right)^2}$$

or

$$pd\left(1 + \frac{1}{pd\eta}\right) = 4.8\sqrt{B} \quad \text{and} \quad pd = 4.8\sqrt{B} - \frac{1}{\eta} \quad .$$

if the mixture is diluted with an inert gas, or:

$$K = \frac{23D_o}{Pd^2} = 2k_2(O_2) = 2k_2N\gamma p \quad .$$

Hence:

$$d^2Pp = d^2p_m(p_A + p_m) = d^2p_m^2\left(1 + \frac{p_A}{p_m}\right) = \frac{23D_o}{2k_2N\gamma} = \text{const.} = d^2p_1^2 \quad (43)$$

where $p_m = p_{H_2} + p_{O_2}$, the partial pressure of the combustible mixture in the presence of inert; p_A the partial pressure of inert gas; p_1 the pressure limit of the $H_2 + O_2$ mixture in the absence of inert gas. Thus

$$\left(\frac{p_1}{p_m}\right)^2 = 1 + \frac{p_A}{p_m}$$

or in more explicit form, taking into account the difference in diffusivities in A and in the $H_2 + O_2$ mixture, we get:

$$\left(\frac{p_1}{p_m}\right)^2 = 1 + \frac{ap_A}{p_m} \quad . \quad (43')$$

This theoretical relation has been repeatedly checked experimentally in this form.

It has been seen that it is practically impossible to determine ϵ from lower limit data when $\epsilon > 10^{-2}$ because p_1 is practically independent of ϵ. A special experiment may be carried out however to determine ϵ in the range 10^{-3} to 1. To this end, in a cylindrical glass or quartz reactor with well treated walls having a value of ϵ around $10^{-4} - 10^{-5}$, a thin probe is introduced made of a material that removes H atoms readily. The lower limit p_1^* is then measured. If $p = p_1$ in the reactor alone is very small, say 0.1 mm but is of the order of 10 mm in a reactor made of the probe material, p_1^* in the system just described will be of the order of a few mm. It may be shown theoretically how p_1^* depends on probe diameter and on ϵ. Therefore data on p_1^* give a way to calculate ϵ for the probe material.

If a gold filament, 0.5 mm in diameter is introduced in a cylindrical vessel at T = 440°C, p_1 increases from 0.94 mm to 8.2 mm.

By means of a magnet, the filament may be introduced or withdrawn. If the mixture is fed into the reactor in the presence of the probe, at a pressure slightly higher than the lower limit, a weak flash occurs but the reaction then subsides. If now the probe is withdrawn from the reactor, the lower limit is lowered and a flash occurs once more.

Nalbandyan and Shubina [17] have measured p_1 in the presence of probes covered with various materials and determined ϵ in this fashion. Large values of ϵ were measured in this way, sometimes close to unity. For instance, a graphite or glass probe covered with a powder of $ZnO.Cr_2O_3$ has a value of $\epsilon = 1$ at $T = 496°C$; a gold probe at 440°G has an ϵ value of 0.1; for platinum, $\epsilon = 0.09$; for tungsten at 540°G, $\epsilon = 6.10^{-3}$ etc. The experiment of Nalbandyan and Shubina strongly reminds us of the control of nuclear reactors by means of cadmium rods. Of course, in both cases, one deals with a branched chain reaction. The active particles (H atoms or neutrons) are easily captured by the rod material.

For $\gamma < 0.5$, in the kinetic region, formula (14') indicates that $p_1\gamma$ = const. In the purely kinetic region, formula (38) gives another relation: $p_1 \sqrt{\gamma}$ = const. where

$$\gamma = \frac{(O_2)}{(O_2) + (H_2)} .$$

What now, are the relations when the mixture is diluted with an inert gas, say nitrogen? Put $p = [O_2] + [H_2]$ and $P = [O_2] + [H_2] + [N_2]$. Equation (43) gives:

$$Pp\gamma = \frac{23D_o}{2k_2Nd^2} .$$

But

$$p\gamma = \frac{p[O_2]}{p} = [O_2] = \gamma' P$$

with

$$\gamma' = \frac{[O_2]}{[O_2] + [H_2] + [N_2]} = \text{mole fraction of oxygen}$$

in the ternary mixture. Hence:

$$P^2\gamma' = \frac{23D_o}{2k_2Nd^2}$$

or

$$P\sqrt{\gamma'} = \text{const.} \tag{44}$$

In the range $5.10^{-3} < \epsilon < 1$, or for η down to $\sim 5.10^{-2}$ where equation (39) is valid, it is easy to show that:

$$P_1 d = \frac{1}{\eta} = 4.8 \sqrt{\frac{D_o}{2k_2 N}} \frac{1}{\sqrt{\gamma'}} \quad ^{*} \qquad (45)$$

or

$$\sqrt{\gamma'} \left(P_1 + \frac{1}{d\eta} \right) = \frac{4.8}{d} \sqrt{\frac{D_o}{2k_2 N}} = \text{const.} \qquad (46)$$

This equation is verified by the data of Baldwin [18] collected in Table 59. The lower limit P_1 (in mm H_2) was measured in reactors with $KC\ell$ treated walls. Their diameters were: 5.1, 3.6, 2.4 and 1.4 cm.

TABLE 59

M_2	O_2	N_2	d=5.1 P_1	d=3.6 P_1	d=2.4 P_1	d=1.4 P_1
0.28	0.72	0	3.28	4.16	5.16	10.15
—"—	0.56	0.16	3.6	4.73	6.12	11.82
—"—	0.28	0.44	5.61	7.11	9.63	19.78
—"—	0.14	0.58	9.07	11.5	16.3	27.74
—"—	0.1	0.62	11.72	15.08	22.01	43.16
—"—	0.07	0.65	15.49	21.15	33.53	——

Baldwin's data are well represented by equation (46) with a value $\eta = 7.10^{-2}$ (i.e., $\epsilon_H \sim 7.10^{-3}$). Indeed, for vessels of diameters $d = 5.1$, 3.6 and 1.4 cm, the quantity $1/d\eta$ is equal respectively to 2.8, 4 and 10.2. The products

---

* Expression (41) for the constant K in the presence of nitrogen may be rewritten as follows:

$$K = \frac{23D}{d^2 \left(1 + \frac{\gamma \lambda_o}{3\epsilon d}\right)^2} = \frac{23D_o}{Pd^2 \left(1 + \frac{\gamma \lambda_o}{3\epsilon dP}\right)^2} = \frac{23D_o}{Pd^2 \left(1 + \frac{1}{Pd\eta}\right)^2}$$

$$2k_2 \gamma Np = 2k_2 \gamma' NP = \frac{23D_o}{Pd^2 \left(1 + \frac{1}{pd\eta}\right)^2}$$

Equation (45) follows directly.

$$\left(P_1 + \frac{1}{d\eta} \right) \sqrt{\gamma'}$$

where P_1 is Baldwin's experimental lower limit, are shown in Tables 60 to 62.

TABLE 60

d=5.1 cm

γ'	P_1	$(P_1 + 2.8)\sqrt{\gamma'}$
0.42	3.28	5.2
0.56	3.6	4.8
0.28	5.6	4.4
0.14	9.1	4.55
0.1	11.7	4.6
0.07	15.5	4.8
		average 4.72

TABLE 61

d=3.6 cm

γ'	P_1	$(P_1 + 4)\sqrt{\gamma'}$
0.72	4.16	6.9
0.56	4.73	6.5
0.28	7.11	6.0
0.14	11.5	5.7
0.1	15.08	6.1
0.07	21.15	6.6
		average 6.3

TABLE 62

d=1.4 cm

γ'	P_1	$(P_1 + 10.2)\sqrt{\gamma'}$
0.72	10.15	17.4
0.56	11.82	16.6
0.28	19.78	16
0.14	27.74	14
0.1	43.16	17
0.07	——	—
		average 16

It can be seen that equation (46) is well obeyed in a very wide interval of values of γ. Other data of Baldwin show this also. He changed the H_2 concentration in a wide range, keeping that of O_2 constant. In the kinetic range, the relation $p_1\gamma$ = const. is obeyed only for $\gamma \leq 0.5$ but in the diffusion range discussed here, the data of Baldwin show that the relation $p\sqrt{\gamma}$ = const. is satisfactorily followed up to $\gamma = 0.9$.*

As can be seen in the tables, the average values of

$$\left(P_1 + \frac{1}{d\eta}\right)\sqrt{\gamma'}$$

for d = 5.1, 3.6 and 1.4 cm are, respectively 4.72, 6.3 and 16. Multiplying these numbers by d, we get

$$\left(P_1 d + \frac{1}{\eta}\right)\sqrt{\gamma'}$$

with values equal to 24, 22.5 and 22.5 respectively. Theoretically [see equation (45)] the same quantity should be equal to

$$4.8\sqrt{\frac{D_0}{2k_2N}} \quad .$$

As was seen already, $D_0 = 7000$ and $k_2N = 1.22 \cdot 10^2$ at 800°K. Hence:

$$4.8\sqrt{\frac{D_0}{2k_2N}} = 25.5 \quad .$$

Calculated and experimental figures agree within 10% and this is excellent agreement since an error can easily be made in the calculation of D_0. Baldwin's data fully support the theory.

§4. The Upper Explosion Limit.

It is an experimental fact that $H_2 + O_2$ mixtures, besides a lower explosion limit p_1, also exhibit an upper limit p_2 above which reaction starts to proceed once more. Hinshelwood [19] discovered this limit in hydrogen-oxygen mixtures. The general trend of the region of chain

---

* This is understandable since in the kinetic range the rate constants a_2' and ϵ_1' are proportional to ϵ_H and ϵ_{OH} and the latter is 10 to 20 times larger than the former. Then, in the kinetic range, the capture of OH radicals on the wall may be neglected up to $\gamma \lesssim 0.5$. In the diffusional range, the corresponding rate constants do not depend on ϵ_H and ϵ_{OH} and they have about the same value. The life time of H atoms in a stoichiometric mixture is 100 times larger than that of OH radicals. It is 10 times larger for $\gamma = 0.9$. Therefore, the disappearance of OH radicals will play no significant role up to $\gamma \sim 0.5$. In the intermediate range, the rate constants will depend on ϵ but not very strongly and consequently the relationships we have derived are valid for $\gamma > 5$. In the case of Baldwin's experiments, they are valid up to $\gamma \sim 0.8 - 0.9$.

explosion is illustrated in Figure 47. [(a) represents lower and upper limits in p - T diagram (20); (b) lower, upper and third limits in a log p - T diagram (21).]

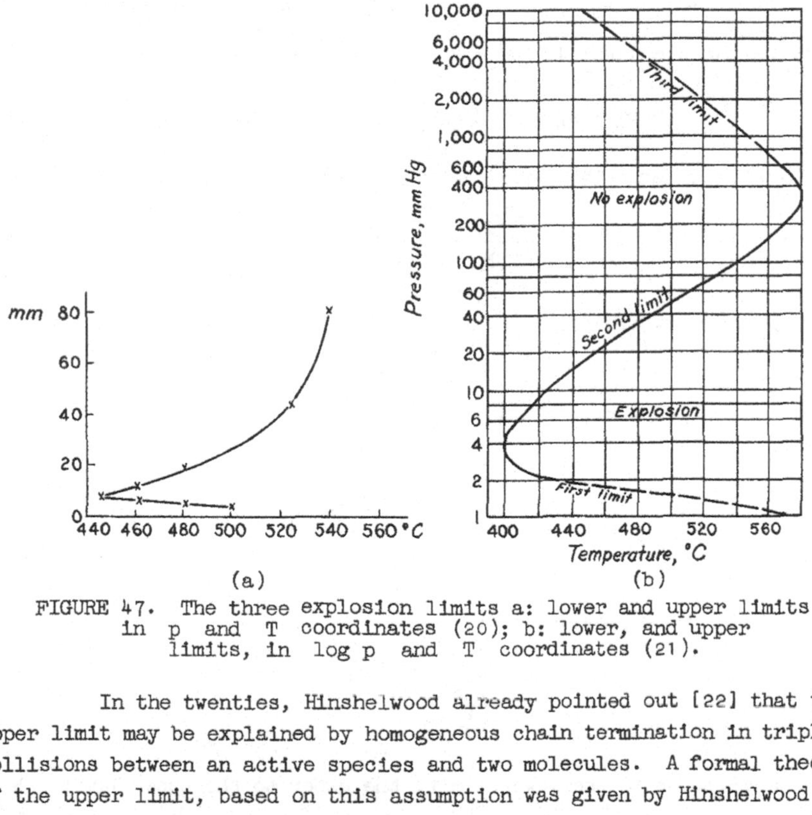

FIGURE 47. The three explosion limits a: lower and upper limits in p and T coordinates (20); b: lower, and upper limits, in log p and T coordinates (21).

In the twenties, Hinshelwood already pointed out [22] that the upper limit may be explained by homogeneous chain termination in triple collisions between an active species and two molecules. A formal theory of the upper limit, based on this assumption was given by Hinshelwood [19] and also by Semenov [14]. This theory delineated a single chain explosion region related to both lower and upper limits. This question was subsequently examined in detail by V. V. Voevodskiĭ [23].

The chemical mechanism of the upper limit was elucidated by Lewis [24] who introduced the assumption that the reaction $H + O_2 + M \longrightarrow HO_2 + M$ is responsible for homogeneous chain termination. In this step, an active H atom is replaced by an inactive HO_2 radical. The latter dies on the walls before it has a chance to regenerate an active H radical in the reaction $H_2 + HO_2 \longrightarrow H_2O_2 + H$.

The analysis of phenomena at higher pressures therefore require the inclusion in scheme II of an additional reaction 6): $H + O_2 + M \longrightarrow HO_2 + M \longrightarrow H_2O_2 + O_2$. This reaction goes at a rate $a_6(H) = k_2(O_2)(M)(H)$

where k_6 is the rate constant for triple collisions and is independent of temperature. This rate contributes an additional term in equation (17):

$$\frac{d(H)}{dt} = n_0 + (2a_2 - a_2' - a_6)(H) = n_0 + \varphi'(H) \quad . \tag{47}$$

The equation for the limit is $\varphi' = 0$ or:

$$2a_2 - a_2' - a_6 + 2k_2(O_2) - a_2' - k_6(O_2)(M) = 0 \quad . \tag{48}$$

Since $(O_2) = p\gamma N$ and $(M) = pN$, this equation can be written in the form:

$$p^2 - \frac{2k_2}{k_6 N} p + \frac{a_2'}{k_6 N^2 \gamma} = 0 \quad . \tag{49}$$

Equation (49) has two roots:

$$p_1 = \frac{k_2}{k_6 N}\left(1 - \sqrt{1 - \frac{a_2' k_6}{k_2^2 \gamma}}\right) \tag{50}$$

$$p_2 = \frac{k_2}{k_6 N}\left(1 + \sqrt{1 - \frac{a_2' k_6}{k_2^2 \gamma}}\right) . \tag{51}$$

There exist two roots provided that

$$\frac{a_2' k_6}{k_2^2 \gamma} < 1 \quad .$$

One root corresponds to the lower limit, the other to the upper limit. When

$$\frac{a_2' k_6}{k_2^2 \gamma} > 1 \quad ,$$

both roots are imaginary and there are no limits. When

$$\frac{a_2' k_6}{k_2^2 \gamma} = 1$$

both roots are the same: this corresponds to the tip of the explosion peninsula. Thus equation (49) describes the experimentally observed explosion peninsula represented on Figure 47.

The quantity:

$$\beta = \frac{a_2' k_6}{k_2^2 \gamma} \tag{51'}$$

does not depend on pressure but is a function of temperature since $k_2^2 \sim \exp(-2E_2/RT)$ and $a_2' \sim \exp(-E/RT)$. Thus β increases rapidly as temperature goes down and at some temperature $T = T_m$, $\beta \approx 1$ (tip of the peninsula). Because of the exponential dependence of β on temperature, when T exceeds T_m by 30 to 50°, $\beta \ll 1$ and expressions (50) and (51) can be expanded in series. To a good approximation, only the first term need be retained and we get:

$$p_1 = \frac{a_2'}{2k_2\gamma N}; \quad p_2 = \frac{2k_2}{k_6 N}\left(1 - \frac{\beta}{4}\right) = \frac{2k_2}{k_6 N} - \frac{a_2'}{2k_2\gamma N}$$

or

$$p_1 + p_2 = \frac{2k_2}{k_6 N} \tag{52}$$

This approximation thus gives the same expression as before for p_1 [equation (14')] although homogeneous chain termination was not previously taken into account. This is not surprising since at such low pressures, the frequency of triple collisions is negligibly small as compared to wall termination.*

When $p_2 \gg p_1$, we have:

$$p_2 = \frac{2k_2}{k_6 N} \tag{52'}$$

so that the upper limit is determined, to a very good approximation, only by homogeneous chain termination. Wall termination of chains is then negligible. Also, p_2 does not depend on vessel diameter at temperatures sufficiently higher than T_m so that $\beta \ll 1$.

* The effect of homogeneous termination on the lower limit can be estimated by keeping the second term in the expansion. Then the expression for p_1 becomes:

$$p_1 = \frac{a_2'}{2k_2\gamma N}\left(1 + \frac{\beta}{4}\right).$$

When $\beta \ll 1$, the additional term is very small.

Thus:

$$p_2 = \frac{2k_2}{k_6 N} = \frac{2A_2}{k_6 N} \exp(-E_2/RT) \qquad (53)$$

with $E_2 = 15100$ cal. The values of p_2 calculated by means of this formula coincide with the experimental ones if $k_6 \sim 10^{-32}$, a figure close to that corresponding to the constant for triple collisions. As observed, p_2 increases rapidly with temperature.

Equation (53) is correct only at temperatures such that $\beta \ll 1$. As the temperature becomes lower and approaches T_m, the upper limit decreases more rapidly with decreasing temperature than according to the law $\exp(-E_2/RT)$. For this very reason, the early determinations of E_2 from the second limit were much too high, around 22 to 26 kcal. Later when investigators started to work with pretreated vessels at lower pressures, the tip of the peninsula was displaced to lower temperatures and more precise observations could be made. There is another cause of error in determinations of E_2 and p_2. The early technique of introducing the mixture in an evacuated vessel at values of the variables within the explosion region, led to partial reaction before explosion. But water vapor strongly catalyzes the formation of HO_2 (25-28) and this leads to low values for the upper limit (see below).

As techniques improved, the value of E_2 became lower. Voevodskiǐ and Nalbandyan [29] estimate $E_2 = 18000$ cal. Lewis and von Elbe [25] find $E_2 = 17000$ cal. The latter figure is only two kcal above the value measured by us [4] in a kinetic study of the activation energy of the reaction $H + O_2 \longrightarrow OH + H$, namely $E_2 = 15100$. Both values coincide within experimental error.

As theory predicts, there is agreement between the experimental values of E_2 obtained from the temperature dependence of the second limit and from reaction kinetics at the lower limit. In the purely diffusional range, the equation for the explosion limits becomes:

$$p^3 - \frac{2k_2}{k_6 N} p^2 + \frac{23D_o}{k_6 d^2 \gamma N^2} = 0 \quad . \qquad (54)$$

This equation also has two positive roots for $T > T_m$ and one for $t < T_m$. At sufficiently high temperatures, the term

$$\frac{23D_o}{k_6 d^2 \gamma}$$

may be neglected. The same value for the limit is obtained as before [equation (52')]:

$$p_2 \approx \frac{2k_2}{k_6 N} .$$

As pointed out above, the position of the upper limit is determined by the reaction $H + O_2 + M \longrightarrow HO_2 + M$ where M is any third body: a molecule of hydrogen or oxygen or an inert gas added to the system. The role of a third body is only to stabilize the HO_2 radical after its formation by removing its excess energy. The upper limit is therefore determined by the total pressure $P = [H_2] + [O_2] + [A]$ and not by the partial pressures $[H_2] + [O_2]$. Consequently a pure $H_2 + O_2$ mixture and one diluted with inert gas must have an identical upper limit with respect to total pressure P. This is also the case, but only approximately. In fact, inert gas dilution sometimes increases or decreases somewhat the upper limit. This is because different molecules have different abilities of removing the excess energy of the radical HO_2 after its formation. This phenomenon is well known from experiments in atom recombination (H, I, Br etc.). Recombination of atoms also requires a third body and is a termolecular process. As it turns out, inert gases modify strongly the rate constant for recombination. Thus, argon is less effective in removing energy than the molecules of the combustible mixture (hydrogen and oxygen). Therefore, P_2 increases when the mixture is diluted with argon. The efficiency of CO_2 molecules is higher than that of H_2 and O_2 so that CO_2 dilution lowers the upper limit. The efficiency of a third body depends on two factors: 1) the collision frequency which increases with molecular velocity (i.e., as molecular weight goes down) and 2) the relative ease of energy transfer. Hydrogen is more effective than oxygen for the first reason and CO_2 is more effective for the second reason.

The efficiency of oxygen and of a hydrogen-oxygen mixture is less than that of hydrogen so that the experimentally observed relation is not

$$p_{H_2} + p_{O_2} = \frac{2k_2}{k_6 N}$$

but:

$$3p_{H_2} + p_{O_2} = 3p(1 - \gamma) + p\gamma = \frac{2k_2}{k_6 N}$$

or

$$p_2 = \frac{2k_2}{k_6(3-2\gamma)} \quad .$$

Consequently, the ratio of upper limits for stoichiometric and equimolar mixtures is:

$$\frac{p_{2_1} \text{ stoich.}}{p_{2_1} \text{ equim.}} = \frac{3 - 1}{3 - \frac{2}{3}} = 0.87 \quad .$$

Water molecules are especially effective as third bodies. Even small quantities of water vapor, a few percent, depress the upper limit appreciably. It is not excluded that this is a case of specific chemical interaction with formation of a complex $H \cdot H_2O$ that gives $HO_2 + H_2O$ in subsequent collisions with O_2. The constant k_6 is strongly increased by this separation into two consecutive collisions, first with H_2O, then with O_2.

Essentially different is the effect on the upper limit of impurities susceptible of reacting with H atoms. As shown by Nalbandyan, a few thousandths of a percent of I_2 are sufficient to decrease p_2 almost by a factor of two. In the reaction $H + I_2 \longrightarrow HI + I$, the active H radical disappears and is replaced by the quite inactive I radical that ultimately recombines to regenerate I_2.

Such impurities thus provide additional homogeneous chain termination steps in binary collisions. The term $- b(I_2)(H)$ has to be added to equation (47). It cuts down branching by decreasing the term $2a_2(H)$ since the latter now becomes $[2a_2 - b(I_2)](H)$. Since practically no activation energy is required for the step $H + I_2 \longrightarrow HI + I$, negligible amounts of iodine are sufficient to depress strongly the upper limit.

Also specific is the effect of the oxides of nitrogen, discovered by Hinshelwood [30] and studied in detail by Norrish [31]. In small concentration, the oxides of nitrogen strongly enhance the upper limit and the concentration of NO_2 is increased at the same time. But with higher concentrations of NO_2, the limit goes down approaching the value it had in the pure mixture (32). Similar antagonistic effects of small and large quantities of NO_2 are also observed in many decomposition reactions of organic compounds, e.g., aldehydes. The mechanism of this phenomenon has not yet been fully elucidated.

As a result of thermal decomposition, NO_2 is in equilibrium with NO. Then NO reacts with HO_2 to give $NO_2 + OH$, an inactive radical HO_2 is replaced by an active one, OH. Of course, this enhances the upper limit. The mechanism of inhibition by larger quantities of NO_2 is more difficult to explain. Maybe NO_2 combines with an H atom to give a molecule of HNO_2 which either decomposes or, in a further collision with NO_2 gives a molecule of nitric acid HNO_3 and NO.

Consider now equation (51) relating p_2 and γ, the quantity characterizing the fraction of oxygen in the system. When γ decreases, β increases and for very small values of γ, $\sqrt{1-\beta}$ becomes imaginary, i.e., the limits disappear. This means that there must exist concentration limits for explosion of $H_2 + O_2$ mixtures. To a certain extent, this has been confirmed experimentally by Chirkov [28, Figure 48].

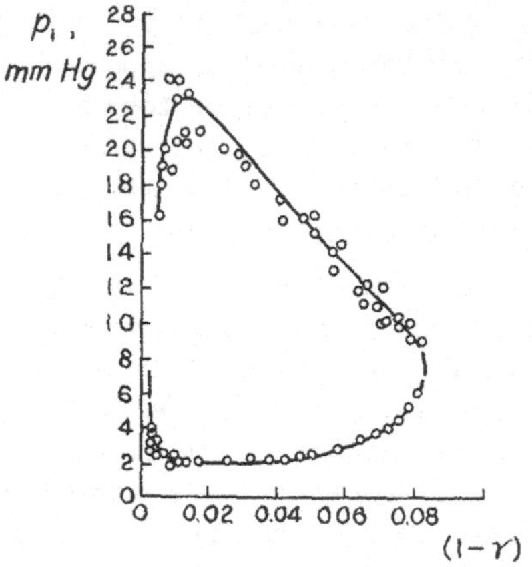

FIGURE 48. Lower and upper limits measured in a durabax vessel, as a function of the hydrogen content of the mixture, at 467°C. Experimental points from (28).

The upper and lower limits are represented as a function of the hydrogen content, i.e., $(1 - \gamma)$.

Over its major part, the curve of upper limits decreases as the hydrogen content increases, according to

$$p_2 = \frac{2k_2}{(k_6(3-2\gamma)} .$$

At the same time, near $1 - \gamma = 0.8$ (i.e., $\gamma = 0.2$), when γ is further reduced, the lower limit starts to go up quite rapidly. At still smaller values of γ, both limits must fuse together. Such a phenomenon, as seen on Figure 48, is observed also at small concentrations of hydrogen but here, the concentration limit seems to correspond to very small quantities of H_2. This limit is due to the fact that, at small H_2 concentrations, the rate constant a_1 of the reaction $OH + H_2 \longrightarrow H_2O + H$ becomes equal to or even smaller than the rate constant of the step $H + O_2 \longrightarrow OH + O$. It becomes then necessary to take into account, not only the wall termination of H atoms but also that of OH radicals [29].

A study of reaction kinetics within the explosion region is extremely difficult since the rate is so fast. Nevertheless, Nalbandyan [26, 33] succeeded in measuring induction periods at temperatures in the vicinity of the tip. Theoretically $\varphi = 2a_2 - a_2' - a_6$. At a pressure p between p_1 and p_2 where p_1 and p_2 are the roots of the equation $2a_2 - a_2' - a_6 = 0$, φ may be written in the form $\varphi = A(p - p_1)(p_2 - p)$. The induction period is related to φ by the relation $\varphi\tau$ = const., or:

$$\tau = \frac{\text{const.}}{\varphi} = \frac{\text{const.}}{(p-p_1)(p_2-p)} \quad . \tag{55}$$

Figure 49 shows one of the curves at $T = 435°C$ ($p_1 = 2.5$ and $p_2 = 17.8$): the points are experimental, the curves theoretical [equation (55)], with an arbitrarily chosen constant. It is seen that theory is well obeyed. Near to the upper or lower limits, the induction periods increase strongly and the rate decreases markedly.

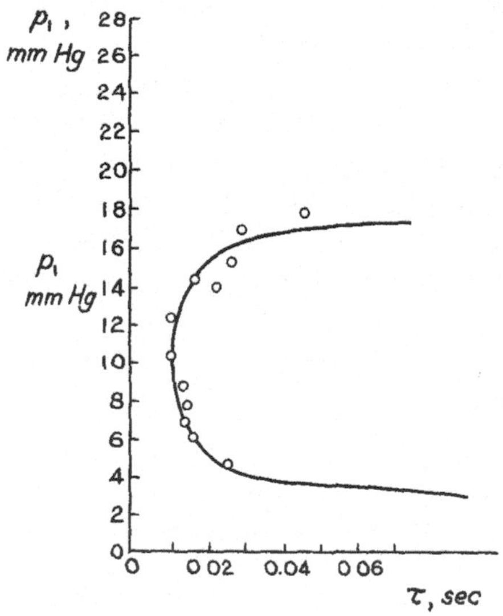

FIGURE 49. Relation between induction period and pressure of $2H_2 + O_2$ mixtures at 435°C. Curves from equation (55). Experimental points from (26) and (33).

As can be seen from the curve of Figure 47, as the temperature is raised to 600°, the upper limit starts to increase abnormally fast with temperature and it disappears altogether at still higher temperatures. The third limit p_3 then appears and it increases as temperature goes down (Figure 47).

An abnormal increase of the upper limit and its disappearance can be understood theoretically. Indeed, at these temperatures and pressures, the rate of the process $HO_2 + H_2 \longrightarrow H_2O_2 + H$ becomes appreciable and the formation of HO_2 does not contribute to chain termination. The very reason for the existence of an upper limit disappears progressively.

As to the third limit, it has a thermal origin but in certain cases (KCl coated walls) it may have a chain character as shown in Chapter VIII.

To sum up, the mechanism of the reaction between hydrogen and oxygen is well understood. The numerous curious phenomena due to the mechanism of this reaction are described quantitatively by theory in complete agreement with experimental facts. This demonstrates the power of the theory of branched chain reactions.

REFERENCES

[1] L. I. Avramenko, Acta Physicochim. URSS, 17, 197 (1942).
[2] N. N. Semenov, Acta Physicochim. URSS, 20, 291 (1945).
[3] A. A. Koval'skiĭ, Phys. Zt. Sow. 4, 723 (1933).

[4] L. V. Karmilova, A. B. Nalbandyan and N. N. Semenov, Zhur. Fiz. Khim., 1958, in press.

[5] P. Harteck and U. Kopsch, Zt. phys. Chem., (B) 12, 327 (1931).

[6] A. E. Biron and A. B. Nalbandyan, Zhur. Fiz. Khim., 9, 132 (1937).

[7] O. A. Ivanov and A. B. Nalbandyan, in preparation.

[8] N. N. Semenov, Acta Physicochim., URSS, 6, 25 (1937).

[9] W. V. Smith, J. Chem. Phys., 11, 110 (1943).

[10] N. N. Semenov, Dok. Akad. Nauk SSSR, 14, 265 (1944).

[11] E. I. Kondrat'eva and V. N. Kondrat'ev, Zhur. Fiz. Khim., 20, 1239, (1946); Acta Physicochim. URSS, 21, 1, 629 (1946).

[12] V. N. Kondrat'ev, 'Spectroscopic Studies of Gaseous Reactions', Moscow, 1944.

[13] V. R. Bursian and V. S. Sorokin, Zt. phys. Chem., (B) 12, 247 (1931).

[14] N. N. Semenov, 'Chain Reactions', Oxford, 1935.

[15] A. B. Nalbandyan, Dok. Akad. Nauk SSSR, 44, 356 (1944).

[16] N. N. Semenov, Acta Physicochim. URSS, 18, 93 (1943).

[17] A. B. Nalbandyan and S. M. Shubina, Zhur. Fiz. Khim., 20, 1249 (1946).

[18] R. R. Baldwin, Trans. Far. Soc., 52, 1344 (1956).

[19] C. N. Hinshelwood and T. W. Williamson, 'The reaction between hydrogen and oxygen', Oxford, 1934.

[20] A. V. Zagulin, A. A. Koval'skiĭ, D. I. Kopp and N. N. Semenov, Zhur. Fiz. Khim., 1, 263 (1930).

[21] B. Lewis and G. von Elbe, 'Combustion, Flames and Explosions of Gases', Academic Press, N. Y., 1951, p. 29.

[22] C. N. Hinshelwood and H. W. Thompson, Proc. Roy. Soc., A 118, 170 (1928); 122, 610 (1929).

[23] V. V. Voevodskiĭ, Dok. Akad. Nauk SSSR, 44, 308 (1944).

[24] B. Lewis and G. von Elbe, 'Combustion, Flames and Explosions of Gases', Cambridge, 1938.

[25] B. Lewis and G. von Elbe, J. Chem. Phys., 10, 366 (1942).

[26] A. B. Nalbandyan, Zhur. Fiz. Khim., 19, 210 (1945).

[27] V. V. Voevodskiĭ and V. L. Tal'roze, Zhur. Fiz. Khim., 22, 1192 (1948).

[28] N. M. Chirkov, Acta Physicochim. URSS, 6, 915 (1937).

[29] A. B. Nalbandyan and V. V. Voevodskiĭ, 'Mechanism of oxidation and combustion of hydrogen', 1948.

[30] G. H. Gibson and C. N. Hinshelwood, Trans. Far. Soc., 24, 559 (1928).

[31] F. S. Dainton and R. G. W. Norrish, Proc. Roy. Soc., 177, 391 (1941).

[32] A. B. Nalbandyan, Zhur. Fiz. Khim., 20, 1238 (1946).

[33] A. B. Nalbandyan, Zhur. Fiz. Khim., 19, 201, 218 (1945).

CHAPTER XI: CHAIN INTERACTION

Until now, attention has been focused on cases where the rates of chain termination and branching were proportional to the concentration of atoms and free radicals formed during the chain reaction. This amounts to assuming that each chain propagates on its own, without interference by the other chains. For this reason, in particular, the quantity n_o, the rate of generation of the primary radicals does not appear in the expression for the limits. There may exist cases, however, where chain termination and branching are due to quadratic branching, i.e., to reactions between the free radicals themselves or to a reaction between a radical and molecules of some intermediate product.

§1. Negative Chain Interaction

The first example of a negative chain interaction is the destruction of radicals by their mutual recombination. This process always takes place but its rate, at low pressures, is usually small as compared to that of wall destruction of free radicals. On the other hand, at high pressures, the chain branching is so fast that the gas temperature increases and the chain explosion becomes a thermal explosion before free radicals have the time to reach high concentrations. Therefore, a quadratic effect can be observed when branching is not too rapid. This question will now be examined theoretically.

The usual equation of chain kinetics for one kind of center reads (see Chapter IX):

$$\frac{dn}{dt} = n_o + (f - g)n \qquad (1)$$

where fn is the branching rate and gn is the termination rate (excluding wall termination). The condition $f - g = 0$ determines the limits of chain ignition.

Destruction of radicals by their mutual recombination leads to an additional term $g_o n^2$ on the right hand side of equation (1). Then:

$$\frac{dn}{dt} = n_o + (f - g)n - g_o n^2 \quad . \qquad (2)$$

Hence:

$$n = \frac{2n_0\left(e^{\sqrt{qt}}-1\right)}{\sqrt{q} + (f-g) + \left\{\sqrt{q} - (f-g)\right\}e^{\sqrt{qt}}} \tag{3}$$

where

$$\sqrt{q} = \sqrt{(f-g)^2 + 4g_0 n_0} \quad . \tag{4}$$

A graph of n versus t is shown on Figure 50.

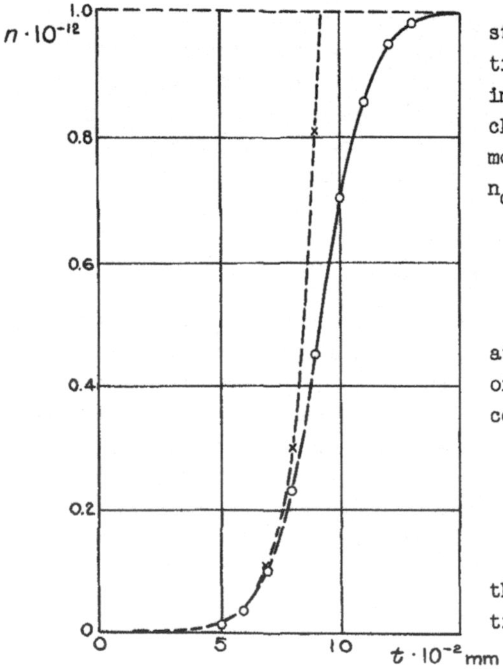

FIGURE 50. Relation between n and t. Full curve from equation (3); dashed curve from equation (5).

In order to define the circumstances under which radical destruction by their mutual recombination is important in the kinetics of branched chain reactions, let us consider the most common case of small values of n_0. Then, equation (3) becomes simply:

$$n = \frac{n_0}{(f-g)}\left(e^{(f-g)t} - 1\right) \tag{5}$$

at small values of t. At large values of t, with $(f - g) > 0$, it becomes:

$$n = \frac{(f-g)}{g_0} \quad . \tag{6}$$

For simplicity, let us now assume that equation (5) is valid up to a time $t = t_1$ such that:

$$\frac{n_0}{(f-g)} e^{(f-g)t_1} = \frac{(f-g)}{g_0} \quad . \tag{7}$$

After t_1, n is constant, equal to

$$\frac{f-g}{g_0} \quad .$$

In words, we replace the continuous curve of Figure 50 by a broken curve (dotted on the Figure). The time t_1 required to build up a constant value n and thus a constant rate $w = an$ can be obtained from equation (7):

$$t_1 = \frac{2.3}{(f-g)} \log \frac{(f-g)^2}{g_0 n_0} \quad . \tag{8}$$

Let $(f - g) = 10^2$; $n_0 = 10^{10}$; $g_0 = 10^{-10}$ on the assumption that radical recombination occurs on every binary collision. Then

$$t_1 = 10^{-2} \cdot 2.3 \log \frac{10^4}{10^{-10} \cdot 10^{10}} \sim 0.1 \text{ sec.}$$

After time t_1, the mixture reacts at a constant rate:

$$w = an = \frac{a(f-g)}{g_0} = \frac{a \cdot 10^2}{10^{-10}} \sim a \cdot 10^{12} \quad .$$

Assume also $a = 10^2$.* Then $w = 10^{14}$. If the reaction is carried out at a pressure of 10 mm, the quantity of reacting molecules is 10^{17}. Therefore the time θ for complete reaction is of the order of 1000 sec. It may well be then that the induction period, lasting in all 0.1 sec will remain unnoticed and that the steady state reaction will be measured from the start. Since f and a are proportional to the concentration of reactant molecules, the reaction rate will be second order. The branched chain reaction may be treated like a simple bimolecular process. One may indeed wonder whether there are not some reactions considered to be bimolecular, that are really branched chain processes with quadratic destruction of free radicals by recombination. Unfortunately, this question first raised by us in 1934 [1] is still unanswered.

It must be noted that n_0 is not in equation (6), so that the most difficult step of usual chain reactions — the primary thermal generation of free radicals is missing. Here, generation is due to branching which is due to the liberation of the chemical energy of the system and requires a small activation energy. If f and a belong to exothermic or thermoneutral processes, the corresponding activation energies will not exceed 10 kcal and the overall activation energy will be less than 20 kcal. In many cases, the branching may be an endothermic step. For instance, with $f = 10^2$ at 600°K, $f = 10^{-10} \exp(-\epsilon/RT) \cdot (N)$ where N is the number of molecules which give way to branching by reaction with a radical. At $p = 10$ mm Hg, $N = 10^{17}$. Hence $\exp(-\epsilon/RT) \sim 10^{-5}$ and

---

* $a = kx$ where x is the component of the mixture with which the active particle reacts. If $x = 10^{17}$ molecules/cm^3 and $k = 10^{-15}$, then $a = 10^2$.

ϵ = 14000 cal. This activation energy corresponds to an endothermicity of 5 kcal (see Polanyi's Relation in Chapter I). Then the activation energy of the overall process will reach say, 25 to 35 kcal.

All that has just been said does not apply to branching ignition processes, e.g., the branched oxidation of hydrogen. There, the H atoms recombine only in triple encounters and this decreases g_0 at a pressure of 10 mm Hg by a factor of 10^5. On the other hand, both f and a are now larger than before ($f = 10^2$) by a factor of 10 to 100. Then the steady state rate, if it could ever be established, would reach the remarkable figure of 10^{20} molecules per second: the reaction would be over in 10^{-3} sec (t_1 in this case is 10^{-2} to 10^{-3} sec). Therefore before the system reaches a constant rate, the reactants have all reacted. Thus in such a case, the reaction occurs in a time during which n and the rate grow according to the usual law [5] and the recombination of atoms may be neglected.

§2. Positive Chain Interaction

The other case, called positive interaction of chains, occurs when collision between two radicals gives branching and thus increases the rate of reaction. This may occur, for instance, when collision between an active radical and an inactive one yields two active radicals. For instance, in the oxidation of hydrogen, chain termination is due to the step $H + O_2 + M \longrightarrow HO_2 + M$. The HO_2 radical has very little activity since the reaction $HO_2 + H_2 \longrightarrow H_2O_2 + H$ may be neglected. In fact, the radical is destroyed on the walls before it has the time to take part in this reaction.

The situation is different when the HO_2 radical collides with H atoms. When these two radicals recombine, a large amount of energy, about 90 kcal is released. In the H_2O_2 molecule just formed, the O - O bond energy is $\sim$ 50 kcal. Therefore, the peroxide decomposes into two OH radicals and the energy release is still 40 kcal: $H + HO_2 \longrightarrow 2OH + 40$ kcal. In this fashion, one active radical, H, is transformed into two active radicals OH that react quickly with H_2: $OH + H_2 \longrightarrow H_2O + H$ and regenerate two new H atoms. We deal here with a typical additional branching step. Of course, the process $OH + HO_2 \longrightarrow H_2O + O_2$ is also possible and it destroys the OH radical. But it must be kept in mind that during hydrogen oxidation, the OH concentration is about 100 times smaller than the H concentration and the latter process will be 100 times less frequent than the former. Therefore, the reaction $OH + HO_2 \longrightarrow H_2O + O_2$ may be omitted.

The upper limit of ignition in $H_2 + O_2$ is determined by the relation $f - g = 0$ where f is the branching rate $2k_2(O_2)$ and g is the termination rate, equal in this case to $10^{-32}(O_2)(M)$. If the concentration of initial centers is raised artificially, the number of branchings increases by the quadratic mechanism just explained and the upper limit should be raised. This is confirmed by experiment.

Let us analyze mathematically the problem of positive chain interaction. For one type of centers, positive interaction gives the equation:

$$\frac{dn}{dt} = n_0 - (g - f)n + bn^2 \quad . \tag{9}$$

Let us try to find a steady-state solution of $n_0 - (g - f)n + bn^2 = 0$ when $g > f$:

$$n = \frac{(g-f) - \sqrt{(g-f)^2 - 4n_0 b}}{2b} \quad .^* \tag{10}$$

If n_0 is very small:

$$n = \frac{n_0}{g - f} \tag{11}$$

and this corresponds to the steady-state solution of the usual equation in the absence of the quadratic term. At the limit $f - g = 0$. It will be assumed that, without additional artificial initiation of chains, n_0 is very small. What happens now if n_0 is increased artificially (say, photochemically)? Then even if $g - f > 0$, a steady-state solution is impossible, if $4n_0 b > (g - f)^2$, since then n has an imaginary value.

This can be explained by interpreting the complete equation (9). As it turns out, as long as $4n_0 b < (g - f)^2$, the number n in time approaches its steady-state value (10). But if $4n_0 b > (g - f)^2$, n will grow continuously as time elapses, without ever reaching a steady-state value. Thus, the condition $4n_0 b = (g - f)^2$ is the condition for chain explosion with quadratic branching. Whereas the condition for chain explosion was $f = g$ at small values of n_0, it now becomes

$$f + \sqrt{4n_0 b} = g \quad .$$

Consequently, explosion occurs in a region where it did not take place before, with given values of f and g. Also then, the boundaries of chain explosion are spaced further apart by this increase of n_0.

---

* The solution with the + sign must be rejected since when n is increased infinitesimally, dn/dt is positive while it is negative if n is decreased. This solution cannot be reached if $n = 0$ at $t = 0$.

Widening of the Explosion Region in the System $H_2 + O_2$ by Short Wave-Length Illumination or Addition of O Atoms

This general discussion can now be applied to an analysis of the displacement of the upper limit in the case of the chain explosion of hydrogen and oxygen, as a result of the deliberate creation of additional O atoms. Two more reactions will be added to the scheme of hydrogen oxidation:

7) $H + HO_2 \longrightarrow 2OH$
8) HO_2 + wall $\longrightarrow$ termination.

The rate of reaction 7) is $k_7(H)(HO_2)$. The constant k_7 is unknown. But evidently it cannot differ very widely from the frequency of binary collisions sustained by a given molecule, namely 10^{-10}. Indeed, as noted already, reaction 7) proceeds via the recombination of H with HO_2 with subsequent rapid break-up of H_2O_2 into two OH radicals. Recombination usually takes place practically on every collision. Nevertheless, when H approaches HO_2 on the H side, recombination cannot take place so that there must exist a steric factor of the order of 0.5. But even upon attack on the O side, decomposition along the O - O bond will proceed with a statistical probability ~ 0.5 since the vibration period of the O - O bond has about the same duration as a collision. Therefore, the full steric factor will be about 0.25. The value 0.1 will be assumed. Then $k_7 \sim 10^{-11}$.

It will be further assumed that chain initiation is due to dissociation of oxygen by absorption of short wave length illumination. The number of quanta absorbed per unit time per unit volume is I_0: these give birth to $2\, I_0$ oxygen atoms per unit time and unit volume. The number n_0, namely the spontaneous generation of H atoms will be neglected as compared to $2\, I_0$.

What is of interest here is the displacement of the upper limit so that, for simplicity, the chain termination by wall destruction of H and OH will be neglected (this leads to an error of less than 10%). The kinetic equations taking all these remarks into consideration are:

$$\frac{d(OH)}{dt} = - a_1(OH) + a_2(H) + a_3(O) + 2k_7(H)(HO_2)$$

$$\frac{d(H)}{dt} = a_1(OH) + a_3(O) - (a_2 + a_6)(H) - k_7(H)(HO_2)$$

$$\frac{d(O)}{dt} = 2I_0 + a_2(H) - a_3(O)$$

$$\frac{d(HO_2)}{dt} = a_6(H) - a_8(HO_2) - k_7(H)(HO_2) .$$

Here

$$a_1 = k_1(H_2), \quad a_2 = k_2(O_2), \quad a_3 = k_3(H_2) ,$$

$$a_6 \sim 10^{-32}(O_2)(M), \quad k_7 \sim 10^{-11}, \quad a_8 = \frac{23D_o}{pd^2}$$

where p is the pressure in mm Hg, d the vessel diameter and D_o is the diffusivity at $p = 1$ mm Hg and the temperature of the experiment.

Since the transition to explosion corresponds to the condition expressing the impossibility of reaching a steady-state solution, we put as before:

$$\frac{d(OH)}{dt} = \frac{d(H)}{dt} = \frac{d(O)}{dt} = \frac{d(HO_2)}{dt} = 0 ,$$

so that we get a system of four algebraic equations. The system will be solved explicitly for the H atom concentration, and remembering that $a_6 > 2a_2$ when $p > p_2$,* we get:

$$2a_2k_7(H)^2 - \left\{(a_6 - 2a_2)a_8 - 4I_ok_7\right\}(H) + 4I_oa_8 = 0 . \tag{12}$$

Hence:

$$(H) = \frac{1}{4a_2k_7}\left[\left\{(a_6 - 2a_2)a_8 - 4I_ok_7\right\} - \sqrt{\left\{(a_6 - 2a_2)a_8 - 4I_ok_7\right\}^2 - 32I_oa_8a_2k_7}\right] . \tag{13}$$

The condition for explosion is:

$$(a_6 - 2a_2)a_8 - 4I_ok_7 = \sqrt{32I_oa_8a_2k_7} . \tag{14}$$

Equation (14) can be rewritten in the form:

---

* p_2 is the upper limit in the absence of illumination, i.e., when $I_o = 0$.

$$\frac{a_6}{2a_2} = 1 + 2\sqrt{\frac{2I_0k_7}{a_2a_8}} + \frac{2I_0k_7}{a_2a_8} \quad . \qquad (15)$$

On the other hand:

$$\frac{a_6}{2a_2} = \frac{k_6N^2\gamma p^2}{2k_2\gamma Np} = \frac{k_6}{2k_2}Np$$

and

$$\frac{2k_2}{k_6N} = p_2$$

where p_2 is the upper limit when $I_0 = 0$ (see Chapter X). Consequently:

$$\frac{a_6}{2a_2} = \frac{p}{p_2} \quad ,$$

$$\frac{k_7}{a_2a_8} = \frac{k_7}{k_2\gamma Np\,\frac{23D_0}{pd^2}} = \frac{k_7d^2}{23D_0k_2N\gamma} = C \quad .$$

Since C does not depend on pressure, the expression giving the upper limit under illumination is:

$$\frac{p_2^*}{p_2} = 1 + 2\sqrt{2I_0C} + 2I_0C \quad . \qquad (16)$$

Nalbandyan [2, 3] has studied the displacement of the upper ignition limit due to oxygen atoms produced photochemically. He found that at $T = 700°K$ and $I_0 = 5 \cdot 10^{13}$ quanta/sec.cm^3, the ratio

$$\frac{p_2^*}{p_2} = 1.6 \quad .$$

Equation (16) permits us to calculate the same ratio. The coefficient C must be calculated first. In Nalbandyan's work, $d = 2.7$ cm and $k_2 = 10^{-10}\exp(-15100/RT)$ according to recent data. At $T = 700°K$, $k_2 = 2 \cdot 10^{-15}$. For k_7, we take 10^{-11}. Further, D_0, the diffusivity of HO_2 radicals must be calculated. Assuming that the diffusivities of HO_2 in H_2 and O_2 are equal to those of O_2 in H_2 and O_2 respectively, we calculate, as we did before in Chapter X, $D_0 = 1400$ at $700°K$ in a stoichiometric mixture $H_2 + O_2$. Introducing all these figures in the expression for C, we get $C \sim 2.5 \cdot 10^{-16}$. Then equation (16) gives:

$$\frac{p_2^*}{p_2} \approx 1.35 \quad .$$

The agreement between calculated and experimental (p_2^*/p_2) values of the ratio is very satisfactory if it is kept in mind that the calculation of k_7 and D_0 is only rough.*

It must be noted that if HO_2 radicals were not destroyed at the wall, the upper limit would disappear altogether. Indeed, all HO_2 radicals would then take part in reaction 7) and the very reason for the existence of an upper limit would vanish. This can be shown easily by solving the original system of equations, putting $a_8 = 0$ and taking into account the wall termination of H atoms. As I_0 is made larger and the upper limit is displaced upward, the chain explosion takes place at higher pressures. But then branching in the mixture occurs faster. On the other hand, at higher pressures, fewer HO_2 radicals reach the wall so that a_8 decreases. Therefore, at these elevated pressures, the HO_2 radical may not reach the wall. Of course, wall termination of HO_2 may then be neglected so that at sufficiently high values of I_0, the upper limit may vanish. Experiment confirms this theoretical prediction. Indeed, Dubovitskiĭ [4] and later Nalbandyan [5] have shown that large concentrations of atomic oxygen produced at low pressure in a discharge can ignite a mixture of H_2 and O_2 with a characteristic sharp lower limit but no upper limit. Figure 51 shows the normal domain of explosion together with that obtained in the presence of a large amount of atomic oxygen.

As was already pointed out, small amounts of NO_2 raise the upper limit sharply [6, 7] or even suppress it completely. Apparently this is due to the interaction of NO (produced by the equilibrium $2NO_2 \longrightarrow 2NO + O_2$) and HO_2: $NO + HO_2 \longrightarrow NO_2 + OH$. The action is then similar to that of atomic oxygen. Figure 52, on the following page, presents the results of Nalbandyan [6].

---

* All known facts concerning recombination and disproportionation of radicals lead us to believe that no large error can be made in estimating the steric factor for which the value 0.1 has been assumed here. Nevertheless, the frequency of binary collisions involving an H atom is higher than the assumed figure, namely 10^{-10} (although it is not higher than 10^{-9}). The true value of k_7 must lie somewhere between 10^{-10} and 10^{-11}. If $k_7 = 10^{-10}$, $p_2^*/p_2 = 2.25$. But the experimental value of p_2^*/p_2 lies precisely in between the two extreme calculated values. The argument can be reversed: from the experimental value $p_2^*/p_2 = 1.6$, the value of k_7 can be calculated. The result is: $k_7 \sim 0.4 \cdot 10^{-10}$.

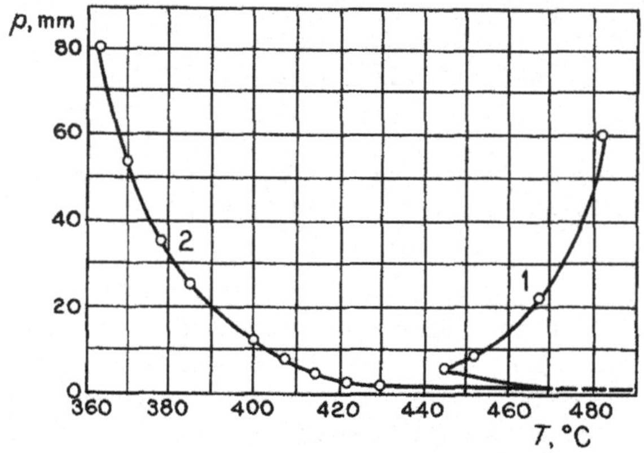

FIGURE 51. Widening of the ignition domain in a $2H_2 + O_2$ mixture (1), due to atomic oxygen. (2)

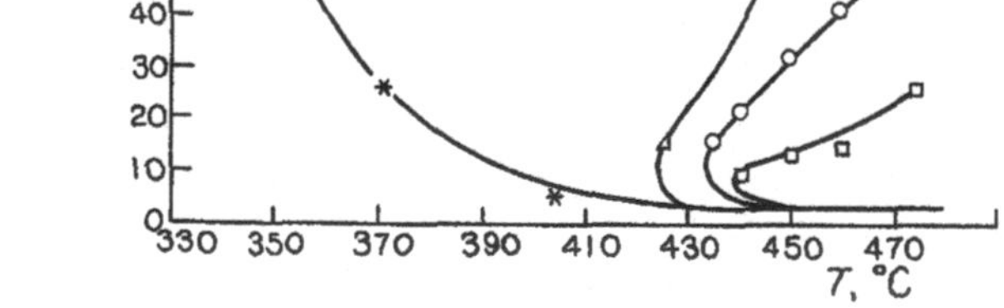

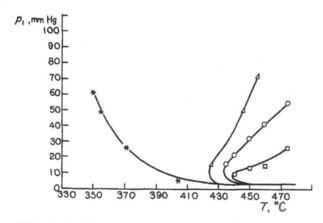

FIGURE 52. Effect of NO_2 on the explosion limits of $2H_2 + O_2$ mixtures
1. Explosion domain in $2H_2 + O_2$ mixtures
2,3,4. In mixtures containing 0.008, 0.026 and 3.4% NO_2. (7).

Similar effects are operative in the work of Lavrov [8] who submitted a mixture of $H_2 + O_2$ to the action of a low intensity spark. At very low intensity, the flame propagates only within certain pressure limits. As the spark intensity increases, the limits go further apart, the upper limit moves upward and disappears ultimately.

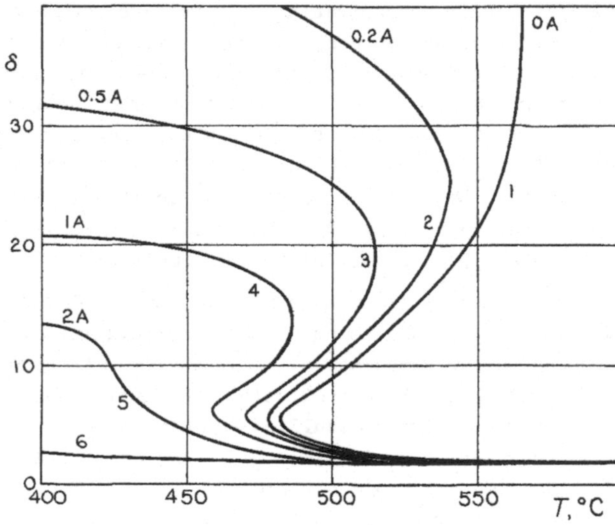

FIGURE 53. Explosion region in $2H_2 + O_2$:
1. Self-ignition
2,3,4,5. As affected by a discharge
(The discharge intensity is measured by the primary current of the transformer; the latter is given on the curves). (8)

Cold Flame Propagation in Lean Oxygen-Carbon Disulfide Mixtures

The propagation of a hot flame is due to heat transfer from hot to cold regions while the reaction takes place at high temperatures. After Lewis [9] formulated the principle of constancy of heat and chemical energy throughout the combustion zone, Zel'dovich [10] and Frank-Kamenetskiĭ [11] put forward their thermal theory of flame propagation, applicable to simple uni- and bi-molecular reactions. In the case of chain reactions, it is necessary to include further the diffusion of free radicals. This, however, affects the results relatively little since the main mechanism of propagation remains the heating of flame elements by heat transfer.

Chain explosion may take place isothermally, the heat evolved being dissipated completely. Without heating of neighboring flame elements, there is no room for flame propagation by chain diffusion if the chain reaction is of the type where termination and branching are linearly related to the concentration of active centers i.e., where each chain propagates independently. The situation is different in the case of positive chain interaction giving additional branching and contributing to the equation expressing the time rate of change of active centers, some new terms related quadratically to their concentration. As we have seen, the artificial increase of original active centers (say by illumination) widens the explosion limits and provokes explosion under conditions where it would not take place normally. In chain combustion, many active centers are produced. They diffuse to adjacent layers where they increase the concentration of centers much above the normal. There, ignition can occur at temperatures less than those required for self-ignition. The situation is quite similar to that created by photochemical initiation. It permits to a cold flame to propagate by a purely diffusional mechanism.

One end of a tube filled with a mixture may be placed in a heater and its temperature raised to the point of chain self-ignition. The rest of the tube is kept at a low temperature. Yet, a flame will propagate in the colder regions. To verify this prediction, we carried out an experiment [12] with a mixture of carbon disulfide and air. The concentration of carbon disulfide was very low (0.03%). The heating of the system due to complete adiabatic combustion did not exceed 15°. In fact, this temperature increase is not reached since the reaction is relatively slow in such lean mixtures and the heat loss in a narrow tube is rather important. One end of the tube was heated to the temperature of chain self-ignition. Then we observed a uniform flame propagation (with a weak luminosity) in the cold part of the tube kept at a temperature 90° lower than that of self-ignition. Figure 54 contains 1) the domain of self-ignition of the mixture investigated and 2) the domain of cold flame propagation.

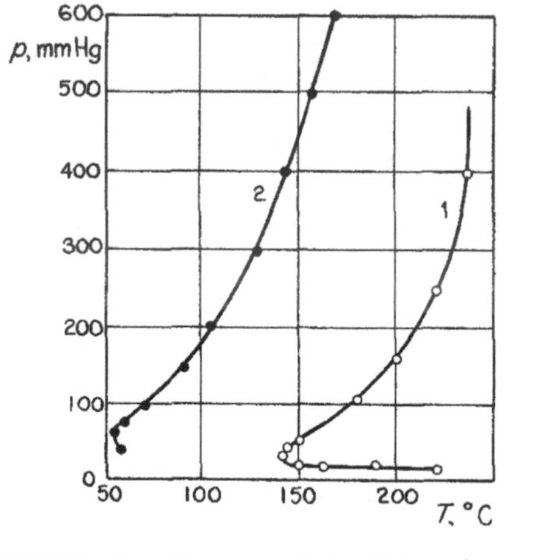

FIGURE 54. Domain of ignition (curve 1) and of flame propagation for an Air mixture with 0.03% CS_2. (12)

This experiment demonstrates that a flame may propagate by a nonthermal mechanism. Measuring the rate of propagation as a function of pressure at various temperatures, we obtained curves, two of which are shown on Figures 55 and 56, on the following page, (curves at the top). We thus obtained an upper limit of flame propagation. At the lower and upper limits, the rate of propagation reaches a minimum value below which propagation becomes impossible.

The rate of propagation is approximately represented by the expression:

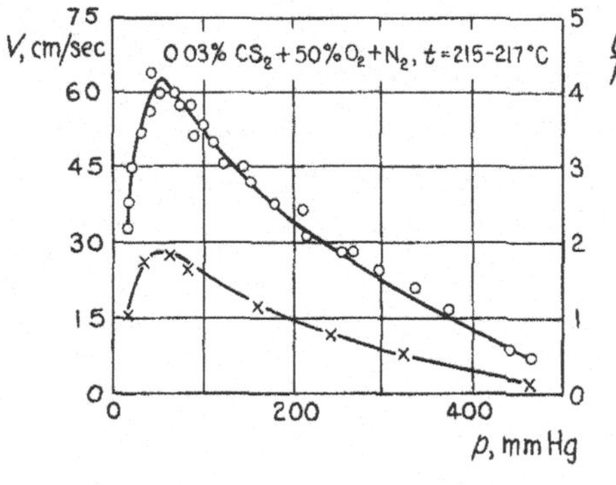

FIGURE 55. Velocity of flame propagation, as a function of mixture pressure p. Upper curve: experimental data. Lower curve: calculated from equation (20). Crosses represent upper curve points, divided by 2.25. (12)

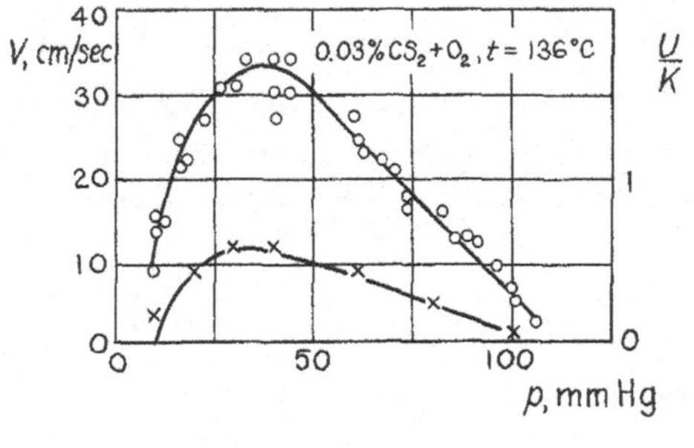

FIGURE 56. Velocity of flame propagation, as a function of mixture pressure p. Upper curve: experimental data. Lower curve: calculated from equation (20). Crosses represent upper curve points, divided by 2.6. (12)

$$V = \sqrt{\frac{bD}{2}}\left(1 - 2\frac{g}{b}\right) \tag{17}$$

where D is the diffusivity of active centers; b and g are coefficients in the expression for the reaction rate:

$$\frac{d\xi}{dt} = -g\xi + b\xi^2 .$$

Here, ξ is the concentration of active centers relative to the initial oxygen concentration.

The dependence of g and b on pressure provides a good theoretical explanation of the curves of V versus p. The cold flame propagation in carbon disulfide-air mixtures offers conclusive proof of the reality of positive chain interaction. The mechanism of oxidation of carbon disulfide is unknown so that the nature of the interaction cannot be explained. For the purpose of illustration, let us consider one of the possible mechanisms:

$$\left.\begin{array}{ll} 1. & S + O_2 \longrightarrow SO + O \\ 2. & CS_2 + O \longrightarrow COS + S \end{array}\right\} \text{chain}$$

3. $SO + SO \longrightarrow SO_2 + S$ quadratic branching

4. $SO + O_2 \longrightarrow SO_3$

5. S + wall wall termination

6. $S + O_2 + M \longrightarrow SO_2 + M$ volume termination .

The kinetic equations can be written down. The chain length is assumed to be much larger than two. It is postulated that S and O react much faster than SO so that

$$\frac{d(S)}{dt} = \frac{d(O)}{dt} = 0 \quad .$$

We get:

$$\frac{d(SO)}{dt} = n_0 - k_4(O_2)(SO) + \frac{k_1k_3(O_2)(SO)^2}{a_5 + x(O_2)(M)} \quad .$$

Let us denote by y the axial distance along the tube. The origin is at $y = 0$. The flame zone moves at a speed V and it is assumed that the process is proceeding at constant velocity. Then:

$$D\,\frac{d^2(SO)}{dy^2} - V\,\frac{d(SO)}{dy} - k_4(O_2)_0(1 - \eta)(SO) + \frac{k_1k_3(O_2)_0(1-\eta)(SO)^2}{a_5 + x(O_2)(M)} = 0$$

where η is the fraction of oxygen reacted.

Put: $(SO)/(O_2)_0 = \xi$. Then:

$$D\,\frac{d^2\xi}{dy^2} - V\,\frac{d\xi}{dy} - k_4(O_2)_0(1 - \eta)\xi + \frac{k_1k_3(O_2)_0^2(1-\eta)\xi^2}{a_5 + x(O_2)(M)} = 0 \quad .$$

It will be assumed that the reaction gives SO + COS in its initial stages, and the final products SO_2 + CO only later. Then, $\xi = \eta$ and:

$$D\,\frac{d^2\xi}{dy^2} - V\,\frac{d\xi}{dy} - g\xi(1 - \xi) + b\xi^2(1 - \xi) = 0 \qquad (18)$$

where $g = k_4(O_2)_0$,

$$b = \frac{k_1k_3(O_2)_0^2}{a_5 + x(O_2)(M)} \quad .$$

In very lean mixtures $(O_2) = (O_2)_o$. With the conditions $\xi = 1$ at $y = +\infty$ and $\xi = 0$ at $y = -\infty$, the equation can now be integrated. A steady-state solution for ξ can be found for a constant value of the parameter V:

$$V = \sqrt{\frac{bD}{2}} \left(1 - 2\frac{g}{b} \right) .$$

The limits of flame propagation p_1 and p_2 are determined by the condition $b = 2g$ or:

$$\frac{k_1 k_3 (O_2)_o^2}{a_5 + x(O_2)_o (M)} = 2k_4 (O_2)_o \quad . \tag{19}$$

Since $(O_2)_o$ and (M) are proportional to the pressure p, we get:

$$\frac{Ap^2}{B + Cp^2} = 2gp \quad \text{or} \quad 2Cp^2 g - Ap + 2gB = 0 \quad .$$

From a well known theorem:

$$\frac{A}{2Cg} = p_1 + p_2$$

and

$$\frac{B}{C} = p_1 p_2$$

where p_1 and p_2 are the roots of this quadratic equation, i.e., the lower and upper limits of flame propagation known experimentally. The expression for the rate of propagation becomes finally:

$$V = k_o \frac{1 - \frac{p_1}{p} - \frac{p_2}{p}}{\sqrt{\frac{p_1}{p} - \frac{p}{p_2}}} \tag{20}$$

where k_o is a constant at a given temperature. As it turns out, this constant retains the same value at various temperatures. In experiments with different amounts (M), (N_2) and (O_2),

$$k_o = k_o' \frac{p_{O_2}}{p}$$

The data are well represented by the formula as shown in Figures 55 and 56 where the lower continuous curves have been calculated by means of equation (20) with a constant value of k_0. The crosses represent points of the upper curves divided by 2.25 (Figure 55) and 2.6 (Figure 56). Discrepancy can be seen only in the sense that at the limits the rate is not equal to zero but to some finite value. Therefore, we correct p_2 by extrapolating the curves to the horizontal axis: the point of intersection gives the true value of p_2. The experimental values of p_1 are very small and uncertain so that an extrapolation is not warranted. Therefore, p_1 has been chosen for each curve so as to obtain the best fit with theoretical curves. Different values of p_1 have thus been used, changing with temperature.

The reaction rate changes with temperature as p_1 and p_2 also change. Knowing p_1 and p_2 as a function of temperature, we can calculate how V changes with tube temperature. The continuous curve in Figure 57 shows the result of such a calculation while the crosses correspond to experimental values of V. The reason why the rate at the limits of propagation retains a finite value is easily understood. It is due to the numerous simplifications of the treatment which ignored the further transformation of the intermediates SO and COS into final products SO_2 and CO.

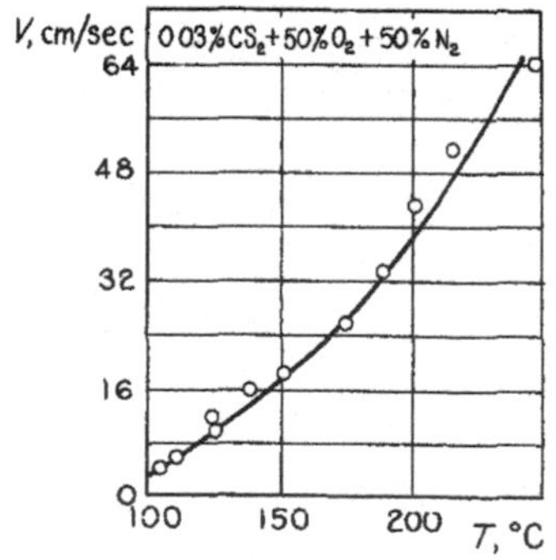

FIGURE 57. Velocity of flame propagation, as a function of temperature. Curves are calculated, points experimental. (12)

REFERENCES

[1] N. N. Semenov, 'Chain Reactions', Oxford, 1935.
[2] A. B. Nalbandyan, Thesis, Kazan', Inst. Chem. Phys., 1942.
[3] A. B. Nalbandyan, Zhur. Fiz. Khim., 20, 1259 (1946).
[4] F. I. Dubovitskiĭ, Acta Physicochim. URSS, 2, 761 (1935).
[5] A. B. Nalbandyan, Acta Physicochim. URSS, 1, 305 (1934).
[6] A. B. Nalbandyan, Zhur. Fiz. Khim., 20, 1283 (1946).
[7] F. S. Dainton and R. G. W. Norrish, Proc. Roy. Soc., A177, 391 (1941).
[8] G. Gorchakov and F. A. Lavrov, Acta Physicochim. URSS, 1, 139 (1934).
[9] B. Lewis and G. von Elbe, Jour. Chem. Phys., 2, 283 (1934).
[10] Ya. B. Zel'dovich and D. A. Frank-Kamenetskiĭ, Zhur. Fiz. Khim., 12, No. 1 (1938); Ya. B. Zel'dovich and N. N. Semenov, Zhur. Tekh. Fiz., 10, No. 12 (1940); Ya. B. Zel'dovich, 'Combustion and Detonation of Gases', Moscow, 1944.

[11] D. A. Frank Kamenetskiĭ, 'Diffusion and Heat Transfer in Chemical Kinetics', Princeton, 1954.

[12] V. G. Voronkov and N. N. Semenov, Zhur. Fiz. Khim., 13, 1695 (1939).

CHAPTER XII: CHAIN REACTIONS WITH DEGENERATE BRANCHING

§1. Introduction

The branched chain reactions considered so far at pressures above the limit exhibit such fast self-acceleration that they have the character of a chain explosion. The reason is that the chain branching as well as the propagation steps are reactions between a radical and a molecule (e.g., $H + O_2 \longrightarrow OH + O$). These processes are quite fast.

Already, between 1930 and 1936, the existence of another class of reactions was recognized. Like branched processes, they are homogeneous and self-accelerating at a pressure or a vessel diameter above a critical value. The acceleration obeys the same law: $w = ae^{\varphi t}$. As a rule, the self-acceleration is not due to catalysis by final products. The acceleration is thousands of times slower than in chain ignition. This actually is the main difference. If the rate increases e times in a fraction of a second in chain ignition, this necessitates minutes or even hours in the reactions of this other type.

At first glance, it seems that the same branching scheme can be applied to these slow self-accelerating reactions. It must be assumed then that the radical of the branching step reacts rather slowly (because, perhaps, of the endothermicity of the process). But every radical will die in the course of time, for example by wall collisions. Suppose that the probability of wall capture is very small, say $\epsilon = 10^{-6}$. Yet, even so, its rate of disappearance will be of the order of

$$\frac{\epsilon v}{4} \sim 10^{-2} \quad .$$

Thus a radical could live during 100 seconds and this is of course a gross exaggeration. Since self-acceleration will occur only when the branching rate is faster than the termination rate, even in this exaggerated case of radical longevity, self-acceleration will occur only if the time required for a radical to enter a branching step, is less than 100 seconds. But in slow oxidation, the time necessary to increase the rate e-times is much longer. In reality, the life time of free radicals considered here does not exceed about one second. Therefore, a radical type branching step is not applicable to reactions with a characteristic time for self-acceleration longer than one second.

Slow self-accelerating reactions were already called, in the thirties, chain reactions with degenerate branching (or reactions with degenerate chain explosion) [1]. To explain them, it was postulated that they do not have branching chains. But, in the course of chain propagation, some molecular intermediate products are produced which may form primary radicals with much more ease than the original molecules. The characteristic time for self-acceleration is then determined by the rate of formation of primary radicals from these intermediate products.

For example, in the oxidation of many organic compounds at temperatures between 50 and 200°C, the following simple scheme seems to apply:

I

1) $R + O_2 \longrightarrow RO_2$

2) $RO_2 + RH \longrightarrow ROOH + R$ etc. .

Here, RH is the oxidized hydrocarbon, ROOH the hydroperoxide formed. This is how the main chain propagates. The first product is the hydroperoxide with a relatively weak $O' - O$ bond ($\sim$ 40 kcal). Therefore, the hydroperoxide can decompose slowly into free radicals:

3) $ROOH \longrightarrow RO + OH$.

This step may be regarded as the process of degenerate branching. The initial substances RH and O_2 are themselves capable of initiating chains, for example, as a result of the step: $RH + O_2 \longrightarrow R + HO_2$. But this is a rare event since the energy involved is usually much larger than that required for step 3). Consequently, at first the reaction proceeds immeasurably slowly. After enough hydroperoxide has accumulated, it accelerates progressively and the rate becomes measurable. Hence, we get long induction periods of several hours. If hydroperoxide is added to the system from the very start, the reaction immediately proceeds at a speed determined by the concentration of added hydroperoxide.

The self-acceleration at the beginning of reaction (in the absence of impurities, e.g., peroxides), follows the law $e^{\varphi t}$. As initial reactants are consumed, it slows down. Then the reaction rate, as measured by oxygen consumption changes with the extent of reaction, as shown by curve 1 of Figure 58 with a maximum close to half conversion. The quantity of oxygen consumed as a function of time is represented by the S-shaped curve 1 of Figure 59. Such cases occur frequently.

The results of Bone, et al, [2] on the oxidation of ethane at

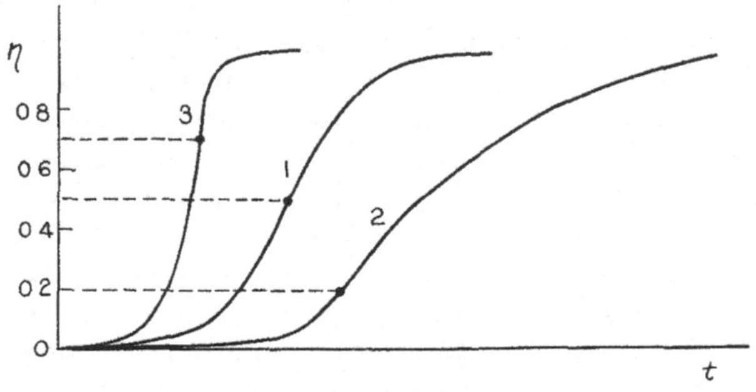

FIGURE 58. Relation between t and η.

318°C and 760 mm Hg are shown in Figure 60.

At high temperatures, branching is not due to hydroperoxide decomposition but to oxidation of aldehydes. At sufficiently high temperatures, the radical RO_2 (See Volume I) can decompose into an alcohol radical (OH in the case of methane) and an aldehyde R'CHO, before it succeeds in reacting following step 2). The radical participates in the main chain while the aldehyde now determines degenerate branching:

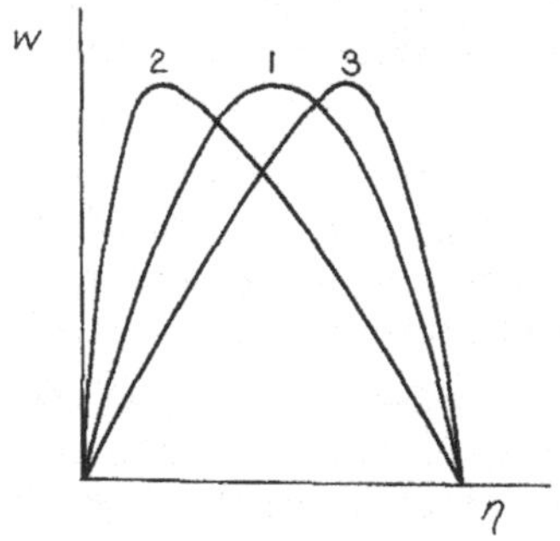

FIGURE 59. Relation between η and w.

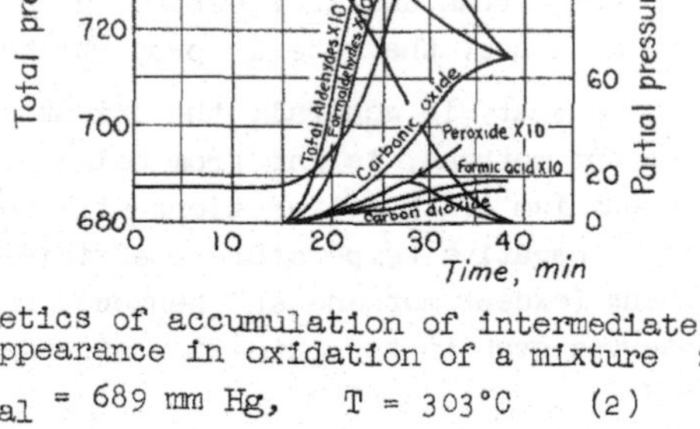

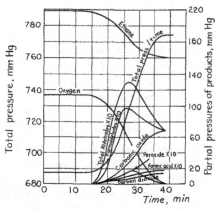

FIGURE 60. Kinetics of accumulation of intermediate products and oxygen disappearance in oxidation of a mixture $2C_2H_6 + O_2$: P_{total} = 689 mm Hg, T = 303°C (2)

4) $\dot{R}'CHO + O_2 \longrightarrow R'CO + HO_2$.

This step is not too slow at temperatures in the vicinity of 300 to 400°C since the C - H bond in aldehydes is relatively weak. Aldehydes accumulate in the system as they are produced by the main chain reaction. But, at the same time, they disappear as they react with the radicals of the main chain (R and the alcohol radicals). A mathematical analysis (3) reveals that the aldehyde disappearance then depends on the second or even third power of the aldehyde concentration. As a result, sometimes at relatively low conversions, the aldehyde concentration reaches a maximum limiting value. Then the reaction rate, proportional to the aldehyde concentration, also reaches a maximum at less than half conversion. It falls off subsequently because of consumption of reactants. The quantity of oxygen reacted plotted versus time will now give a non-symmetrical S-shaped curve with a maximum rate near the origin (see curves 2 on Figures 58 and 59). Such curves are often found. A good example is the oxidation of methane (see below).

The aldehyde concentration at the maximum is determined by the ratio of rate constants for the reactions between radicals and hydrocarbons or aldehydes respectively. This ratio will also determine the position of the maximum rate, between the start of the reaction and half conversion.

Sometimes, the maximum rate is found at conversions higher than 0.5, (see curves 3, Figures 58 and 59). An example is the oxidation of ethane at high temperatures (about 600°C) where the maximum lies at 66% conversion. As shown by Chirkov [4], the reaction rate can be expressed as a function of η, the fraction of oxygen reacted:

$$w = w_0 + k\eta^2(1 - \eta) \quad .$$

The curve of η versus time has now an unsymmetrical S-shape with an inflection point beyond half conversion. In this case, it corresponds to $\eta \sim 0.66$ (Figure 61, on the following page). The mechanism is not clear. Formally, it would seem that, as reaction proceeds, some other intermediate product is formed which reacts with aldehydes and so gives a branching step easier than that corresponding to the reaction between aldehyde and oxygen (hence, the rate is proportional to η^2).

It must be noted that, in general, the oxidation kinetics of all gaseous hydrocarbons except methane, is far from being elucidated. It is our opinion that the mechanism of these reactions will be explained only after the phenomenon of a negative temperature coefficient in the oxidation of most hydrocarbons (except methane and benzene) has been elucidated. This phenomenon is observed even in the oxidation of ethane and is quite sharp for higher hydrocarbons. As temperature is raised, a certain temperature range is soon reached where the time for reaction does not decrease, as it usually does, with further increase in temperature but rather starts increasing. At still higher temperatures, the time for reaction returns to its normal behavior and decreases as temperature goes up. In ethane oxidation, this interval of negative temperature coefficient, determined by the change in induction period, lies between 300 and 350°C (in a $2C_2H_6 + O_2$ mixture, at $p \sim 600$ mm Hg) (4).

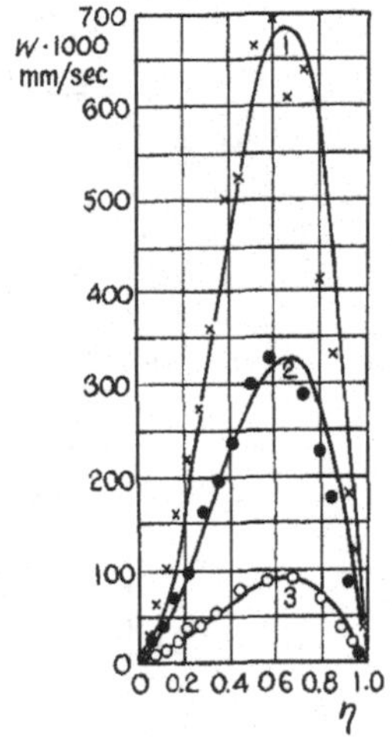

FIGURE 61. Reaction rate versus conversion in oxidation of a mixture $C_2H_6 + 3.5\ O_2$. T = 546°C. Circles and crosses are experimental points. (4)

	p_0 mm Hg	w	k
1	134	0.013	4.6
2	104.5	0.01	2.16
3	64.5	0.044	0.58

For the majority of the other hydrocarbons, the interval of negative temperature coefficient, between 350 and 450°C is determined by the change of the maximum rate with temperature. Only in the cases of butane (5) and propane (6) have parallel measurements been made of the change in $w_{max.}$ together with the induction period. It seems that the behavior of both $w_{max.}$ and $1/\tau$ with respect to temperature is the same (Figures 62 and 63). The abnormal behavior of the curves of induction periods (they go up as temperature increases) would seem to indicate that branching is now caused by reaction between molecules of two intermediate products and that the concentration of both or of only one of them becomes smaller as temperature increases.

It is known, for instance, that in propane oxidation (7), the concentration of active intermediate CH_3CHO falls, by a factor of four to five, as temperature is increased from 300°C (the cold flame region) to 387°, the region of the negative temperature coefficient.

It can be assumed that the oxidation reaction follows scheme I at temperatures below inflammation. Branching is then due to the decomposition of ROOH into free radicals, and the induction period is shorter, the higher the temperature. At higher temperatures, aldehydes become important and part of the reaction follows the aldehyde scheme. Enikopolyan [8] has postulated the following sequence of steps:

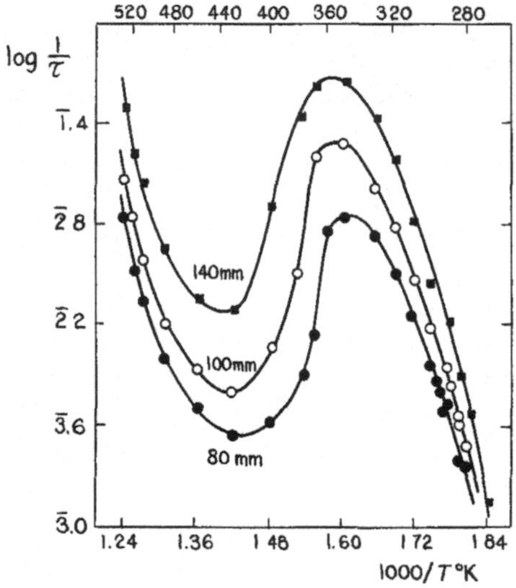

FIGURE 62. Inverse of induction period as a function of temperature for an equimolar mixture $C_4H_{10} + O_2$, p_{total} = 200 mm Hg. (5)

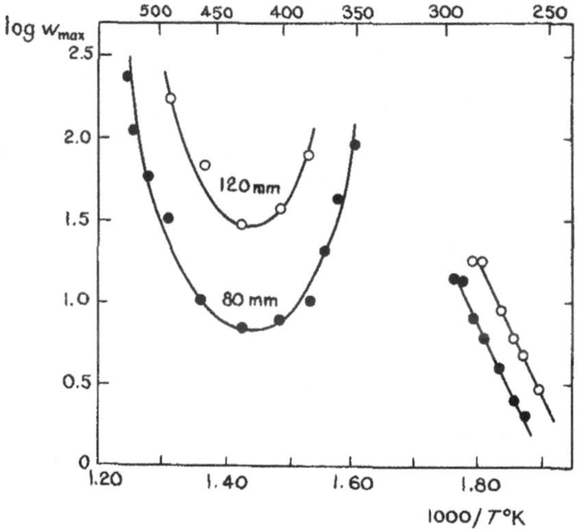

FIGURE 63. Reaction rate versus temperature for an equimolar mixture $C_4H_{10} + O_2$, p_{total} = 200 mm Hg. (5)

a) $RO_2 + CH_3CHO \longrightarrow ROOH + CH_3CO$

b) $CH_3CO + O_2 \longrightarrow CH_3C(=O){-}O-O$

c) $CH_3C(=O){-}O-O + RH \longrightarrow CH_3C(=O)OOH + R$.

The peroxyacid formed is known to be an oxidation product of the aldehyde at temperatures in the neighborhood of 100°C. At 300°C, this peroxyacid decomposes readily. The strength of its O - O bond is weaker than in hydroperoxides and according to Enikopolyan, this peroxyacid is responsible for branching at that temperature level. At still higher temperatures, the radicals

$$RO_2,\ RO,\ CH_3C(=O)OO,\ CH_3CO$$

start to decompose extensively, for example:

$$CH_3CH_2OO \longrightarrow CH_2O + CH_3O; \quad CH_3CO \longrightarrow CH_3 + CO \ .$$

Then the quantity of RO_2 available to enter in step a) becomes smaller and the branching process consequently suffers. This is precisely what causes the negative temperature coefficient in hydrocarbon oxidation.

It must be recalled that Shtern [9], in a special study of propane oxidation, has shown CH_3CHO to be the active intermediate responsible for degenerate branching. Thus, at 300 and 350°G, addition of CH_3CHO to a $C_3H_8 + O_2$ mixture, shortens or suppresses completely the induction period (suppression occurs when CH_3CHO is added in amounts corresponding to its maximum concentration). But if CH_3CHO is added at 387°C (where a zero temperature coefficient is observed), it has no effect on the reaction. These observations support the view that the occurrence of a negative temperature coefficient is related to a decrease in branching.

As to a branching step of the type: $HCHO + O_2 \longrightarrow HCO + O_2$, its rate must be small at these temperatures because of an appreciable activation energy (32 to 35 kcal).

Yet, at still higher temperatures, branching steps of the type $HCHO + O_2 \longrightarrow HCO + HO_2$ become fast enough and at some temperature level, the region of a negative temperature coefficient is overcome. The reaction rate now increases normally with temperature and induction periods become shorter. The chain is now propagated via an aldehyde scheme:

$$C_2H_5 + O_2 \longrightarrow C_2H_5OO$$

$$C_2H_5OO \longrightarrow CH_3CHO + OH$$

$$OH + C_2H_6 \longrightarrow H_2O + C_2H_5$$

$$CH_3CHO + O_2 \longrightarrow CH_3CO + HO_2 \quad .$$

The scheme of Enikopolyan has not been sufficiently checked experimentally and remains largely hypothetical. Nevertheless, it shows how a negative temperature coefficient can arise in hydrocarbon oxidation.

In oxidation of higher hydrocarbons, the region of negative temperature coefficients borders the cold flame region.

The phenomenon of cold flame oxidation consists in the following. In a certain temperature range (usually between 260 and 320°C) the oxidation reaction is accompanied by a blue flash with a simultaneous pressure jump in the system. After the explosion has subsided (it lasts about one second), the pressure also falls. There may be one to five explosions of this kind. The region of cold flame (a) is represented on Figure 64 as well as the cold flame oxidation (b) of $C_3H_8 + O_2$ mixtures (10).

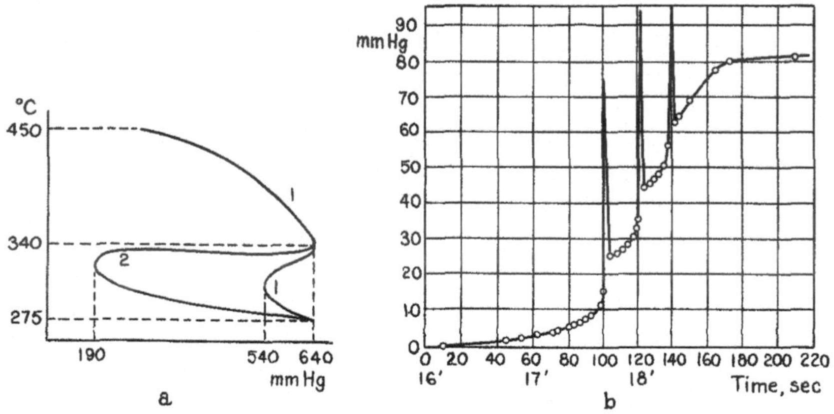

FIGURE 64. a- Region of inflammation (curve 1) and of cold flames (curve 2); $C_3H_8 + O_2$ mixture. (10)
b- Cold flame oxidation of a $C_3H_8 + O_2$ mixture $T = 280°C$, $p_o = 420$ mm Hg. (10)

It is now established that cold flame oxidation exists for saturated hydrocarbons (from ethane up), unsaturated hydrocarbons (starting with propylene), aldehydes and ethers. As hydrocarbon oxygen mixtures are made leaner, cold flames become weaker and disappear ultimately. The studies of Kondrat'ev [11] and Emeleus [12] have proved that the spectrum of cold flames is the fluorescence spectrum of formaldehyde. Cold flame oxidation is frequently characterized by long induction periods (~10 minutes) while the interval between flames is of the order of 6 to 10 seconds. These long induction periods may be eliminated completely by addition of intermediates (e.g., acetaldehyde in the case of propane and propylene). The cold flame induction period is related to temperature and pressure changes. In the oxidation of $C_4H_{10} + O_2$ mixtures, Neĭman [13] has found the following relation:

$$\tau p^{1.2} \exp(-23000/T) = \text{const.}$$

As to the mechanism responsible for cold flames, there are two common hypotheses at the present time. The first one is that of M. B. Neĭman [14]: the cold flame consists in an explosion due to peroxides building up to a critical concentration. In such an explosion, according to Neĭman, a large quantity of radicals are formed and these attack the hydrocarbons, oxidizing them violently and incompletely. This violent reaction is the cold flame.

The second hypothesis is that of Pease [15] and Norrish [16]: the cold flame explosion has a thermal origin and is due to the upsetting of the heat balance of a thermally unstable system which is being oxidized by a degenerate branching mechanism.

It seems most likely that a cold flame is a chain ignition in which branching is determined by positive chain interaction. This viewpoint is supported by the phenomenon of propagation of cold flames.

According to Enikopolyan, Polyak and Shtern [17], a cold flame is a chain explosion due to branching that occurs when a radical reacts with an active intermediate, e.g.:

1) $RO_2 + CH_3CHO \longrightarrow RO + OH + CH_3CO$.*

Such a branching step is possible only at the temperature level corresponding to cold flame regions. Indeed, at lower temperatures, the RO_2 radical does not decompose and therefore aldehydes are not formed. At considerably higher temperatures, RO_2 will decompose so quickly that it has no time to react with aldehydes as postulated above.

---

* Such a branching step was proposed by V. V. Voevodskiĭ and V. I. Vedeneev [18].

With this branching step, the concentration of active centers is given by:

$$n = \frac{w_o}{g - fx}$$

where x is the concentration of active intermediate, e.g., the aldehyde. When CH_3CHO reaches a limiting value $(CH_3CHO)_{max}$, $g = f_{max}$ and a chain explosion takes place in the form of a cold flame.

A study of existing theories reveals that none is free from objections and the question of the mechanism of cold flames may not be considered as solved. In our opinion, the most likely mechanism is that of a chain explosion.

Reactions with degenerate branching are by no means restricted to hydrocarbon oxidation. The oxidation of hydrogen sulfide follows a mechanism similar to that of methane oxidation, but at lower temperatures (about 200°C). The oxidation of arsine exhibits typical behavior of a degenerate branching explosion (the reaction lasts several days) at temperatures only slightly above room temperature.

Mechanisms of degenerate explosion are not limited to oxidation reactions. An analysis (1) of various data on the decomposition of $C\ell_2O$ reveals that a similar mechanism must also operate in this case. The final decomposition products are $C\ell_2$ and O_2, none of which exerts any catalytic effect. Nevertheless, the curve of reaction versus time has a typical S-shape and the initial self-acceleration obeys the law: $A(e^{\varphi t} - 1)$. Direct experiments where fresh gas was introduced in a vessel where $C\ell_2O$ decomposition took place, show that some active substance must be formed during reaction and be responsible for the self-acceleration. At the very end, this substance disappears in the explosion. But it must be relatively stable. If the reacting mixture is cooled down to 0°C so that reaction stops and then reheated to the original temperature, the reaction starts immediately at a rate equal to that before quenching.

Slow chain reactions with degenerate branching, just like fast reactions with chain ignition, exhibit the phenomenon of lower limit with respect to pressure and critical diameter. This was vividly demonstrated by Shantarovich [19] by means of the oxidation of arsine. At 50°C, a mixture $AsH_3 + O_2$ was introduced in three spherical reactors with diameters $d = 5$, 10 and 15 cm, at a pressure of about 150 mm Hg. In the first reactor ($d = 5$ cm), the mixture stayed unchanged during several days. In the second reactor ($d = 10$ cm), reaction was observed and a slight deposit of the reaction product As_2O_5 appeared on the walls. But reaction soon stopped. In the third reactor ($d = 15$ cm), reaction started and continued during several days while the walls were covered with a thick deposit of As_2O_5. Therefore, in the conditions of this work, the critical diameter is approximately $d_c = 10$ cm.

The lower pressure limit was determined in the following way. The reactor was filled with a mixture of arsine and oxygen at a certain pressure p. Reaction started, continued for a while but not to the end, then stopped when the pressure reached a value p_1. The extent of reaction $\Delta p = p - p_1$ was measured. It was found that Δp increases proportionally with $p - p_1$. This means that p_1 is the lower pressure limit. When the reactor was filled with a mixture at a pressure slightly below p_1, no sight of reaction could be detected, confirming the conclusion that p_1 is a lower pressure limit. It was found to depend on diameter, according to the already familiar law: p_1d = const. Thus at 50°C, $p_1 = 94$ mm Hg, d = 15 cm and $p_1d = 1400$. With $p_1 = 143$ mm Hg, d = 10 cm, $p_1d = 1430$. With $p_1 = 271$ mm Hg, d = 5 cm, $p_1d = 1360$. The relation p_1d = const. is thus well obeyed. As in the case of the ignition limit of hydrogen, the limit of the slow oxidation of arsine decreases with temperature according to the law $p_1 = c \exp(E/RT)$ where E is about 5 kcal. The reaction rate, in the initial stage of self-acceleration, obeys the law $e^{\varphi t}$ but rapidly reaches a maximum at a low conversion. It then falls off. Apparently, the reaction is inhibited by the wall deposit of As_2O_5, presumably because of a more effective chain termination.

Similar limits are also observed in hydrocarbon oxidation. However, they are not so sharp, apparently because of heterogeneous wall reactions taking place at high temperatures.

In a study of acetylene oxidation at 320°C and p = 400 mm Hg, Spence [20] has shown that when the reactor diameter is less than 4 mm, the reaction rate drops off sharply.

Sadovnikov [21], in a study of ethane oxidation between 600 and 790°C, at p between 30 and 180 mm Hg, observed a sudden increase in reaction time when the reactor diameter was reduced to 5 mm.

A similar investigation of the effect of reactor diameter on the rate of oxidation of a series of hydrocarbons (CH_4, C_2H_6, C_3H_8, C_2H_4, C_3H_6, C_2H_2) was conducted by Norrish [22] who showed that upon reducing the diameter below a certain definite value, the reaction rate, in every case, becomes extremely small.

The mechanism of oxidation of the simplest hydrocarbon – methane – will now be considered.

§2. Oxidation of Methane

In the oxidation of the simplest hydrocarbons, aldehydes are among the main intermediate products (formaldehyde in the case of methane). Additional free radicals, as was proposed in the preceding section, are produced in a degenerate branched chain reaction between aldehyde and oxygen:

$$HCHO + O_2 \longrightarrow HCOOH + O$$

or $\qquad\qquad\qquad\qquad\qquad\qquad\qquad\qquad\qquad\qquad\qquad\qquad$ (a)

$$\searrow H_2O + CO + O \quad .$$

In subsequent sections, degenerate branching will be ascribed to peroxide decomposition, e.g.:

$$ROOH \longrightarrow RO + OH \quad (35 \text{ to } 40 \text{ kcal}) \quad . \qquad (b)$$

This is based on many experimental data showing the initiation due to peroxides that also appear as intermediate products in hydrocarbon oxidation.

When a sufficient amount of organic peroxides is formed by the propagation of the main chain as it frequently happens in liquid phase oxidation of organic compounds, the decomposition of peroxides is thus responsible for branching. But in gas phase oxidation of hydrocarbons, usually at temperatures above 300°C, aldehydes are apparently primary products (see Chapter II) which may cause degenerate branching, for example by reactions of type (a). Nevertheless, our theoretical analysis (see Chapter VII) shows convincingly that the following step is more likely:

$$3) \quad HCHO + O_2 \longrightarrow HO_2 + HCO - Q \quad .$$

In our analysis of methane oxidation, this step will be used as degenerate branching reaction. Indeed,

$$Q = - Q_{HCO-H} + Q_{H-O_2} = - 79 + 47 = - 32 \text{ kcal} \quad .$$

Since the reverse process is a radical recombination step with a low activation barrier (see Chapter VII), the activation energy of step 3) must be close to its endothermicity. The primary generation of free radicals from reactants: $CH_4 + O_2 \longrightarrow CH_3 + HO_2 - 55$ kcal is considerably more difficult. At the start of reaction, with little aldehyde present, the rate of initiation is:

$$w_{CH_4} = k_0(CH_4)(O_2) = 10^{-10} \exp(- 55000/RT) \cdot (CH_4)(O_2) \quad .$$

As aldehydes accumulate, radical initiation also occurs by means of step 3) at a rate:

$$w_{CH_2O} = k_3(HCHO)(O_2) \approx 10^{-10} \exp(-32000/RT) \cdot (HCHO)(O_2) \quad .$$

The ratio

$$\left(w_{CH_2O}/w_{CH_4} \right) = \alpha$$

is then:

$$\alpha = \frac{k_3(HCHO)(O_2)}{k_4(CH_4)(O_2)} = \exp(23000/RT) \cdot \frac{(HCHO)}{(CH_4)} \quad .$$

At 700°K,

$$\alpha = 10^{7.15} \frac{(HCHO)}{(CH_4)} \quad .$$

Thus, at 423°C, in an original mixture of methane and oxygen, generation of primary radicals is 14000 times slower than in the same mixture containing 0.1% HCHO and 1400 times slower than in a mixture with 0.01% HCHO.

Consequently, as reaction proceeds and aldehydes accumulate, the rate, which is at first immeasurably small, accelerates progressively because of step 3). This determines the induction period.

The chain propagation in methane oxidation above 300 - 350°C may be represented in the following scheme:

1) $CH_3 + O_2 \longrightarrow CH_3OO + q_1$

1') $CH_3OO \longrightarrow CH_2O + OH + q_1'$

2) $OH + CH_4 \longrightarrow CH_3 + H_2O + 15$ kcal .

According to several investigations [23, 24], step 1) proceeds at a rate $\sim 10^{-13}(O_2)(CH_3) = a_1(CH_3)$. At a total pressure of 235 mm Hg and $T = 700$°K, $a_1 = 2.2 \cdot 10^5$ in a stoichiometric mixture. Apparently, no activation energy is required for step 1). It is not clear whether it takes place on every triple collision [25] or following binary collisions with a steric factor of 10^{-3}.

According to Avramenko's data [26], the rate of step 2) is equal to: $10^{-10} \exp(-8500/RT) \cdot (CH_4)(OH) = a_2(OH)$. The activation energy can be estimated from the empirical relation (Chapter I): $\epsilon_0 = 11.5 - 0.25q = 11.5 - (0.25 \cdot 15) \cong 8$ kcal, in close agreement with the experimental result. At $T = 700°K$, the rate of step 2) is equal to:

$$10^{-10} \exp(-8500/1400) \cdot (CH_4)(OH) = 2.2 \cdot 10^{-13}(CH_4)(OH) = a_2(OH) \quad .$$

At $p = 235$ mm Hg, in a stoichiometric mixture $(CH_4) = 10^{18}$ molecules/cm^3. Hence $a_2 = 2.2 \cdot 10^5$.

Before a radical CH_3OO is able to decompose following step 1'), it must first isomerize: $CH_3O\dot{O} \longrightarrow \dot{C}H_2 - O - OH$. Isomerization is undoubtedly associated with a large activation energy. A rough estimate indicates that $\epsilon_1' = 20000$ cal. Then the rate of reaction 1') is equal to: $10^{13} \exp(-20000/RT) \cdot (CH_3OO)$. At 700°K: $10^{13} \cdot 6.3 \cdot 10^{-7}(CH_3OO) = 6.3 \cdot 10^6(CH_3OO) = a_1'(CH_3OO)$ where $a_1' = 6.3 \cdot 10^6$. As can be seen, a_1' is 30 times larger than a_1 and a_2. Therefore at high temperatures at which methane oxidation takes place, reaction 1') presumably follows reaction 1) fast enough so that the formation of formaldehyde may be pictured by means of a single step:

$$CH_3 + O_2 \longrightarrow CH_2O + OH \quad .$$

Consider now chain termination. It may occur when OH or CH_3 radicals get to the wall. In the cases considered here, a_1 is about equal to a_2 so that the concentrations of radicals OH and CH_3 may be considered of the same order of magnitude.

It is known [27] that the wall capture of a OH radical is more likely than that of H atoms. There are no data available on the wall capture probability of CH_3 radicals. Nevertheless, it may be assumed (if, for instance, the kinetics of cracking reactions is kept in mind) that the wall capture probability of a CH_3 radical is smaller than that of H atoms and therefore much smaller than that of OH radicals. Consequently, the wall termination rate of OH radicals, namely $a_6(OH)$ will be taken as considerably larger than the rate of termination of CH_3, viz., $a_6'(CH_3)$. The latter will be neglected.

In the initial stages of the process, when the aldehyde concentration is low, the reaction between aldehydes and chain radicals will proceed at such a slow rate that it can be overlooked. Initially, only the aldehyde reaction leading to branching must be considered since even at very low aldehyde concentrations, the rate of generation of primary radicals by that process, far exceeds that of the reaction between methane and oxygen.

Therefore, for the early stages of methane oxidation, the following scheme may be written:

0) $CH_4 + O_2 \longrightarrow CH_3 + HO_2$

1) $CH_3 + O_2 \longrightarrow CH_2O + OH$

2) $OH + CH_4 \longrightarrow CH_3 + H_2O$

3) $CH_2O + O_2 \longrightarrow HCO + HO_2$

4) $HCO + O_2 \longrightarrow CO + HO_2$

5) $HO_2 + CH_4 \longrightarrow H_2O_2 + CH_3$

6) $OH \xrightarrow{\text{wall}}$

8) $CH_2O \xrightarrow{\text{wall}}$.

The last step has been added to include the possibility of heterogeneous oxidation of aldehyde on the walls of the reactor.

The rate of change of radical concentration is given by the equations:

$$\frac{d(CH_3)}{dt} = n_0 + a_2(OH) + a_5(HO_2) - a_1(CH_3)$$

$$\frac{d(OH)}{dt} = a_1(CH_3) - a_2(OH) - a_6(OH) = a_1(CH_3) - (a_2 + a_6)(OH)$$

$$\frac{d(HCO)}{dt} = a_3(CH_2O) - a_4(HCO) \quad \text{where} \quad a_3 = k_3(O_2)$$

$$\frac{d(HO_2)}{dt} = n_0 + a_3(CH_2O) + a_4(HCO) - a_5(HO_2)$$

$$\frac{d(CH_2O)}{dt} = a_1(CH_3) - a_3(CH_2O) - a_8(CH_2O) \quad .$$

In all these equations, a has the dimension of inverse time.

Since the time required to propagate the chain by means of the radicals OH, CH_3, HCO and HO_2 is much less than the time $1/a_3$ required for branching (as in the mechanism of the $H_2 + O_2$ reaction), it may be said that the radical concentration is completely determined by the aldehyde concentration.

Therefore:

$$\frac{d(CH_3)}{dt} = \frac{d(OH)}{dt} = \frac{d(HCO)}{dt} = \frac{d(HO_2)}{dt} = 0 \quad .$$

This algebraic system of equations can be solved and the concentration of all active centers may be expressed in terms of the concentration of CH_2O:

$$(OH) = \frac{2[n_o + a_3(CH_2O)]}{a_6} \quad \text{and} \quad (HO_2) = \frac{n_o}{a_5}$$

Then:

$$\frac{d(CH_2O)}{dt} = \left(\frac{a_2}{a_6} + 1\right)\left[2n_o + 2a_3(CH_2O)\right] - a_3(CH_2O) - a_8(CH_2O) =$$

$$2n_o\left(\frac{a_2}{a_6} + 1\right) + 2a_3(CH_2O)\left(\frac{a_2}{a_6} - \frac{1}{2}\right) - a_8(CH_2O) \quad . \tag{1}$$

The ratio a_2/a_6 of propagation and termination rates is the chain length ν. Nalbandyan [28] has found that at room temperature, the quantum yield is equal to 0.55. It is equal to 19 at 393°. But at room temperature even in the absence of chains, the quantum yield should be equal to two. Therefore, at least 75% of the quanta absrobed by the system are not used to form radicals. If it is assumed that this fraction of unused quanta does not increase with temperature, the quantum yields of Nalbandyan should be multiplied by four. Thus the quantum yields impose a lower limit to chain length. Without committing a large error, it may be assumed that at 423°, $\nu = a_2/a_6 \cong 100$. But then both 1 and 1/2 may be neglected as compared to a_2/a_6 and the previous expression becomes:

$$\frac{d(CH_2O)}{dt} = 2n_o \frac{a_2}{a_6} + 2a_3 \frac{a_2}{a_6}(CH_2O) - a_8(CH_2O)$$

$$= 2n_o(a_2/a_6) + \left[2a_3 \frac{a_2}{a_6} - a_8\right](CH_2O) \quad . \tag{2}$$

Integration of equation (2) with $(CH_2O) = 0$ at $t = 0$, gives:

$$(CH_2O) = \frac{2n_o(a_2/a_6)}{\varphi}(e^{\varphi t} - 1) \tag{3}$$

where

$$\varphi = 2a_3 \frac{a_2}{a_6} - a_8 \quad . \tag{4}$$

The rate of methane oxidation with this scheme is then:

$$-\frac{d(CH_4)}{dt} = a_2 \frac{2n_o + 2a_3(CH_2O)}{a_6} \quad . \tag{5}$$

Substituting the value of (CH_2O) from (3), we get:

$$-\frac{d(CH_4)}{dt} = \frac{2n_0 a_2}{a_6} + \frac{4a_3 n_0 a_2^2/a_6^2}{\varphi}(e^{\varphi t} - 1) =$$

$$2n_0 \frac{a_2}{a_6}\left[1 + \frac{2a_3 \cdot a_2/a_6}{\varphi}(e^{\varphi t} - 1)\right] . \qquad (6)$$

In equation (4), the term

$$2a_3 \frac{a_2}{a_6}$$

is proportional to at least the square of the pressure while the term a_8 apparently does not depend so strongly on pressure. Therefore φ depending on pressure may be positive or negative. If φ is positive, the rate is given by equation (6). If φ is negative, the following equation obtains:

$$-\frac{d(CH_4)}{dt} = 2n_0 \cdot a_2/a_6\left[1 + 2a_3 \frac{a_2}{a_6}(1 - e^{-\varphi t})\right] \qquad (7)$$

where

$$\varphi = a_8 - 2a_3 \frac{a_2}{a_6} = \left| 2a_3 \frac{a_2}{a_6} - a_8 \right| .$$

The condition $\varphi = 0$, or

$$2a_3 \frac{a_2}{a_6} = a_8 \qquad (8)$$

determines the critical pressure and the critical diameter of the slowly accelerating reaction. The product a_3a_2 is proportional to p^2, $a_6 \propto 1/d$, $a_8 \propto 1/d$. If it is assumed that a_8 does not depend on pressure, the condition for passing from a slow reaction at constant rate to an accelerating one, is given by:

$$p^2d^2 = a_8 . \qquad (9)$$

When the quantity

$$2a_3 \frac{a_2}{a_6}$$

and consequently also φ are small, the transition from a very slow to a rather fast reaction, is not very sharp and extends over a certain range of values of φ. This is what happens in the oxidation of methane.

At pressures considerably exceeding the critical, the rate of heterogeneous reaction of CH_2O becomes quite small as compared to the term $2a_3 \cdot a_2/a_6$ and therefore a_8 may be neglected. Then

$$\varphi = 2a_3 \frac{a_2}{a_6}$$

and equation (6) takes the form:

$$-\frac{d(CH_4)}{dt} = 2n_o \nu e^{\varphi t} \tag{10}$$

where ν is the length of the main chain

$$\nu = \frac{a_2}{a_6}$$

and $\varphi = 2a_3\nu$.

Integration of equation (10) gives, if unity is neglected in front of $\exp(\varphi t)$:

$$\Delta(CH_4) = \frac{2n_o \nu e^{\varphi t}}{\varphi} = \frac{n_o}{a_3} e^{\varphi t} = Ce^{\varphi t} \tag{11}$$

where $\Delta(CH_4) = (CH_4)_o - (CH_4)_t$ is equal to the quantity of methane consumed at time t.

This is the expression used in our 1934 book [1] for the disappearance of original reactant or the accumulation of intermediates. It was there applied to the oxidation of several hydrocarbons. This law is valid only at the very start of reaction. However, the method used to detect concentration changes has a limit of sensitivity $a = \Delta'(CH_4)$. Then, as long as $\Delta(CH_4) < \Delta'(CH_4)$, reaction will not be detected. The time required to decrease the methane concentration by the amount $\Delta'(CH_4)$ is called the induction period. The latter is determined directly. It may be shown that the induction period is determined by equation (11) and is not due to any secondary phenomenon (for example, the presence of inhibiting impurities). Indeed, when φ is determined independently from the curve $\log \Delta(CH_4)$ versus t, at times greater than τ (the induction period), at various pressures and temperatures, it is found that $\tau\varphi = \text{const}$. This relation immediately follows from equation (11). Consider the end of the induction period when $\Delta(CH_4) = \Delta'(CH_4)$. Then $Ce^{\varphi\tau} = \Delta'(CH_4)$ or $\varphi\tau = \log \Delta'(CH_4) - \log C$, or $\varphi\tau + \log C = \text{const.}$ Hence

$$\tau = \frac{\text{const.} - \log C}{\varphi},$$

or

$$\tau\varphi = \text{const.} - \log G \quad . \tag{11'}$$

In a restricted interval of temperatures and pressures, the quantity C varies somewhat. However, since C is under the logarithmic sign in equation (11'), this variation is small and can be ignored. Therefore, the product $\tau\varphi$ should be a constant, as found actually.

The induction period may be reduced or even eliminated by addition of good chain initiators (e.g., peroxides, NO_2 etc.). On the contrary, it may be increased by means of inhibitors that terminate chains readily.

Equation (11), derived for the early stages of the reaction is well observed with all hydrocarbons studied. During the course of the process, the equation ceases to apply at higher conversions. In particular, according to Enikopolyan's study of methane oxidation [29], the law ceases to be obeyed at about 10% conversion. In this conversion level, self-acceleration tapers down and the rate becomes constant.

It is now recalled that in the scheme of methane oxidation considered thus far, the reactions between aldehyde and chain radicals were not included (e.g., $OH + CH_2O \longrightarrow H_2O + HCO$). This can be justified only at the very beginning of reaction. But as aldehyde accumulates during oxidation, these reactions start playing an essential role. For higher conversions, it is then necessary to include reactions between aldehyde and OH radicals, also HCO and HO_2 radicals that are found in the steps: $OH + CH_2O \longrightarrow H_2O + HCO$; $HCO + O_2 \longrightarrow CO + HO_2$.

A new scheme can then be written, taking into account reactions between aldehyde and radicals:

0) $CH_4 + O_2 \longrightarrow CH_3 + HO_2$

1) $CH_3 + O_2 \longrightarrow CH_2O + OH$

2) $OH + CH_4 \longrightarrow H_2O + CH_3$

2') $OH + CH_2O \longrightarrow H_2O + HCO$

3) $CH_2O + O_2 \longrightarrow HCO + HO_2$

4) $HCO + O_2 \longrightarrow CO + HO_2$ (II)

5) $HO_2 + CH_4 \longrightarrow H_2O_2 + CH_3$

5') $HO_2 + CH_2O \longrightarrow H_2O_2 + HCO$

6) $OH \xrightarrow{wall}$

8) $CH_2O \xrightarrow{wall}$.

The following steps have not been included: 1) a reaction between HCO and methane;* 2) the step: $CH_3 + CH_2O \longrightarrow CH_4 + HCO$; 3) the wall termination of HO_2. This is justified as follows: 1) the reaction $HCO + CH_4 \longrightarrow CH_2O + CH_3$ is endothermic (20 - 25 kcal) while the reaction between HCO and O_2 is exothermic (20 kcal) so that the former can be neglected; 2) it may be shown that the rate constants of $CH_3 + CH_2O$ and $CH_3 + O_2$ differ by a factor of ten while the maximum concentration of CH_2O is about 400 times smaller than that of O_2. Therefore, the rate of $CH_3 + O_2$ is 40 times larger than that of $CH_3 + CH_2O$, so that the latter can be neglected; 3) the OH radical, as is well known, is very active and is easily destroyed at the wall. Insofar as the HO_2 radical is much less reactive than OH, it is probable that the reaction of HO_2 with the wall will also be slower than that of OH. This point will be justified more thoroughly in what follows.

The relevant kinetic equations can now be written. If further:

$$\frac{d(CH_3)}{dt} = \frac{d(OH)}{dt} = \frac{d(HCO)}{dt} = \frac{d(HO_2)}{dt} = 0 \quad ,$$

the concentrations of all active centers can be expressed in terms of (CH_2O). Then, with $a_2 >> a_6$ and $n_0 << a_3(CH_2O)$, the reaction rate becomes:

$$\frac{d(CH_2O)}{dt} = \left[\frac{2a_2a_3}{a_6} (CH_2O) - a_3(CH_2O) - a_8(CH_2O) \right] - \frac{2a_3k_5'}{a_5} (CH_2O)^2 - \frac{2k_2'k_5'a_3}{a_5a_6} (CH_2O)^3 \quad . \qquad (12)$$

The expression between square brackets is that which was found above from scheme (I) for

$$\frac{d(CH_2O)}{dt} \quad .$$

* The reaction $HCO + CH_2O$ has then of course been omitted as well.

The second and third terms correspond to the reactions between aldehyde and radicals (OH, HCO, HO_2). Since methane oxidation will be treated here only at pressures far in excess of the critical, the term $a_8(CH_2O)$ can be neglected. The second term of the expression between square brackets, $a_3(CH_2O)$ is about 200 times smaller than the first term since a_2/a_6, according to Nalbandyan [28] is of the order of 100 so that $a_3(CH_2O)$ may also be neglected. The above expression takes the simpler form:

$$\frac{d(CH_2O)}{dt} = \frac{2a_2a_3}{a_6}(CH_2O) - \frac{2a_3k_5'}{a_5}(CH_2O)^2 - \frac{2k_2'k_5'a_3}{a_5a_6}(CH_2O)^3 =$$

$$\frac{2a_2a_3}{a_6}(CH_2O)\left[1 - \frac{k_5'a_6}{k_5a_2}\frac{(CH_2O)}{(CH_4)}\left\{1 + \frac{k_2'a_2}{k_2a_6}\frac{(CH_2O)}{(CH_4)}\right\}\right] . \qquad (12')$$

The ratios k_5'/k_5 and k_2'/k_2 can now be estimated at 427°C = 700°K.

The first ratio is determined by the competition between formaldehyde and methane for HO_2 radicals. The second ratio depends on the same competition for OH radicals. Let us assume that the steric factors for reactions 5) and 5') are almost the same, and similarly for reactions 2) and 2'). Then the ratios of rate constants are determined only by differences in activation energies. Let us estimate these differences $\epsilon_5 - \epsilon_5'$ and $\epsilon_2 - \epsilon_2'$. Both reactions 2) and 2') are exothermic (q_2 = 15 kcal; q_2' = 37 kcal). From the relation $\epsilon_0 = 11.5 - 0.25q$, we get $\epsilon_2 - \epsilon_2'$ = 6 kcal. The exothermicity of reaction 5') is about 12 kcal while reaction 5) is endothermic (11 kcal). Then, $\epsilon_5 - \epsilon_5' \sim 11$ kcal. At T = 700°K, the ratios of rate constants are:

$$k_5'/k_5 = \exp(11000/RT) = 2.5 \cdot 10^3; \quad k_2'/k_2 = \exp(6000/RT) \sim 70 .$$

Then:

$$\frac{d(CH_2O)}{dt} = \frac{2a_2a_3}{a_6}(CH_2O)\left[1 - \frac{2.5 \cdot 10^3}{100}\alpha\{1 + 70 \cdot 100\alpha\}\right]$$

where

$$\alpha = \frac{(CH_2O)}{(CH_4)} \quad \text{and} \quad a_2/a_6 \sim 100 .$$

Thus:

$$\frac{d(CH_2O)}{dt} = \frac{2a_2a_3}{a_6} (CH_2O) \left[1 - 25\alpha \; (1 + 7000\alpha) \right] .$$

With a dimensionless time

$$\tau = \frac{ta_6}{2a_2a_3} :$$

$$\frac{1}{(CH_2O)} \cdot \frac{d(CH_2O)}{d\tau} = [1 - 25\alpha(1 + 7000\alpha)] = [1 - f(\alpha)] . \qquad (13)$$

The right hand side of this equation can be calculated for various values of α. The corresponding curve is shown on Figure 65. It can be seen that as long as $\alpha \leq 0.05\%$, $1 - f(\alpha) \sim 1$ with an error of 5 or 6%. The second term in the square bracket of the equation can thus be neglected as compared to unity. This means that equation (2) derived from scheme (I) can be used to determine the concentration (CH_2O) at pressures above the critical. As α increases, $1 - f(\alpha)$ decreases rapidly. The second term in the square bracket grows and at $\alpha = 0.234\%$, $1 - f(\alpha) = 0$, or $f(\alpha) = 1$. Then

$$\frac{d(CH_2O)}{dt}$$

becomes equal to zero. It can be concluded that (CH_2O) after reaching the value $(CH_2O)_{max}$ does not change any more. This is actually observed. From this moment on, the rate of formation of CH_2O is equal to its rate of disappearance. The data of Enikpolyan [30] show that, at 423°C, the stationary concentration of aldehyde corresponds to 0.51% of the methane content of the mixture. This is twice as large as the value just calculated (0.234%) from the postulated reaction mechanism. The discrepancy between calculated and experimental values of $(CH_2O)_{max}$ varies little with temperature. Thus, at 423°C,

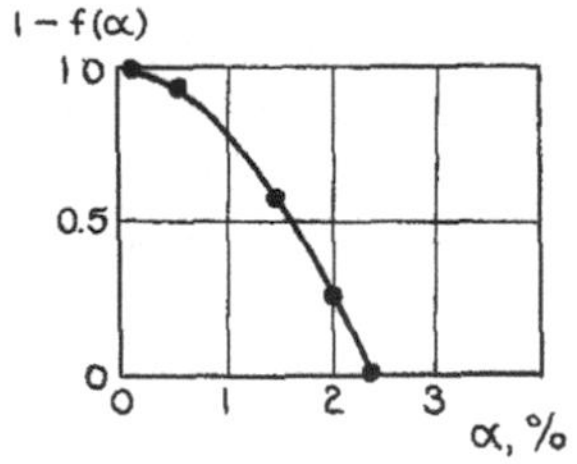

FIGURE 65. $\alpha = \frac{(CH_2O)}{(CH_4)}$ versus $1 - f(\alpha)$ where $f(\alpha) = 25\alpha(1 + 7000\alpha)$

$$\frac{(CH_2O)_{exp.}}{(CH_2O)_{theo.}} = 2$$

while at $T = 491°C$,

$$\frac{(CH_2O)_{exp.}}{(CH_2O)_{theo.}} = 2.8 \quad (31) \ .$$

Since the theoretical figure $(CH_2O)_{max.} = 0.234\%$ was obtained from an independent estimate of absolute values of a few constants, a discrepancy by a factor of two can be considered as an argument in support of theory.

According to our analysis, when $\alpha = 2.34 \cdot 10^{-3}$, the second term in the square bracket of (13) is equal to $(1 + 7000\alpha) = 1 + 16.3$. Thus, in equation (12'), unity may be neglected in front of

$$\frac{k_2'}{k_2} \frac{a_2}{a_6} \frac{(CH_2O)}{(CH_4)}$$

and in subsequent calculations a simplified expression shall be used:

$$\frac{d(CH_2O)}{dt} = \frac{2k_2k_3}{a_6} (CH_4)(O_2)(CH_2O) \left[1 - \frac{k_2'k_5'}{k_2k_5} \frac{(CH_2O)^2}{(CH_4)^2} \right] . \qquad (14)$$

When the maximum concentration of CH_2O has been reached, the expression between brackets in equation (14) becomes equal to zero and:

$$(CH_2O)_{max.} = \sqrt{\frac{k_5}{k_5} \cdot \frac{k_2}{k_2'}} (CH_4) \ . \qquad (15)$$

Experiments [31] actually show that $(CH_2O)_{max.}$ is proportional to (CH_4) as shown on Figure 66, on the following page, and does not depend on (O_2). The expression under the square root sign can be calculated since:

$$\frac{k_5}{k_5'} = e^{-11000/RT}; \quad \frac{k_2}{k_2'} = e^{-6000/RT} \quad \text{and} \quad \frac{k_5k_2}{k_5'k_2'} = e^{-17000/RT} \ .$$

Hence:

$$(CH_2O)_{max.} = \sqrt{e^{-17000/RT}} (CH_4) = \exp(-8500/RT) \cdot (CH_4) \ . \qquad (15')$$

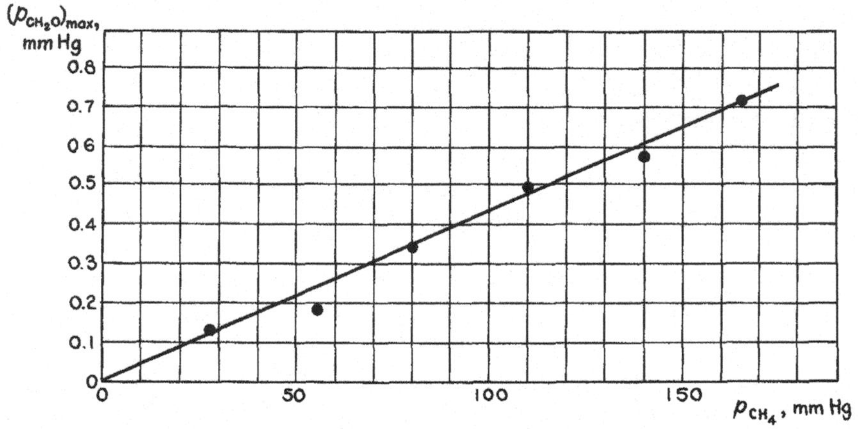

FIGURE 66. Maximum value of (CH_2O) versus (CH_4) in mixture, $T = 491.5°C$, $p_o = 235$ mm Hg. (31)

Theory thus tells us that $(CH_2O)_{max.}$ depends on temperature exponentially. Experimentally, Enikopolyan [32] found:

$$(CH_2O)_{max.} \propto \exp(-10000/RT) \quad .$$

This is in excellent agreement with theory, especially if it is remembered that several activation energies were estimated by an approximate formula.

The rate of methane disappearance will now be calculated. Scheme (II) gives:

$$-\frac{d(CH_4)}{dt} = a_2(OH) + a_5(HO_2)$$

$$= \left[k_2'(CH_2O) + k_2 \right] 2 \left\{ \frac{n_0 + a_3(CH_2O)}{a_6} \right\} + n_0 + 2a_3(CH_2O).^{*} \tag{16}$$

* At the beginning of reaction, when the concentration of CH_2O is small, $k_2'(CH_2O) < a_2$ and therefore the term $k_2'(CH_2O)$ may be neglected in front of a_2. Equation (5) is then obtained, valid for methane oxidation at low conversions:

$$-\frac{d(CH_4)}{dt} = \frac{a_2[2n_0 + 2a_3(CH_2O)]}{a_6}$$

Since $n_o << a_3(CH_2O)$ and $1 << a_2/a_6$, we get:

$$-\frac{d(CH_4)}{dt} = \frac{2}{a_6}\left[k_2(CH_4) + k_2'(CH_2O)\right]\left[k_3(CH_2O)(O_2)\right]. \quad (17)$$

When (CH_2O) becomes equal to $(CH_2O)_{max.}$, the rate of reaction

$$-\frac{d(CH_4)}{dt}$$

reaches a maximum constant value.

Let us determine this maximum rate. From equation (15), we get $(CH_2O)_{max.}$. Substituting this value into equation (17), we have:

$$-\frac{d(CH_4)}{dt} = \frac{2}{a_6}\left[k_2 + k_2'\left(\frac{k_2k_5}{k_2'k_5'}\right)^{1/2}\right]\left[k_3\left(\frac{k_2k_5}{k_2'k_5'}\right)^{1/2}(O_2)(CH_4)^2\right]. \quad (18)$$

But

$$\frac{k_2}{k_2'} = \exp(-\ 6000/RT)$$

and

$$\left(\frac{k_2k_5}{k_2'k_5'}\right)^{1/2} \sim \exp(-\ 8500/RT)\ .$$

Thus, the second term in the square bracket of equation (18) is about six times smaller than the first term. For a rough approximation, it may be neglected. Equation (18) then becomes:

$$-\frac{d(CH_4)}{dt} = \frac{2k_2k_3}{a_6}\left(\frac{k_2k_5}{k_2'k_5'}\right)^{1/2}(CH_4)^2(O_2)\ .^* \quad (19)$$

Norrish [33], in a study of methane oxidation, found a relation

* This calculation may be compared to a twenty-year old calculation by Norrish [33]. In order to obtain a constant aldehyde concentration at low conversions, Norrish introduced a ter-molecular reaction for aldehyde consumption: $2HCHO + O_2 \longrightarrow 2CO + 2H_2O$. This gave him a quadratic term analogous to the second term in the square bracket of equation (14). Norrish did not include a chain mechanism for aldehyde oxidation. This is not in accord with present views. But we have just seen that a chain mechanism for aldehyde oxidation precisely gives the needed second term in the square bracket of equation (14).

of this type. There are some indications in the literature that sometimes the reaction rate is proportional to the oxygen concentration to a power less than one. In our selection of a mechanism for methane oxidation, we assumed that the branching step $HCHO + O_2 \longrightarrow HCO + HO_2 - 32$ kcal proceeds homogeneously with an activation energy of 32 kcal. As a result, the linear dependence of reaction rate on oxygen concentration is obtained. It is not excluded, however, that branching occurs in part on reactor walls. In such a case, the rate of branching may be independent of oxygen concentration so that the overall rate depends only on methane concentration. With different wall activities, branching may thus be either homogeneous or heterogeneous and this will modify the relation between rate and oxygen concentration.

The wall termination constant a_6 must decrease with increasing diameter. Correspondingly, the reaction rate must become larger. There are many reports in the literature, confirming this point in many cases of hydrocarbon oxidation. Yet, Sadovnikov [21] has found in a study of ethane oxidation, that with proper pre-treatment of the walls, the reaction rate ceases to depend on reactor diameter. Norrish [22] has also shown that, in the oxidation of various hydrocarbons, the dependence of rate on diameter disappears, or becomes very weak, when a rather small diameter is reached. This may well be due also to the possibility of branching at the wall so that a_3 would then decrease as reactor size increases. The ratio a_3/a_6 would then become independent of reactor size.

The absolute reaction rate of methane oxidation will now be calculated for scheme (II) from equation (19). We have already: $a_2/a_6 = \nu \sim 100$ (at $T = 423°C$)

$$\left(\frac{k_2 k_5}{k_2' k_5'}\right)^{1/2} \sim \exp(-\,8500/RT) = 2.3 \cdot 10^{-3}$$

$k_3 = f_3 \cdot 10^{-10} \exp(-\,32000/RT)$, or at 423°C: $k_3 = f_3 \cdot 10^{-10} \cdot 10^{-10} = f_3 \cdot 10^{-20}$.

The methane concentration at 423°C, at a total pressure $p = 235$ mm Hg, in a stoichiometric mixture is equal to:

$$(CH_4) = \frac{0.97 \cdot 10^{19} \cdot 78}{700} = 1.1 \cdot 10^{18} \text{ molecules/cm}^3 \quad .$$

Substituting these values in equation (19), we get:

$$-\frac{d(CH_4)}{dt}\,\frac{1}{(O_2)} = 2 \cdot 100 \cdot f_3 \cdot 10^{-20} \cdot 2.3 \cdot 10^{-3} \cdot 1.1 \cdot 10^{18}$$
$$= f_3 \cdot 4.8 \cdot 10^{-3} \quad .$$

Experimentally, Enikopolyan [30] found $w_{max.} = 4.5 \cdot 10^{-5}$ sec.$^{-1}$. To get agreement between theoretical and experimental figures, f_3 should be of the order of 10^{-2}. It must be remembered that the estimation of $(CH_2O)_{max.}$ was off by a factor of two, probably because some errors in the values of the constants were used in the calculation. Then the steric factor should be 1/200. Such a value for f_3 is reasonable since the branching step must require a relatively specific orientation of HCHO and O_2 molecules before reaction can take place.

The activation energy predicted by the postulated mechanism will be calculated next. According to Enikopolyan [31, 32], the activation energy is 44 to 46 kcal.*

Equation (19) shows that the overall activation energy is made of the activation energy of step 3), 32 kcal; that of step 2), 8.5 kcal and the temperature coefficient of the group

$$\left(\frac{k_2 k_5}{k_2' k_5'} \right)^{1/2},$$

namely 8.5 kcal. Summing, we find $E \sim 49$ kcal if a_6 is assumed to be temperature independent. In fact, data show that the probability of wall capture is a function of temperature with an activation energy between 4 and 9 kcal. Taking into account the variation of a_6 with temperature, we get $E \sim 46$ kcal.

In this entire analysis, the wall termination of HO_2 radicals has been neglected. Changes in the results obtained, due to inclusion of a step 7) $H_2O \xrightarrow{\text{wall}}$, will now be examined. The rate of this wall process is $a_7(HO_2)$. With this new reaction, the value of (OH) becomes:

$$(OH) = \frac{2n_0 + 2a_3(CH_2O)}{a_6\left[1 + \frac{a_7}{a_6}\frac{k_2'}{k_5}\frac{(CH_2O)}{(CH_4)}\right]} = \frac{(OH)^x}{1 + \gamma} \tag{20}$$

where

---

* The activation energy E for methane oxidation differs widely from one paper to another. This is probably due to poor reproducibility. It also depends strongly on wall conditions and mixture composition. According to Walsh [34], $E = 40$ kcal at $T = 475°$ for the mixture $2CH_4 : O_2$.

Enikopolyan took special precautions to get reproducible results. His data give $E = 44$ to 46 kcal. This is for a stoichiometric mixture in a HF washed reactor after a long series of runs. Complete reproducibility was achieved.

$$\gamma = \frac{a_7}{a_6} \cdot \frac{k_2'}{k_5} \frac{(CH_2O)}{(CH_4)}$$

and

$$(OH)^x = \frac{2n_0 + 2a_3(CH_2O)}{a_6} .$$

Thus $(OH)^x$ is identical to the value found above in the absence of step 7). Nalbandyan and Voevodskiĭ [27] have calculated that, at temperatures above 400°C, in reactors that were not pre-treated in any special way, the probability of wall destruction is 10^3 times smaller for a radical HO_2 than for a radical OH, i.e.,

$$\frac{a_7}{a_6} \sim 10^{-3} .$$

Since

$$\frac{(CH_2O)_{max.}}{(CH_4)} = \frac{0.3}{78} = 3.8 \cdot 10^{-3}, \qquad \gamma = 3.8 \cdot 10^{-6} \frac{k_2'}{k_5} .$$

At 423°C:

$$\frac{k_2'}{k_5} = \frac{f_2'}{f_5} \cdot 1.9 \cdot 10^5 .$$

Then

$$\gamma_{423} = \frac{k_2'}{k_5} \cdot 3.8 \cdot 10^{-6} = 1.9 \cdot 10^5 \cdot 3.8 \cdot 10^{-6} \frac{f_2'}{f_5} = 0.72 \frac{f_2'}{f_5} .$$

Consequently, it appears that if the steric factors f_2' and f_5 were identical, the concentration (OH) would change by no more than 70% following inclusion of step 7). It would remain of the same order of magnitude. But there are reasons to believe that the steric factor for a radical-aldehyde reaction is smaller than for a radical-aldehyde reaction. As is well known, the activation energy is also smaller for the radical-aldehyde reaction [26, 35]. Therefore, γ should be substantially smaller than 0.72 and it is permitted to ignore wall destruction of HO_2 radicals, in first approximation.

In order to confirm the validity of the postulated scheme, it would be essential to show that aldehyde oxidation in the absence of hydrocarbons proceeds via the steps postulated in the scheme. Also, the proposed constants and the homogeneous character of the branching step $a_3(CH_2O)$ should be checked. This reaction is the chain initiation step in oxidation of pure aldehyde. Unfortunately, data on aldehyde oxidation are scarce, contradictory and they pertain to much higher pressures of CH_2O than the ones encountered in the oxidation of methane. Moreover, aldehyde oxidation has usually been studied at temperatures lower than in hydrocarbon oxidation. Data give conflicting values of the activation energy, between 17 and 39 kcal in the oxidation of formaldehyde.

The theoretical scheme of methane oxidation has given values of

$$(CH_2O)_{max.} \quad \text{and} \quad \left(\frac{d(CH_4)}{dt}\right)_{max.}$$

in good agreement with experiment, as well as the dependence of these quantities on temperature and CH_4 and O_2 concentrations. All this was done without taking into account the exhaustion of reactants. In order to compare reaction rates over the entire range of conversion to experimental values, it is necessary to integrate equation (14). If, at time $t = 0$, $(CH_2O) = 0$ (in a mixture without addition of CH_2O), the expression for (CH_2O) as a function of time is the following:

$$\xi = \frac{e^{\alpha\tau}\sqrt{e^{2\alpha\tau} + \alpha^2} - \alpha}{e^{2\alpha\tau} + \alpha^2} \tag{21}$$

where

$$\xi = \left(\frac{k_2'k_5'}{k_2k_5}\right)^{1/2} \frac{(CH_2O)}{(CH_4)}$$

and

$$\alpha = \frac{k_3(O_2)}{n_0} \frac{(CH_4)}{\left[\dfrac{k_2'k_5'}{k_2k_5}\right]^{1/2}}, \qquad \tau = \frac{2k_2n_0}{k_6}\left[\frac{k_2'k_5'}{k_2k_5}\right]^{1/2} t .$$

Curve 1 on Figure 67, on the following page, is drawn according to equation (21).

One point, on this curve, designated by A, has been chosen to determine α or, which is the same thing, a value of n_0. Other points are experimental data.

If, at $t = 0$, $(CH_2O) \neq 0$, integration of equation (14) gives:

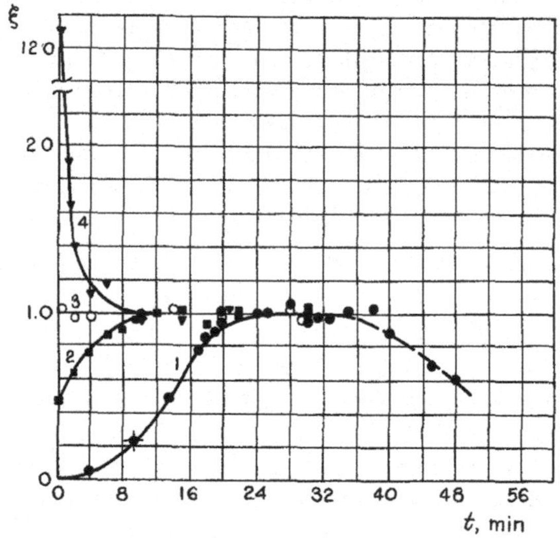

FIGURE 67. Relation between ξ and t. Continuous curve (1) calculated from equation (21). The points are experimental. Curves 2, 3 and 4 calculated from equation (22). Reference [30]

$$\xi = \frac{e^{2\alpha\tau}\sqrt{\theta_0 e^{2\alpha\tau} + \alpha^2\theta_0 - \theta_0} - \alpha}{\theta_0 e^{2\alpha\tau} + \alpha^2} \tag{22}$$

where

$$\theta_0 = \frac{(1+\alpha\,\xi)^2}{1 - \xi^2}$$

and ξ_0 is the dimensionless concentration of formaldehyde added to the original mixture.

Curves 2, 3, 4 on Figure 67 were drawn according to equation (22), points are experimental data.

The rate of disappearance of methane in dimensionless variables is:

$$\frac{d\zeta}{d\tau} = 1 + \alpha\xi \; ^{*} \tag{23}$$

* In (23, the value of ξ as a function of time was obtained from the equation

$$\frac{d\xi}{d\tau} = (1 + \alpha\xi)(1 - \xi) \quad . \tag{24}$$

where

$$\zeta = \frac{p_0 - p_{CH_4}}{p_0\left(\frac{k_2k_5}{k_2'k_5'}\right)^{1/2}}$$

(with p_0: the initial concentration of methane).

Curves on Figure 68 represent the change of methane concentration with time for different amounts of CH_2O added to the mixture. Points are experimental.**

In spite of the good agreement between experimental and calculated values, it must be noted that the theoretical curve taking into account consumption of reactants starts to fall off at large reaction times corresponding to the maximum rate. On the other hand, experimental curves for methane disappearance and formaldehyde accumulation, retain a constant value too long (Figure 69, on the following page). This might be explained by an unknown catalytic effect of water vapor which would compensate the decrease in rate

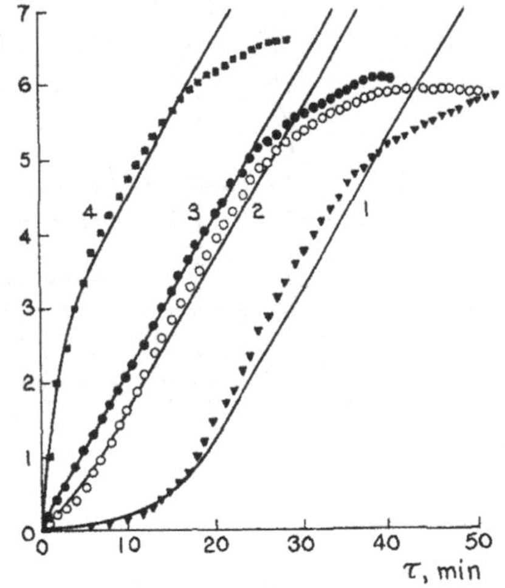

FIGURE 68. Theoretical curves of methane disappearance in oxidation of mixtures $CH_4 : O_2$ = 1 : 2, containing various amounts of CH_2O [4] 1) none; 2) 0.5; 3) 12.6; 4) 12.6. T = 423°C, p_0 = 235 mm Hg. Points are experimental [30].

---

* (cont.

The theoretical expression for $\frac{d\xi}{d\tau}$ was

$$\frac{d\xi}{d\tau} = (1 + \alpha\xi)(1 - \xi)^2 \quad . \tag{25}$$

Since $0 < \xi < 1$, equation (25) may be rewritten as:

$$\frac{d\xi}{d\tau} = \frac{3}{2}(1 + \alpha\xi)(1 - \xi) \quad . \tag{26}$$

Equation (26) would necessitate a small correction to the initial stages of curves 1 and 2, Figure 67.

** For long times of reaction, experimental and theoretical values diverge since the latter were calculated without taking into account consumption of reactants.

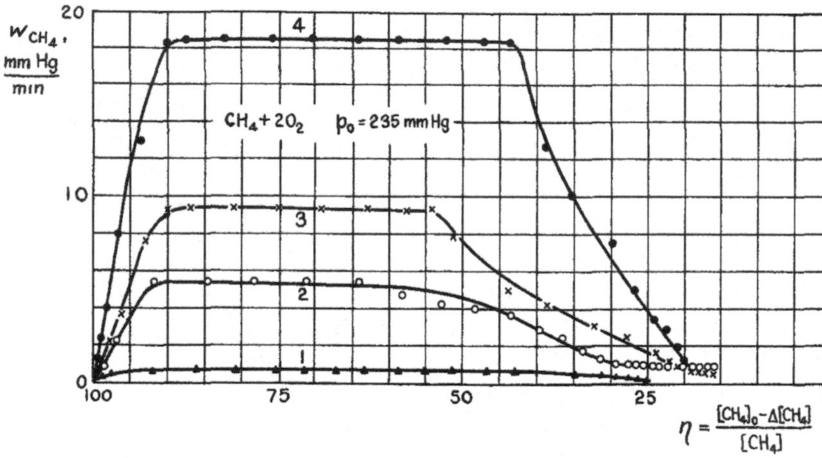

FIGURE 69. Relation between w and η for a mixture $CH_4 + 2O_2$; $p = 235$ mm Hg, $T = 423$°C.

due to methane disappearance. It is a little hard to see why the compensation should be so exact, or why the experimental values would stay so constant while the decrease in rate is a cubic function of conversion. This constancy is also observed in other oxidation reactions (CH_2O, CH_3CHO, acetylene and ethylene). It is known in the case of CH_2O oxidation that water does not accelerate the process.

I think that this curious effect deserves a thorough study since its explanation might lead to new concepts in chemical kinetics. The mechanism of methane oxidation considered here cannot be applied to large conversions until this phenomenon is understood.

Nevertheless, in the initial stages of reaction, the mechanism of methane oxidation may be considered as well established on the whole, although some obscurities remain unexplained.

This was used as an example to illustrate the application of chain theory to reactions with degenerate branching.

§3. Kinetics of Oxidation of Hydrogen Sulfide

The slow oxidation of hydrogen sulfide belongs to the class of reactions with degenerate branching. It was studied by Emanuel' [36] between 250 and 300°C. The kinetic curves (amount reacted as a function of time) measured by pressure change have an S-shape with a definite induction period and an inflection point at 10 to 15% conversion [36]. Since final products — SO_2 and water — are practically without effect on the kinetics of the process, the self-acceleration cannot be explained by a catalytic action of final products. On the other hand, two ignition limits, even though within a small temperature range (350 to 400°C) establish with certitude the chain character of the oxidation of hydrogen sulfide.

Self-acceleration in chain reactions is usually due to chain branching. In this case, the acceleration is much too slow to be due to branching caused by active free radicals. Obviously, a relatively stable intermediate is formed in the system and it is able to produce free radicals, i.e., to cause degenerate branching. Nevertheless, the oxidation of hydrogen sulfide is relatively fast (reaction is complete in two minutes) at low temperatures. It must be assumed, therefore, that the intermediate is more active and consequently less stable than aldehydes or even peroxides in hydrocarbon oxidation.

To investigate the kinetic characteristics of the intermediate that catalyzes the oxidation of hydrogen sulfide, we developed together with N. M. Emanuel' [37, 38] and applied a kinetic method.

The basic idea of this kinetic method is that the kinetics of the process under standard conditions is used to estimate the concentration of intermediate. The natural assumption is made at a given pressure, temperature and mixture composition, the kinetic behavior of the system is determined uniquely by the concentration of intermediate.

The apparatus used in the application of this kinetic method to hydrogen sulfide oxidation, consisted of three reactors. At any time, the mixture could be transferred from one reactor to another, either directly or through the third one that was kept at room temperature.

Preliminary experiments showed that a mixture kept in the first reactor during a time to the induction period, would react, after transfer to the second reactor, at a rate corresponding to the pressure and temperature prevailing there. This means that the intermediate fully survives the transfer.

If a mixture is transferred from the first reactor to the intermediate reactor and kept there for various lengths of time, the rate of destruction of the intermediate at cold temperatures may be determined. As it turned out, the intermediate is relatively stable and disappears completely only after several hours.

A mixture submitted to preliminary reaction may be introduced in the intermediate reactor where it is diluted with a fresh mixture of hydrogen sulfide and oxygen. The concentration of intermediate is thus diminished when the new mixture is transferred to the second reactor, all other conditions remaining equal. As a result of the diminished concentration of intermediate, reaction in the second reactor does not start at once but after a certain time. This time is longer when the dilution of the prereacted mixture is more pronounced. There is a one to one correspondence between the induction period in the second reactor and the degree of dilution, i.e., the relative concentration of intermediate. Conversely, therefore, from a measurement of induction period in the second reactor, one can determine a relative concentration of intermediate in the mixture at the moment of its transfer. Unit concentration is defined as that which corresponds to the end of induction period. After the end of the induction period, at the beginning of the reaction period, the concentration of intermediate is larger than unity and in order to measure it by the kinetic method, it is necessary to dilute the mixture under study in the intermediate reactor.

By means of the kinetic method, the concentration of intermediate during hydrogen sulfide oxidation was determined (Figure 70). In agreement with the requirements of chain theory, the concentration of intermediate increases exponentially at first. Subsequently, it reaches a maximum and falls off. The time corresponding to the maximum of the kinetic curve of the intermediate, coincides with the inflection point on the kinetic curve obtained by measuring pressure change in the system.

The kinetic method for the study of intermediates, even when their nature is not known, yields valuable information on several of their important characteristics, such as the kinetics of their accumulation, their decomposition at low temperatures, their reaction with other substances. Emanuel' has found that the decomposition of the intermediate formed during hydrogen sulfide oxidation follows a first order rate law.

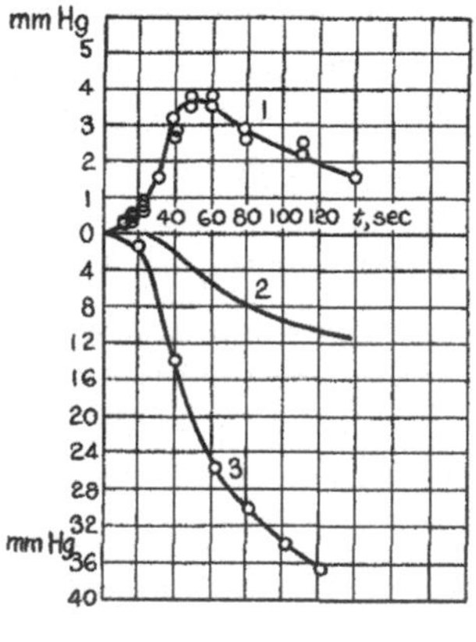

FIGURE 70. Kinetic curves of accumulation of (SO_2): Curve (1), according to kinetic data. Pressure drop: Curve (2). Decrease of partial pressure of O_2: Curve (3) calculated by means of the relation: $\Delta p_{O_2} = 3\Delta p + p_{SO}$.

Oxidation of a mixture: 40 mm Hg H_2S + 60 mm Hg O_2. T = 270°C [38].

He also studied the reaction of the intermediate with water vapor. In the presence of a definite amount of water in the intermediate reactor, the active intermediate decomposes much more rapidly, in a matter of 10 or 15 minutes instead of hours in the absence of water.

The rate of disappearance of sulfur monoxide reacting with water, obeys the equation:

$$-\frac{dc}{dt} = kc^{1.5}\gamma$$

where c is the relative concentration of sulfur monoxide and γ is a dimensionless water concentration. The reaction is obviously a complex one. As the temperature becomes higher, the rate of this reaction decreases so that its temperature coefficient is negative.

The kinetic method has later been applied successfully to the study of intermediates in the oxidation of acetaldehyde [39], propylene [40, 41], isobutane [41], propane [42], and also, in part, in a study of the mechanism of propane oxidation catalyzed by hydrogen bromide [43] as will be explained in the last section of this Chapter.

However, by itself the kinetic method is unable to lead to identification of the intermediate. Such identification was achieved by N. M. Emanuel' [36] by means of absorption spectra of reacting mixtures $H_2S + O_2$. Emanuel' quenched samples of the mixture, taken during oxidation and observed bands characteristic of the spectrum of sulfur monoxide as it appears in a discharge [44]. He also showed that the kinetic curves of accumulation of sulfur monoxide as determined by the change in intensity of its absorption bands during oxidation, coincide with the curves of accumulation of active intermediate obtained by the kinetic method. It was thus established that sulfur monoxide is the intermediate responsible for the autocatalytic character of hydrogen sulfide oxidation.

This conclusion was confirmed in a study of $H_2S + O_2$ mixtures to which was added sulfur monoxide produced by an electric discharge in a mixture of SO_2 and sulfur vapors. The reaction rate in an artificial mixture of this kind containing a certain amount of sulfur monoxide is equal to that observed in an ordinary $H_2S + O_2$ mixture in which sulfur monoxide has reached the same concentration.

In the same series of investigations [36], it was found that SO exists as a monomer only at temperatures around 250-300°C. Upon cooling, dimerization to S_2O_2 occurs. This was shown by measuring the decrease in pressure in a mixture containing SO cooled to room temperature. Thus also, the absorption spectra at room temperature, must be assigned to and not to SO.* The dimerization of SO at low temperatures indicates that it behaves like a diradical. But it must be a relatively stable one, not like oxygèn atoms responsible for branching in the oxidation of hydrogen. Its stability can be judged from the fact that it accumulates in considerable quantities at the maximum of the rate. At that moment, a pressure of 10 mm Hg of SO has been observed, i.e., 20% of the original H_2S or 75% of the H_2S reacted. At the same time it is considerably less stable than aldehydes and hydroperoxides which are responsible for degenerate branching in hydrocarbon oxidation. Therefore, hydrogen sulfide oxidation may be regarded as an intermediate case between pure branched processes and processes with degenerate branching.

Prediction of definite properties of the active particle SO is not yet possible because there are no data on the bond energy of SO, nor on the heat of the reaction $SO + O_2$.

The kinetic method together with spectroscopic studies have nevertheless established the kinetics of accumulation of SO during the course of the reaction.

Figure 70 [36] shows the kinetics of accumulation of SO during the oxidation of H_2S at 270°C (curve 1). Curve 2 on the same Figure represents pressure change of the mixture. It can be seen that SO buildup starts in the very first seconds of the process without any detectable induction period. This does not appear on the kinetic curves of pressure change since the process $H_2S + O_2 \longrightarrow H_2O + SO$ occurs without a pressure change. The end of the induction period measured on the pressure curves must therefore correspond to the appearance of SO_2, the final oxidation product. Knowing Δp and the SO pressure, we may calculate oxygen consumption $\Delta p_{O_2} = 3\Delta p + p_{SO}$. The consumption of oxygen calculated in this way is shown by curve 3.

The mechanism of hydrogen sulfide oxidation is unknown. The experiments of Emanuel' lead only to the conclusion that SO is a long lived radical responsible for branching. One of the possible schemes that can be postulated is the following:

0) $H_2S + O_2 \longrightarrow HS + HO_2$

1) $HS + O_2 \longrightarrow SO + OH$

---

* A recent mass-spectrometric study of the product obtained in a discharge in S vapors + SO_2 has revealed masses 80 (S_2O) and 64 (SO_2) but not 96 (S_2O_2). It was concluded that S_2O and SO_2 are present but not SO. However, as was noted already by the authors of this work, this may be due to the transformation of $S_2O_2^+$ into S_2O^+ in the ionization chamber.

2) $OH + H_2S \longrightarrow H_2O + HS$

2') $OH + SO \longrightarrow SO_2 + H$

3) $SO + O_2 \longrightarrow SO_2 + O$

3') $O + H_2S \longrightarrow HS + OH$

4) $H + H_2S \longrightarrow H_2 + HS$

4') $H + SO \longrightarrow HSO$

5) $HSO + O_2 \longrightarrow SO_2 + HS$

6) $HS \longrightarrow$ termination.

As in the mechanism of methane oxidation, we have included in the scheme a reaction between the active intermediate SO and active centers (steps 2' and 4') leading to the oxidation of SO to SO_2.

Unlike the oxidation of methane where the limiting step is the reaction of a radical with the hydrocarbon and where the rate depends to the square of the methane concentration, we assume here that reaction 1) is the slowest one. The rate is then proportional to the square of the oxygen concentration, as observed. This assumption is based on the fact that in spite of its exothermicity, step 1) must proceed through a four-center activated complex requiring, it would seem, an appreciable activation barrier. Since step 1) is the slowest, chain termination is determined by the destruction of HS radicals.

As can be seen, degenerate branching is due to step 3), a reaction between SO and oxygen. The diradical O thus formed, as in the $H_2 + O_2$ reaction, is rapidly replaced by two monoradicals produced in 3').

This mechanism gives for the steady-state concentration of HS radicals:

$$(HS) = \frac{1}{k_6} \cdot \left[2n_0 + 2k_3(O_2)(SO) \right] \tag{27}$$

where n_0 is the rate of spontaneous generation of active centers.

If $a_4' > a_4$, the rates of O_2 disappearance and SO accumulation are respectively:

$$-\frac{d(O_2)}{dt} = \frac{2k_1(O_2)}{k_6} \left[1 + \frac{k_2'}{k_2} \frac{(SO)}{(H_2S)} \right] \left[n_0 + k_3(O_2)(SO) \right] \tag{28}$$

and

$$\frac{d(SO)}{dt} = \frac{2k_1(O_2)}{k_6} \left[1 - \frac{k_2'}{k_2} \frac{(SO)}{(H_2S)} \right] \left[n_0 + k_3(O_2)(SO) \right] . \tag{29}$$

Since the maximum concentration of SO follows from the condition

$$\frac{d(SO)}{dt} = 0,$$

it is determined by means of equation (29):

$$(SO)_{max.} \approx \frac{k_2}{k_2'} (H_2S) \quad . \tag{30}$$

In fact, Emanuel' has shown that the maximum concentration of SO is related linearly to the total pressure of the mixture.

Some time after the start of reaction, when a sufficient quantity of SO has been formed, the formation of active centers is essentially determined by step 3). The quantity n_0 may then be neglected as compared to $k_3(O_2)(SO)$.

With new variables:

$$\xi = \frac{[SO]}{[SO]_0} \quad \text{and} \quad \eta' = \frac{[O_2]_0 - [O_2]}{[O_2]_0}$$

equations (28) and (29) take the form:

$$\frac{d\xi}{dt} = C\xi(1 - \eta')^2 - B\xi^2(1 - \eta') \tag{31}$$

and

$$\frac{d\eta'}{dt} = C\xi(1 - \eta')^2 + B\xi^2(1 - \eta') \tag{32}$$

with

$$C = \frac{2k_2k_3}{k_6} [O_2]_0^2 \quad \text{and} \quad B = \frac{4k_2'k_1k_3}{k_2k_6} [O_2]_0^2 \quad .$$

Equations (31) and (32) coincide with the equations derived by Emanuel' [36, 38] by means of his reaction scheme. In this scheme, Emanuel' did not consider the chain disappearance of SO but assumed that the oxidation of H_2S followed a chain mechanism only up to the appearance of SO. Subsequently then, SO would disappear in a triple collision with O_2: $SO + SO + O_2 \longrightarrow 2SO_2$. This is analogous to the postulated scheme of Norrish in methane oxidation. Truly, this quadratic disappearance of SO permits Emanuel' to obtain equations similar to (31) and (32). But in both examples, the intermediate reacts in chain fashion.

As Emanuel' has shown, these equations fit relatively well the curves of SO accumulation and O_2 disappearance as a function of time and conversion.

After the maximum concentration has been reached

$$\left(\frac{d(SO)}{dt} = 0\right) \quad ,$$

the reaction rate for O_2 disappearance is given by:

$$-\frac{d(O_2)}{dt} = \frac{2k_1k_3}{k_6} \cdot \frac{k_2}{k_2'} (O_2)^2(H_2S) \quad . \tag{33}$$

This equation may be rewritten in the form:

$$\frac{d\eta}{dt} = k_W \frac{p_0^2}{(1+\mu)^2} \left(1 - \frac{3}{2}\mu\eta\right)^2 (1 - \eta) \tag{34}$$

where

$$k_W = \frac{2k_1k_2k_3}{k_6k_2'}$$

the effective reaction rate constant,

$$\eta = \frac{\Delta p}{\Delta p_\infty}$$

with Δp, the pressure change at time t, and Δp_∞ the final pressure change;

$$\mu = \frac{[H_2S]_0}{[O_2]_0} \quad ,$$

p_0 is the total initial pressure.

Equation (34) corresponds reasonably well to the empirical equations used by Emanuel' to represent the kinetics of oxidation of H_2S after the maximum of the rate. The main discrepancy consists in the fact that the initial pressure appears to the second power in equation (34) while it appears linearly in Emanuel's empirical equation. An analysis of Emanuel's kinetic curve reveals, however, that the exponent of p_0 lies between one and two and is close to two.

With the condition $(SO) = 0$ at $t = 0$, the simultaneous solution of equations (28) and (29) gives:

$$[SO] = \frac{[H_2S]}{\alpha}\left[1 - \left(\frac{[H_2S]}{[H_2S]_0}\right)^2\right] \tag{35}$$

where

$$\alpha = \frac{k_2'}{k_2} .$$

Figure 71 shows a curve calculated by means of equation (35). The points are Emanuel's data. It is seen that the agreement is quite satisfactory.

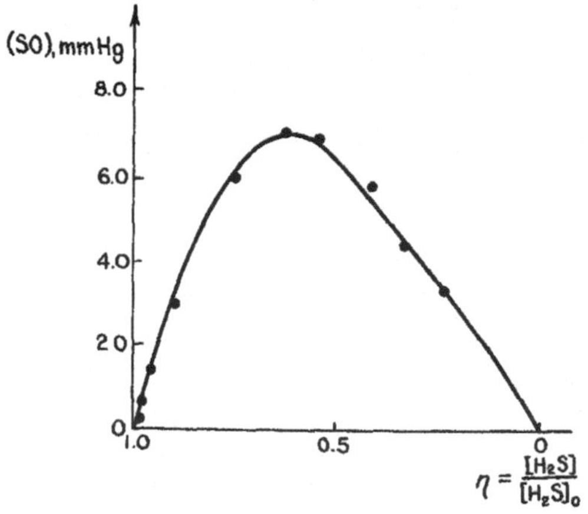

FIGURE 71. Relation between (SO) and $\eta = \frac{[H_2S]}{[H_2S]_0}$.

Curve calculated from equation (35). Points are experimental. T = 270°C; p = 100 mm Hg, $\mu = 2/3$, $\alpha = 3.64$. [36, 38]

§4. Kinetics of Liquid Phase Oxidation of Hydrocarbons

Liquid phase oxidation of hydrocarbons proceeds entirely via the formation of hydroperoxides which are the first stable intermediate products. A large number of olefinic and alkylaromatic hydrocarbons have been studied and it appears that the unique oxidation product at not too high temperature and moderate conversions is the corresponding hydroperoxide [46-51]. In the case of saturated compounds, even in the early stages of the reaction, other oxidation products appear besides the hydroperoxides: alcohols, carbonyl compounds, acids. But here also, the first oxidation product is the hydroperoxide and all other compounds are due to its further transformation. This has been shown by N. M. Emanuel' et al. for the oxidation of cyclohexane [52] and n-decane [53] by a comparison of the rate of decomposition of the hydroperoxide and the rate of formation of the remaining products.

In the case of cyclohexane, the rate of formation of cyclohexanol and cyclohexanone (except the fraction of the latter which is formed by the oxidation of cyclohexanol) was determined by tracer techniques and found to be proportional to the concentration of cyclohexylhydroperoxide. Thus both products of controlled oxidation are produced directly from the hydroperoxide. In the case of decane, a special kinetic method was used. Decane oxidation was interrupted at a given stage, oxygen was carefully removed from the system and the rate of decomposition of the hydroperoxide was determined in a nitrogen atmosphere. In this fashion, the rate of hydroperoxide disappearance could be measured at any moment during the oxidation of n-decane. This rate was found to be almost completely equal to the rate of formation of all stable products, during the entire course of the process. Thus, all these products are formed as a result of the decomposition of the hydroperoxide.

Liquid phase oxidation is considerably easier than the gas phase oxidation of the same molecule. Apparently, this is due to a number of factors: the high concentration of the oxidized material, the less pronounced rate of termination since radicals have a harder time diffusing to the walls, favorable conditions permitting the accumulation of hydroperoxide since its heterogeneous decomposition is minimized.

Liquid phase oxidation of hydrocarbons is accelerated by addition of compounds dissociating readily into free radicals — peroxides [46, 47, 54-58], azoisobutyric acid dinitrile [59], lead tetraacetate [60], benzoyldiazoacetate [61, 62]. This shows the chain nature of the process and in particular of its first stage, the formation of the hydroperoxide. Further proofs of the chain mechanism of oxidation are quantum yields much higher than unity in photochemical oxidation [63] and the strong inhibiting effect of small quantities of several compounds, especially phenols and aromatic amines [64]. A common mechanism of chain propagation in agreement with all observed kinetic relations, consists in the alternation of two elementary steps:

1. $R + O_2 \longrightarrow RO_2$
2. $RO_2 + RH \longrightarrow ROOH + R$

where RH is the original hydrocarbon.

<u>Rates of Chain Propagation and Termination</u>. In the liquid phase, the diffusion of free radicals is difficult and linear termination

of chains at the wall, as in gas phase reactions, is not expected. The most natural termination is by recombination of radicals RO_2 and R. This is indeed the case in the photochemical oxidation of tetraline [65] and the oxidation of ethyllinoleate, initiated by benzoyl peroxide [66]. In both examples, the oxidation rate is proportional to the square root of the rate of initiation, namely in the first example to the square root of light intensity and in the second example to the square root of the concentration of benzoyl peroxide. (The latter breaks up monomolecularly along the O - O bond to give two free radicals.)

This demonstrates the quadratic nature of chain termination in oxidation.

When the oxygen pressure is high enough, termination occurs practically exclusively by recombination of the RO_2 radicals. Let n_0 be the rate of initiation, k_1 and k_2 the rate constants of steps 1) and 2), k_3 the rate constant for recombination of RO_2 radicals. With the steady state assumption, the rate w of oxidation is:

$$w = \frac{k_2}{\sqrt{k_3}} [RH] \sqrt{n_0} \quad . \tag{36}$$

If the initiation rate is known and the rate of oxidation measured, the ratio $k_2/\sqrt{k_3}$ can be determined. This ratio, of course, should not depend on the initiating agent but only on the nature of the hydrocarbon.

The temperature dependence of the ratio gives the difference

$$E_2 - \frac{1}{2} E_3 \quad .$$

Since E_3 is the activation energy for recombination and should be close to zero, the activation energy E_2 for chain propagation can thus be determined. In this way, activation energies have been obtained for a large variety of hydrocarbons [55, 63, 66-71]. It is interesting to compare these activation energies to those calculated from the formula (see Chapter I):

$$\epsilon_0 = 11.5 - 0.25 \; |q| \quad . \tag{37}$$

The heat of reaction, in chain propagation can be estimated from two assumptions: the O - H bond energy of the hydroperoxide is 90 kcal, i.e., equal to the O - H bond energy in hydrogen peroxide and the C - H bond energy depends only on the class of the hydrocarbon considered. In first approximation, we will take for q_{C-H} the C - H bond energy in α position to a double bond, the figure of 77 kcal for all olefins, i.e., the same value as in propylene. We take $q_{C-H} = 75$ kcal for all alkylaromatics by analogy with ethylbenzene and 94 kcal for paraffins, as in propane. For a C - H bond in a CH_2 group in α position to two double bonds there are no data but a value smaller than that for a single double bond is expected. We shall take 65 kcal.

Table 63, on the following page, contains experimental and calculated [equation (37)] values of E_2 for a series of hydrocarbons. It must be kept in mind that the heat of reaction for $RO_2 + RH$ could be estimated only very roughly. Indeed, the heat of reaction is determined not only by the position of the C - H bond with respect to the double bond or the aromatic ring but also by other structural details. Therefore the calculated values of E_2 must be compared to the average values of E_2 measured for hydrocarbons of four different types. Then the agreement is not too bad and it is particularly important to note that the trends are the same for the calculated values and the average experimental values of E_2.

Kinetic data are not capable of giving a separate value for both k_2 and k_3. This gap can be filled by photochemical data. The rotating sector method permits us to evaluate the rate constant for recombination of RO_2 radicals. Then k_2 can be obtained since the ratio $k_2/\sqrt{k_3}$ follows from kinetic data.

Absolute values of rate constants can also be obtained by a method using photochemical after-effects. Here the after-effect is measured, i.e., the quantity of oxygen consumed in the oxidation process after switching off the light, i.e., after cessation of initiation:

$$\Delta[O_2] = \int_0^\infty w\,dt \quad .$$

Strictly speaking, because of quadratic termination, the process should continue till exhaustion of the hydrocarbon since, as the radical concentration goes down, the chain length increases without limit. In reality, of course, when the concentration of radicals is quite small, linear chain termination starts to operate by diffusion of the radicals to the wall or by their reaction with traces of any inhibitor present in the system. The measurement of $\Delta[O_2]$ must therefore be continued until a certain rate w_1 is reached, at the limit of experimental detection. The rate of oxidation:

$$w = k_2[RH][RO_2]$$

$$\frac{d[RO_2]}{dt} = -k_3[RO_2]^2 \qquad (38)$$

TABLE 63

Activation energy for chain propagation: $RO_2 + RH$ in oxidation of hydrocarbons and their derivatives [72].

Hydrocarbon	E_2 Experimental		E_2 Calculated	Reference
Ethyllinoleate	5.0	Average 5.0	5.25	[68]
Ethyllinoleate	4.2			[66]
	5.0			[67]
	5.7			[63]
Tetraline	4.3	Average 6.8	7.75	[73]
	9.3			[68]
Squalene	6.8	Average 8.9	8.25	[70]
Octene-1	7			[69]
	11.3			[68]
Hexadecene-1	12.1			[68]
4-methylpentene-3	8.1			[68]
2,2,4-trimethylpentene-1	12.5			[68]
Cyclohexene	8.5			[68]
1-methylcyclohexene	8.0			[68]
1,2-dimethylcyclohexene	7.5			[68]
1,3,5-trimethylcyclohexene	7.0			[68]
Methyloleate	8.0			[55]
Decane	11.5		14.5	[71]

At $t = 0$:

$$[RO_2] = [RO_2]_L = \frac{w_L}{k_2[RH]}$$

where w_L is the rate of photochemical oxidation. Hence:

$$\Delta[O_2] = \int_0^{t_1} k_2[RH] \frac{dt}{\dfrac{1}{[RO_2]_L} + k_3 t} = \frac{k_2[RH]}{k_3} \ln \frac{1/[RO_2]_1}{1/[RO_2]_L}$$

where $[RO_2]_1$ and t_1 are radical concentration and time corresponding to rate w_1 at which measurements are terminated. Finally:

$$\Delta[O_2] = \frac{k_2[RH]}{k_3} \ln \frac{w_L}{w_1} \quad . \tag{39}$$

A measurement of the after-effect gives therefore the ratio k_2/k_3. Together with $k_2/\sqrt{k_3}$ determined from kinetic data, it gives k_2 and k_3 separately. It is also necessary to take into account the initiation in the dark and thus the oxidation in the dark, proceeding at a rate w_D. Thus:

$$\Delta[O_2] = \int_0^\infty (w - w_D)\, dt \quad .$$

After integration:

$$\Delta[O_2] = \frac{k_2}{k_3} [RH] \, \ell n \, \frac{w_D + w_L}{2w_D} \quad .$$

The ratio k_2/k_3 can also be calculated from this expression.

Both techniques — the rotating sector and the photochemical after-effect — have been used to determine absolute values of the rate constants k_2 and k_3 [65, 69, 73, 74]. Together with the measured values of E_2, this now permits us to write k_2 as an Arrhenius function of temperature.

Values of k_2 and k_3 for several hydrocarbons have been collected in Table 64. The striking feature there is that the pre-exponential factors are 5 to 7 orders of magnitude smaller than the collision factors. In the gas phase, the rate constants for recombination of radicals are close to 10^{-10}. This peculiar feature of liquid phase oxidation has not received any adequate explanation. Maybe this is related to the so-called cage effect.

TABLE 64

Rate constants for chain propagation k_2 and radical recombination

Hydrocarbon	k_2 cm^3/sec.	k_3 cm^3/sec.	Reference
Octene-1	$10^{-17}\ e^{-7300/RT}$	5.10^{-16}	[69]
Cyclohexene	$2.8.10^{-15}\ e^{-8500/RT}$	$1.6.10^{-15}$	[73]
1-Methylcyclohexene	$2.3.10^{-15}\ e^{-8000/RT}$	8.10^{-16}	[73]
Ethyllinoleate	$2.8.10^{-16}\ e^{-5700/RT}$	5.10^{-16}	[69]
Tetraline	$4.5.10^{-17}\ e^{-4300/RT}$	$6.8.10^{-14}$	[68]

Chain Initiation and Degenerate Branching. The characteristic feature of both gas and liquid phase oxidation of hydrocarbons is the autoacceleration of the reaction. This is quite natural since hydroperoxides accumulate during the liquid phase oxidation of hydrocarbons. But peroxide compounds have the ability to initiate the process. In this sense, liquid phase oxidation of hydrocarbons is a process with degenerate chain branching and the branching agents are the hydroperoxides.

In many investigations, a direct proportionality has been observed between the reaction rate and the concentration of accumulated hydroperoxide [55, 75-77]. When benzoyl peroxide is added from the very start of the reaction, the rate is proportional to the square root of the benzoyl peroxide concentration. On the other hand, the rate, in many cases, depends linearly on the hydroperoxide formed during the reaction itself. With quadratic branching, this indicates that the rate of initiation is proportional to the square of the concentration of hydroperoxide.

According to Bateman [76], at not too small concentrations of hydroperoxide, its decomposition into free radicals occurs primarily not by a unimolecular but by a bimolecular process:

$$2ROOH \longrightarrow RO_2 + RO + H_2O \quad .$$

With an O - O bond energy of $\sim$ 40 kcal and an O - H bond energy of 90 kcal, this bimolecular reaction necessitates an expenditure of $\sim$ 14 kcal, versus 40 kcal required to break to O - O bond unimolecularly. Such a process may be made easier by the association of two hydroperoxide molecules, hydrogen bonded as follows:

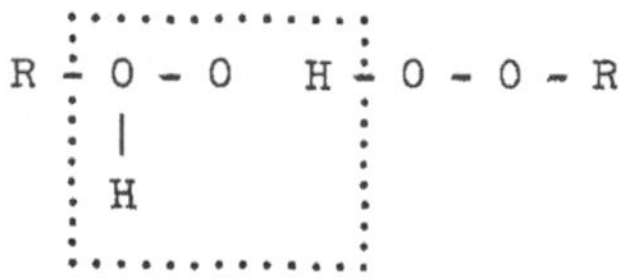

This picture receives support in the fact that in the presence of molecules able to join to hydroperoxide molecules by means of a hydrogen bond and thus opposing complexing between two hydroperoxide molecules, the rate of oxidation is proportional to the square root of the hydroperoxide concentration. The decomposition of hydroperoxide must then be unimolecular.

At small hydroperoxide concentrations, the number of binary collisions decreases and thus also the concentration of dimers so that the bimolecular reaction ought to be quite slow. Then a first order process with respect to the hydroperoxide become important. This explains why, in the initial stages of the oxidation, w is observed to depend linearly on (ROOH).

It is commonly believed that this process is a unimolecular decomposition of the hydroperoxide along the O - O bond. With a few hydrocarbons, this reaction has been studied directly by measuring the rate of hydroperoxide disappearance in a nitrogen atmosphere, i.e., under conditions precluding its formation. Since free radicals are formed following hydroperoxide decomposition, an induced decomposition of the hydroperoxide may take place. Therefore, the measured rate constants are extrapolated to infinite dilution. The limiting value is then taken as the true rate constant. Values of these rate constants are collected in Table 65.

TABLE 65

Rate constants for hydroperoxide decomposition reactions in dilute solutions

Hydroperoxide	Rate Constant	Reference
Cumyl	$2.7 \cdot 10^{12}\ e^{-30400/RT}$	[79]
Sec.-decyl	$0.8 \cdot 10^{12}\ e^{-31700/RT}$	[80]
Cyclohexyl	$1.2 \cdot 10^{13}\ e^{-34000/RT}$	[81]
1,4 Dimethylcyclohexyl	$1.6 \cdot 10^{13}\ e^{-32800/RT}$	[82]
Decalyl	$8.5 \cdot 10^{13}\ e^{-32100/RT}$	[83]
Tert.-butyl	$1.2 \cdot 10^{15}\ e^{-39000/RT}$	[84]

Activation energies should correspond to O - O bond energies. These data would indicate that the O - O bond energy is about 30 to 34 kcal (to the exception of tert.-butylhydroperoxide). But such a low value appears improbable since the O - O bond energy in hydrogen peroxide is 48 kcal. It seems more likely that, in the liquid phase, a bimolecular reaction takes place between hydroperoxide and solvent:

$$R - O \vdots \underset{\displaystyle H}{\overset{}{O}} \quad + \quad H \vdots R \longrightarrow RO + H_2O + R$$

Even if the O - O bond energy is as low as 40 kcal, this bimolecular process involving a n-paraffin with a C - H bond energy of 94 kcal, would be endothermic only to the extent of (94 + 40 - 116) = 18 kcal. It would thus be more favorable than a unimolecular decomposition along the O - O bond. Another argument in favor of this view is the strong dependence of rates of hydroperoxide decomposition on the nature of the solvent. Furthermore, in most cases, the decomposition is easier in solvents with a less tightly bound H atom [85].

In the general case, the expression for the oxidation rate has the form:

$$w = \frac{k_2}{\sqrt{k_3}} [RH] \sqrt{k'[ROOH] + k''[ROOH]^2} \quad . \tag{40}$$

It must be kept in mind that k' may be an effective constant containing a constant solvent concentration.

As shown by Bateman [76], this relation represents adequately the variation of w with [ROOH] for a number of hydrocarbons, in a very wide range of hydroperoxide concentrations.

Bateman has also measured the bimolecular rate constants for hydroperoxide decomposition in a variety of cases. In the case of ethyl-linoleate, the activation energy for the bimolecular decomposition was also measured and its value is 26 kcal [55]. The rate constant is then:

$$k'' = 1.6 \cdot 10^{-10} \exp(-26000/RT) \frac{cm^3}{sec} \quad .$$

Consequently, so far as the pre-exponential factor is concerned, the reaction may be considered as a straight bimolecular process without the preliminary formation of a complex.

The most likely chain initiation step in the absence of peroxides (i.e., in the very early stages of the process) is the reaction:

$$RH + O_2 \longrightarrow R + HO_2 \quad .$$

The energy required here is the difference between the C - H bond energy q_{C-H} and the $H - O_2$ bond energy (47 kcal). Since the activated complex has a linear configuration, the activation barrier ought to be small and the reaction rate can be written as:

$$w_o = k_4[RH][O_2] = f \cdot 10^{-10} \exp\left[-(q_{C-H} - 47000)/RT\right] \cdot [RH][O_2]$$

where f is a steric factor.

The concentration of dissolved oxygen is usually of the order of 10^{18} per cm^3. Then:

$$w_o \approx f \cdot 10^{8} \exp\left[-(q_{C-H} - 47000)/RT\right] \cdot [RH]$$

$$= f \cdot 10^{29} \exp\left[-(q_{C-H} - 47000)/RT\right] \quad .$$

Let us estimate w_o for three different molecules: an olefin, a branched paraffin and a straight chain paraffin. For an olefin, q_{C-H} = 77 kcal and:

$$w_o = f \cdot 10^{29} \exp(-30000/RT) \quad .$$

At 65° (a temperature commonly used in olefin oxidation), this gives:

$$w_o \sim f \cdot 10^{10} \leq 10^{10} \quad .$$

For a tertiary C - H bond in paraffins, q_{C-H} = 89 kcal and

$$w_o = f \cdot 10^{29} \exp(-42000/RT) \quad .$$

For a secondary C - H bond in paraffins, q_{C-H} = 94 kcal and $w_o = f \cdot 10^{29} \exp(-47000/RT)$.

At 127° (a temperature at which paraffin oxidation proceeds at an appreciable rate), we get $w_o \sim f \cdot 10^6$ in the case of branched paraffins and $w_o \sim f \cdot 10^3$ (a very small value) in the case of straight chain paraffins.

<u>The Induction Period and the Kinetics of the Reaction</u>. The rate w_o largely determines the length of the induction period. In order to determine the latter, it is necessary to know the kinetic equation of the oxidation process.

The system of differential equations describing the kinetics of the initial stages of the oxidation process, as long as the decomposition of the hydroperoxide can be neglected, can be written as follows:

$$\frac{d[R]}{dt} = w_o - k_1[R][O_2] + k_2[RO_2][RH] + k_4\,[ROOH]$$

$$\frac{d[RO_2]}{dt} = k_1[R][O_2] - k_2[RO_2][RH] - k_3[RO_2]^2 \qquad \text{(I)}$$

$$\frac{d[ROOH]}{dt} = k_2[RO_2][RH] \quad .$$

At the oxygen pressures normally employed (a few hundreds mm Hg), the radicals R react much faster than the radicals RO_2 and the concentration of R may be treated as quasi-steady. This reduces the number of equations to two:

$$\frac{d[RO_2]}{dt} = w_o + k_4[ROOH] - k_3[RO_2]^2$$

$$\frac{d[ROOH]}{dt} = k_2[RO_2][RH] \quad .$$

This system may be written in terms of dimensionless variables:

$$\frac{d\xi}{d\tau} = w_o' + \eta - \xi^2$$

$$\frac{d\eta}{d\tau} = \xi$$

where

$$\xi = \frac{k_3[RO_2]}{\sqrt{k_2k_4[RH]}} , \qquad \tau = \sqrt{k_2k_4[RH]}t, \qquad \eta = \frac{k_3[ROOH]}{k_2[RH]} ,$$

$$w_o' = \frac{w_o k_3}{k_2k_4[RH]} \quad .$$

This system may now be integrated:

$$\int_0^{\eta} \frac{d\eta}{\sqrt{\eta - (\frac{1}{2} - w_o')(1 - e^{-2\eta})}} = \tau \quad . \qquad (41)$$

The induction period will be defined as the time required to build up a concentration of hydroperoxide equal to $10^{-4}\%$ (the threshhold of analytical detection).

For cyclohexene at 65°, $k_3/k_2 = 10^5$ and therefore, at the end of the induction period $\eta = 0.1$.

Independently of the hydrocarbon type, the rate constants for hydroperoxide decomposition have about the same value. To the exception of tert.-butylhydroperoxide, the pre-exponential factors are equal to $10^{12} - 10^{13}$ sec^{-1} and the activation energies vary between 31 and 34 kcal. For cyclohexenylhydroperoxide, we shall take:

$$k_4 = 10^{13} \exp(- 32000/RT) \ sec^{-1} \quad .$$

This gives:

$$w_o = \frac{f \cdot 10^{10} \cdot 10^5}{10^{13} \exp(-32000/338 \cdot R) \cdot 10^{21}} = f \cdot 10^2 \quad .$$

Putting $f = 1/200$, we get $w_o = 0.5$. Although f has never been determined, it is probably substantially lower than unity. The value of $f = 0.005$ simplifies the calculation a great deal, without affecting the functional dependence of η on τ.

The induction period becomes:

$$\tau = \int_0^{0.1} \frac{d\eta}{\sqrt{\eta}} = 2\sqrt{0.1} \approx 0.33 \quad .$$

The time elapsed between the end of the induction period and the moment at which 1% conversion is reached (this interval will be called 'reaction period'), is equal to:

$$\tau_r \quad \int_{0.1}^{10^3} \frac{d\eta}{\sqrt{\eta}} = 2\sqrt{10^3} - \sqrt{0.1} = 62.6; \qquad \frac{\tau_r}{\tau} = 209 \quad .$$

Therefore, the induction period is very small as compared to the reaction period.

The kinetic curve of hydroperoxide accumulation or, which is the same, of oxygen absorption has a parabolic shape characteristic of a relatively weak auto-acceleration. This agrees with kinetic data on olefin oxidation. As an illustration, the kinetic curve of cyclohexene oxidation at 65° is reproduced on Figure 72 [86]. It can be seen that auto-acceleration is not very pronounced. After 12 hours, conversion reaches 1.2%, i.e., $\eta = 1000$. Note that one unit of dimensionless time corresponds to:

$$t = \frac{1}{\sqrt{k_2 k_4 [RH]}} = \frac{1}{\sqrt{10^{-20} \cdot 10^{-8} \cdot 10^{21}}}$$

$$= 3 \cdot 10^3 \text{ sec} \sim 1 \text{ hour} \quad .$$

Thus, the calculated time of conversion is equal to about 60 hours, of the same order of magnitude as the time observed.

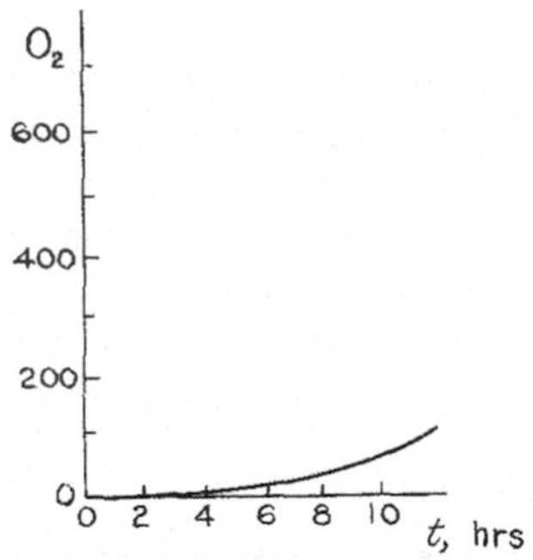

FIGURE 72. Kinetic curve of oxygen absorption in oxidation of cyclohexene at $T = 65°C$ [86].

In the case of branched paraffins,

w_0 is considerably smaller. The ratio k_3/k_2 on the other hand is not much larger. As shown in Table 64, the pre-exponential factor of k_2 differs little from that of k_3 and therefore $k_3/k_2 = \exp(E_2/RT)$. For branched paraffins, E_2 should be a little larger than for cyclohexene but since the reaction is carried out at a slightly higher temperature, it may be assumed that k_3/k_2 will be of the same order of magnitude. Then:

$$w_0 = \frac{f \cdot 10^6 \cdot 10^5}{10^{13} \exp(-32000/400R) \cdot 10^{21}} \approx 10^{-8} \quad .$$

For $\eta < 0.1$, the expression under the square root sign in the integral may be expanded in series and only the first two terms of the expansion need be retained:

$$\tau = \int_0^{0.1} \frac{d\eta}{\sqrt{w_0 + \eta + \eta^2}} = \cosh^{-1} \frac{0.1}{w_0} \approx \ln \frac{0.2}{w_0} = 16.8$$

Also:

$$\tau_r = \int_{0.1}^{10^3} \frac{d\eta}{\sqrt{\eta - (\frac{1}{2} - w_0)(1 - e^{-2\eta})}} = 64.4$$

so that:

$$\frac{\tau_r}{\tau} = 3.8 \quad .$$

Therefore, a distinct induction period must be observed. Figure 73 on the following page, shows the kinetic curve for isodecane oxidation. There, the ratio of reaction period to induction period is equal to $\tau_r/\tau = 3.3$.

The kinetic curve for n-paraffin oxidation does not exhibit an induction period so distinctly. As an illustration, Figure 74, also on the following page, shows the kinetic curve for oxidation of n-decane [72]. The reason is that, for straight chain paraffins, w_0 is further reduced and moreover the ratio k_3/k_2 becomes larger. Thus, for n-decane, the activation energy of chain propagation is 11.5 kcal and this means that k_3/k_2 is about 10^6 at 127°. Then, the ratio τ_r/τ will be larger. For instance, with $k_3/k_2 = 10^6$:

$$\tau = \int_0^1 \frac{d\eta}{\sqrt{\eta - (\frac{1}{2} - w_0)(1 - e^{-2\eta})}} = 26.4 \quad \tau_r = \int_1^{10^4} \frac{d\eta}{\sqrt{\eta - (\frac{1}{2} - w_0)(1 - e^{-2\eta})}} =$$

$$2 \cdot (100 - 2.7) = 194$$

so that $\tau_r / \tau = 7.3$.

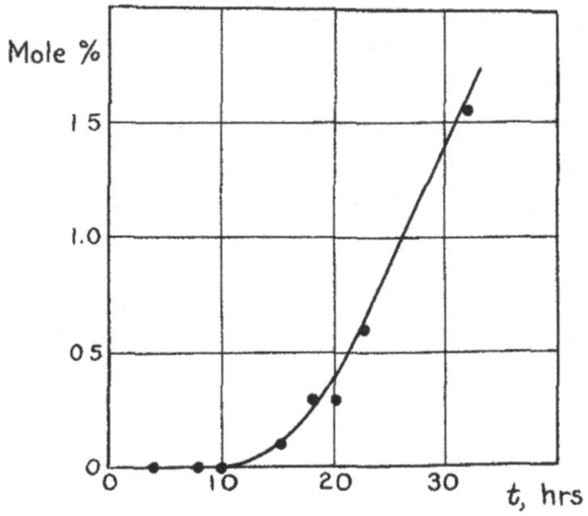

FIGURE 73. Kinetic curve of accumulation of hydroperoxide in oxidation of isodecane at 120°C.

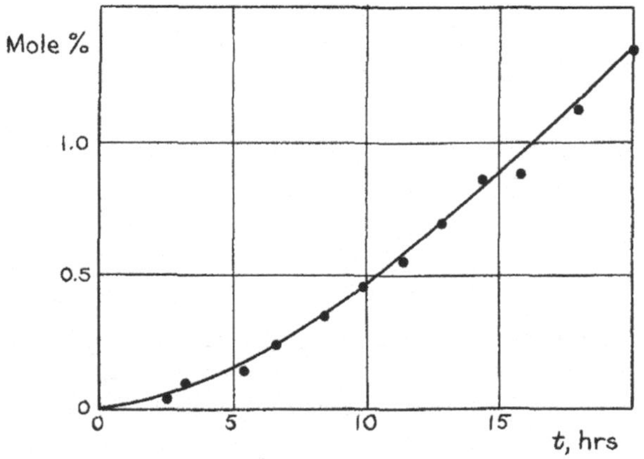

FIGURE 74. Kinetic curve of accumulation of hydroperoxide in oxidation of n-decane at 120°C [72].

If the ratio k_3/k_2 is larger yet, τ_r and also the ratio τ_r/τ will become even larger. As a result, in the case of straight chain paraffins, the induction period will be short as compared to the reaction period.

It must be noted that in the preceding calculations, the reaction $RH + O_2$ has been assumed to be the unique source of free radicals at zero time. This is apparently correct in the case of olefins for which the rate of initiation is not too small. But with branched and especially straight chain paraffins, w_o is so small that it cannot be the main contributor to chain initiation. Negligible amounts of hydroperoxides or some wall reaction will generate free radicals faster than can be done by means of the process $RH + O_2$. In effect, this shortens the induction period calculated above and increases the ratio τ_r/τ.

The existence of long induction periods when w_o is small is due to the exponential build-up of hydroperoxides in the early stages:

$$\eta = \cosh w_o\tau - 1 \approx \frac{1}{2} e^{w_o\tau} .$$

On the other hand, if upon integration of the system of equation, the concentration RO_2 or ξ is assumed to be quasi-steady, the following expression is obtained:

$$\eta = \frac{\tau^2}{4} + \sqrt{w_o}\tau .$$

This is a parabolic expression. When system (I) is solved rigorously, it is seen that such an expression is valid only when w_o is sufficiently large. In other words, when the rate of initiation is small, the quasi-steady state approximation with respect to RO_2 radicals, is not applicable. This is due to the quadratic nature of chain termination. Indeed, the time required to reach the quasi-steady concentration

$$\frac{1}{\sqrt{w_o k_3}}$$

depends not only on the rate of radical destruction but also on the rate of initiation. In oxidation, in spite of the large value of the rate constant for termination, the time required to reach the steady state concentration may be long. For example, with $w_o = 10^3$, this time reaches 10^6 sec so that a steady state concentration cannot be attained [87].

In connection with induction periods, it must be kept in mind that in many oxidation reactions, the induction period may be affected by inhibitors present in the oxidized substance. In the presence of inhibitors, the system of differential equations describing the kinetics of the process, takes the form:

$$\frac{d[RO_2]}{dt} = w_o + k_4[ROOH] - k_1[RO_2][I]$$

$$\frac{d[ROOH]}{dt} = k_2[RO_2][RH] - k_4[ROOH]$$

since now the main termination step occurs by reaction of the RO_2 radical with a molecule of inhibitor. The character of kinetic curves is determined by the sign of the roots λ of the characteristic equation of this system:

$$\begin{vmatrix} -k_1[I] - \lambda & k_4 \\ k_2[RH] & -k_4 - \lambda \end{vmatrix} = 0$$

or:

$$\lambda^2 + (k_1[I] + k_4)\lambda + k_4(k_1[I] - k_2[RH]) = 0 \quad .$$

If both roots are negative, as is the case when $k_1[I] > k_2[RH]$, the process will go at a constant rate. If $k_1[I] < k_2[RH]$, the process will be slow but autoaccelerating and non-stationary. Thus, in the presence of potent inhibitors, the reaction will practically stop until the inhibitor has been consumed. For example, when n-decane is oxidized in the presence of α-naphthol, the oxidation does not start until all the α-naphthol has disappeared. On the contrary, with a less effective inhibitor, the process is slowed down but retains its self-accelerating character. This is observed when isopropylbenzene is oxidized in the presence of phenol [88] (Figure 75).

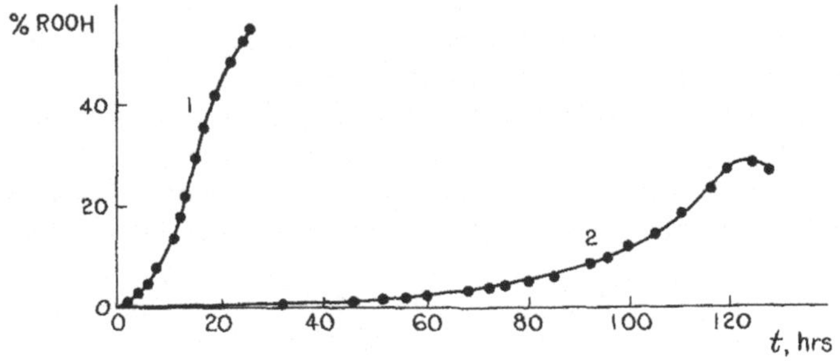

FIGURE 75. Kinetic curves of accumulation of hydroperoxide in oxidation of cumene. 1. Without inhibitor. 2. In the presence of 0.1% phenol; T = 105°C [88].

These considerations indicate that there should exist a critical concentration of inhibitor above which the rate of the process stays stationary, i.e., very slow and below which it is self-accelerating. This phenomenon is observed in n-decane oxidation catalyzed by copper stearate. Here the Cu^{2+} ion catalyzing the process also acts as an inhibitor, apparently reacting with RO_2 radicals. It has been found that above a certain critical concentration of stearate, the process starts only after several hours [77]. On the other hand, with concentrations of stearate below the critical, the self-accelerating oxidation process starts after a moderate induction period (less than one or two hours).

Critical concentrations must exist also with other inhibitors and this question deserves a serious experimental study.

Competing with the radical initiation by means of the process $RH + O_2 \longrightarrow R + HO_2$, there is an alternative mode of initiation by degenerate branching. This becomes important already in the early stages of the reaction. The rate of this process may be estimated roughly:

$$w_b = 10^{13} \exp(-32000/RT) \cdot [ROOH] \quad .$$

Therefore:

$$\frac{w_o}{w_b} = f \cdot 10^{-5} \exp\left[-(q_{C-H} - 79000)/RT\right] \cdot \frac{[RH]}{[ROOH]} \quad .$$

For olefins, $q_{C-H} = 77000$ cal. and:

$$\frac{w_o}{w_b} = f \cdot 10^{-4} \frac{[RH]}{[ROOH]} \quad .$$

Thus, even if $f = 1$, the rates of initiation and of degenerate branching become comparable in the very early stages of the process, when $[ROOH]/[RH] \sim 10^{-4}$ i.e., at 0.01% conversion. Since f is probably substantially lower than unity by analogy with steric factors of other liquid phase reactions, the radical initiation step ceases to play any role at even earlier stages of the process. This argument is even stronger in the case of paraffins with $q_{C-H} = 94000$ cal. Here, at 127°C:

$$\frac{w_o}{w_b} = f \cdot 10^{-13} \frac{[RH]}{[ROOH]} \quad .$$

On the contrary, in the case of hydrocarbons with bond energies substantially less than 77 kcal, the initiating non-branching step $RH + O_2 \longrightarrow R + HO_2$ will be observed during the initial phases of the reaction.

This is actually observed in the oxidation of ethyllinoleate. Here the oxidized CH_2 group is in α-position to two double bonds:

$$CH_3 - (CH_2)_4 - CH = CH - \underset{\displaystyle H}{\underset{|}{CH}} - CH = CH - (CH_2)_7 - COOC_2H_5 \quad .$$

It must therefore be weaker than in olefins. The kinetics of oxidation, as long as the rate of degenerate branching may be neglected, is represented by the equation (55):

$$w = k[RH]^{3/2}[O_2]^{1/2} \quad .$$

As indicated by equation (36), this means that the rate of initiation is given by:

$$w_0 = k'[RH][O_2]$$

so that the initiation is indeed due to a bimolecular process between RH and O_2.

Accumulation of Hydroperoxides in Liquid Phase Oxidation. In a reacting system, hydroperoxides not only decompose but they also attack RO_2 radicals. The most likely reaction is the abstraction of a hydrogen atom from the carbon bound to the peroxidic group:

$$R' - \underset{\displaystyle OOH}{\underset{|}{CH}} - R'' + RO_2 \longrightarrow R' - \underset{\displaystyle OOH}{\underset{|}{\dot{C}}} - R'' + ROOH \quad .$$

The C - H bond in question must be somewhat weaker than normally. This also happend in aldehyde where the C - H bond of the carbonyl group is weakened by the oxygen. But the weakening in the case of hydroperoxides is not so marked since the carbon-oxygen bond is only a single bond and not an aldehydic double bond. It then seems that the radical

$$R' - \underset{\displaystyle OOH}{\underset{|}{\dot{C}}} - R''$$

readily breaks down to a ketone and a OH radical.

Consequently, the hydroperoxide takes part in the chain process. If the disappearance of the hydroperoxide is mainly due to the chain process, its rate of accumulation will be:

$$\frac{d[ROOH]}{dt} = k_2[RO_2][RH] - k_5[RO_2][ROOH] \quad .$$

The maximum concentration of hydroperoxide is equal to:

$$[ROOH]_{max.} = [RH] \frac{k_2}{k_5} .$$

Since k_5 differs but little from k_2, the hydroperoxide will accumulate in sizeable amounts, in agreement with observation.

This is the fundamental difference between this process and the oxidation of methane, where the intermediate product — formaldehyde — reacts with free radicals much more readily than the original hydrocarbon. Indeed, there is a difference of 8 kcal between the activation energies for chain propagation in methane and formaldehyde oxidation. As a result, in methane oxidation, the intermediate product may accumulate only in small quantities.

The difference between intermediate hydrocarbon oxidation products in the liquid and gaseous phases is particularly evident in studies of the behavior of these intermediates in the absence of the original hydrocarbon. In the gas phase, the chain oxidation of formaldehyde can be propagated at substantially lower temperatures than the oxidation of methane because the propagation steps are easier in the former case. On the contrary, hydroperoxides, as a rule, will decompose appreciably at temperatures equal to or even higher than the temperature at which oxidation of the corresponding hydrocarbon takes place. Hydroperoxide decomposition is apparently a radical chain process with very short chains. For instance, hydroperoxide decomposition cannot successfully be initiated by initiators such as azoisobutyric acid dinitrile.

It is precisely because of this difficulty of participating in a chain process that hydroperoxides can be accumulated in large quantities during the course of liquid phase oxidation of hydrocarbons.

§5. Some New Phenomena in the Oxidation of Hydrocarbons and Aldehydes

Acceleration of Liquid Phase Oxidation by Additives and Exciting Radiation. Recontly, N. M. Emanuel' [89] has attracted attention to the fact that in liquid phase oxidation, because of the degenerate branching, initiation of the process by initiators or by radiation ought to have a substantial effect only in the early stages of the process.

After accumulation of the branching agent in sufficient quantities, the formation of radicals by degenerate branching becomes larger than that due to initiation. The initiator then ceases to play any important role of the kinetics of the process.

This proposition has been verified by using chemically active gases as initiators: NO_2, HBr (gaseous initiation), γ-radiation from Co^{60} and α-particles from radon. Oxidation was carried out in a flow system. In the case of NO_2, HBr and radon, initiation was accomplished by mixing the initiator with the incoming air or oxygen during some period of time. In this fashion it was easy to interrupt initiation without perturbing the oxidation process, simply by stopping the addition of initiating gas. In the case of γ rays from Co^{60}, initiation was interrupted by removing the Co^{60} source.

In the oxidation of n-decane, it was found that the process is strongly accelerated by addition of HBr. But the kinetic curves of accumulation of carbonyl compounds and acids (Figure 76) are identical whether HBr is used as an initiator all the time or only during ten minutes [90]. The same result was obtained in a study of cetane oxidation

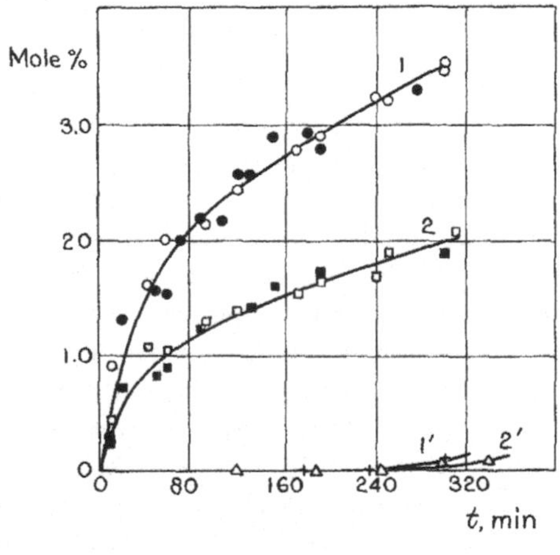

FIGURE 76. Kinetic curves for the formation of carbonyl compounds (curve 1) and acids (curve 2) in oxidation of n-decane at 110°C. a,b with 17% HBR; c,d after a 10-minute preliminary initiation of the reaction by means of oxygen with 17% HBR. 1' and 2': kinetic curves for the formation of ketones and acids in uncatalyzed oxidation [90].

initiated by NO_2 addition. Here 10% NO_2 in the incoming air during one minute cut down the induction period from 360 to 8-10 hours. But continuous feeding of NO_2 containing air resulted in a strong inhibition of the process in the early stages of conversion. Thus a long exposure to the initiator is harmful to the continuation of the process, apparently because of some secondary reactions [91]. A shorter initiation at the beginning (1.5 hours) also proved to be more effective than a continuous initiation in the oxidation of liquefied butane in the presence of NO_2 [92]. A sizeable decrease of the induction period was similarly observed in the oxidation of cetane initiated during 70 minutes by γ-rays from Co^{60} [93] and in the oxidation of isodecane initiated by radon [94].

To sum up, an initiator exerts an accelerating effect on oxidation only in the early stages of the process. Later on, either it is without effect on the reaction rate or it is harmful because of undesirable secondary reactions.

Stages in Oxidation Processes. The examples just quoted indicate that the kinetic characteristics of oxidation may change sharply during the course of the reaction. This is due to a change in the relative rates of the various elementary steps: stages that play an important role at the start, become secondary later on and vice versa. Emanuel' has shown that such stages also exist in gas phase oxidation. Changes may sometimes be so sharp that one may speak of several macroscopic stages of the complex process, each stage having its own elementary mechanism [95].

Thus, in the gas phase oxidation of acetaldehyde, an autocatalytic reaction develops at first, accompanied by a decrease of pressure and the formation of acetyl hydroperoxide. Then this stage subsides and is replaced by a second stage during which acetaldehyde is mainly oxidized by the acetyl hydroperoxide formed during the first stage [96-98]. A similar picture is observed in the gas phase oxidation of isobutane in the presence of HBr. At the beginning, the main process is the formation of tertiary butylhydroperoxide. After a while, this is replaced by oxidation due to the hydroperoxide formed [99]. A remarkable feature of both reactions just mentioned is that, during the second stage oxygen ceases to be consumed in spite of the presence in the system of considerable quantities of the original reactants. A similar interruption in the process is also observed in other cases. Thus in the oxidation of propane in the presence of HBr [43] or Cl_2 [100] and in the oxidation of ethane with HBr added [101], oxidation stops at a moment when the three components, hydrocarbon, oxygen and initiator are still present in the system in abundant quantities. An analogous phenomenon takes place also in uninitiated processes. Thus a sharp decline in the rate of oxidation is observed in the gas phase oxidation of propylene [102] and the liquid phase oxidation of n-decane [103].*

* It must be noted that an interruption of the process before exhaustion of reactants is not a necessary feature of oxidation processes. There are studies reported where oxidation proceeded to complete exhaustion of one

A detailed study of the macroscopic stages of a complex oxidation process has been carried out by Z. K. Maĭzus and N. M. Emanuel'. They investigated the oxidation of propane in the presence of HBr. In this system, propane oxidation proceeds selectively to acetone, the main product of the reaction. But production of acetone stops at a moment when the system still contains large amounts of all components of the reaction. Although the process involves three components, the kinetic curves of acetone accumulation can be represented by a simple equation:

$$p_{ac.} = (p_{ac.})_\infty (1 - e^{-kt}) \quad .$$

Thus the reaction obeys a quasi-unimolecular law if the concentrations refer to the part of the mixture that has reacted. Here, the partial pressures of acetone at time t and at the end of the process are denoted by $p_{ac.}$ and $(p_{ac.})_\infty$ respectively [110]. The authors were thus led to believe that, during the first few seconds, a fast reaction occurs in the system. This, the authors called the initial initiating stage during which some intermediate initiator is produced. The initiator then decomposes unimolecularly and initiates a straight chain oxidation of propane to acetone. The rate of that process is:

$$\frac{dp_{ac.}}{dt} = w_0 \nu$$

where w_0 is the rate of decomposition of the intermediate catalyst and ν is the chain length of the oxidation process. Denoting the concentration of intermediate catalyst by I, we get:

$$\frac{dI}{dt} = - kI; \quad I = I_0 e^{-kt} \quad .$$

Hence:

$$p_{ac.} = I_0 \nu (1 - e^{-kt}) = (p_{ac.})_\infty (1 - e^{-kt}) \quad .$$

* (cont.)
of the components. An example is the oxidation of diethyl ether. It consists of two stages: the oxidation by molecular oxygen and the oxidation by the hydroperoxide formed in the first stage. Yet it continues until all the oxygen available has been consumed [104]. In the oxidation of isobutane [105], with a system containing 80 mm Hg C_4H_{10} and 350 mm Hg O_2, reaction practically stops after the partial pressure of hydrocarbon has reached 20 mm Hg. A large excess of oxygen is still available. But with a system containing 100 mm Hg C_4H_{10} and 200 mm Hg O_2, oxidation proceeds to the end, i.e., to complete exhaustion of the available oxygen. Complete consumption of oxygen has also been reported in uninitiated oxidation of methane [106], ethane [102], ethylene [107] and propane [108, 109].

Thus effectively, the kinetics of acetone accumulation follows a quasi-unimolecular decomposition of initiator.

According to this interpretation, the interruption of the process is due to the exhaustion of intermediate initiator. In order to confirm this view, the authors made use of their observation of the inhibiting action of acetone on the oxidation process [111]. They found that a concentration of acetone above a critical limit will suppress the oxidation completely when it is added to the starting mixture. But when the same critical amount is added to the reacting system 10 or 15 seconds after the reaction has started, the reaction proceeds at rate which is identical to that in the absence of acetone. This can be understood if acetone suppresses the initial initiating process taking place during the first few seconds but has no effect on the second straight chain process.

Since the maximum concentration of acetone produced in the reaction is equal to $I_0\nu$, at a given temperature, i.e., at a given value of ν, it is possible to infer from the maximum concentration of acetone, the concentration of intermediate initiator produced during the initial initiating stage.

The author then studied the kinetics of the initial initiating stage [112]. To do this, the reaction was carried out at a certain temperature T and at a given moment t, the initiating step was interrupted by addition of a critical amount of acetone. Then the reactor was heated to a temperature of 215°C and the quantity of acetone formed $(\Delta p_{ac.})_\infty$ was measured. Since the second stage was always carried out at the same temperature so that ν was constant, the value $(\Delta p_{ac.})_\infty = I_\nu$ gave a relative measure of the concentration of intermediate initiator at time t and temperature T. Thus were obtained kinetic curves of accumulation of intermediate initiator at various temperatures. The curves are S-shaped. This indicates that the formation of intermediate initiator is a branched chain process. The formation of intermediate initiator is interrupted when the system still contains large quantities of the three initial reactants: propane, oxygen and HBr. Consequently the initial initiating stage is self-inhibiting. The overall activation energy of the initial initiating stage is 13.5 kcal. This figure is quite different from the activation energy of the reaction leading to acetone, namely 25 kcal. This is another indication of the distinct nature of both processes.

The existence during propane oxidation with HBr of two macroscopic stages was also confirmed by measuring the heating due to the process, by means of sensitive thermocouples [113]. The curves representing temperature change versus time are shown on Figure 77, on the following page. They have two maxima indicating two different stages of

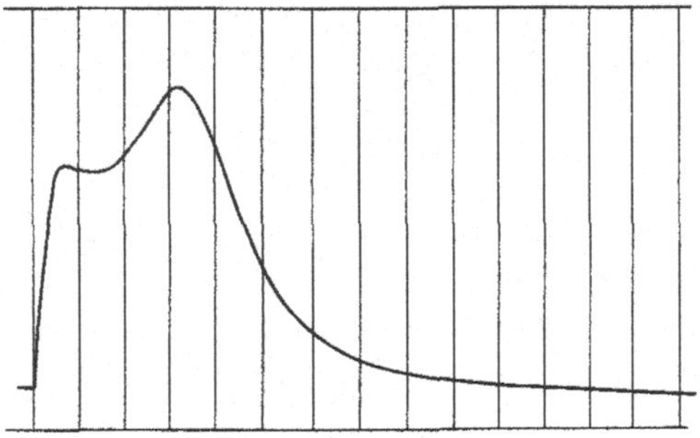

FIGURE 77. Photographic record of the temperature increase [113] in a mixture $2C_3H_8 + 2O_2 + 1HBr$ at 220°C, $p_{total} = 250$ mm Hg.

that release and consequently, two different macroscopic stages. The falling branch of the temperature curve is well represented by the expression:

$$\Delta T = Ae^{-kt}$$

and the constant k coincides with the constant calculated from the kinetic curves of acetone accumulation under the same conditions. Since ΔT is proportional to the reaction rate, the height of the first maximum must be proportional to the maximum rate of the first stage of the process. The temperature dependence of the height of this maximum gives the activation energy of the initial initiating stage, 13.8 kcal, a figure in complete agreement with that obtained independently from kinetic measurements.

Subsequently, the same thermocouple technique revealed the existence of separate macroscopic stage in the oxidation of ethane with addition of HBr [101], the oxidation of propane with addition of NO_2 [114], $NOC\ell$ [115], $C\ell_2$ [100] and in the oxidation of methane with addition of NO_2 [116]. In the last instance, it was possible to elucidate the nature of the initial macroscopic stage; it was observed that the first temperature maximum corresponded to the formation of nitromethane which is the initiator of the second stage in methane oxidation.

The existence in oxidation processes of several macroscopic stages separated in time, reopens the question of the selectivity of chemical changes. A process consisting of several macroscopic stages must not be carried out under constant conditions since the optimum conditions may well vary from one stage to another. The best illustration of this situation is provided by the gas phase initiated oxidation processes considered above, where the same initiator accelerates the process in its early stages but inhibits it in its later stages.

REFERENCES

[1] N. N. Semenov, 'Chain Reactions', Oxford, 1935.

[2] W. A. Bone and S. G. Hill, Proc. Roy. Soc., A 129, 434 (1930).

[3] N. S. Enikopolyan, Zhur. Fiz. Khim., 30, 769 (1956).

[4] N. M. Chirkov and S. G. Entelis, 'Kinetics of Ethane Oxidation' in 'Kinetics of Chain Oxidation', Moscow, 1950, p. 118.

[5] Nan-Chiang Wu Shu and J. Bardwell, Canad. J. Chem., 33, 1415 (1955).

[6] D. M. Newitt and L. S. Thornes, J. Chem. Soc., 1669 (1937).

[7] N. Ya. Chernyak, V. L. Antonovskiĭ, A. F. Revzin and V. Ya. Shtern, Zhur. Fiz. Khim., 28, 240 (1954).

[8] N. S. Enikopolyan, Dok. Akad. Nauk SSSR, 1958, in press.

[9] V. Ya. Shtern, Zhur. Fiz. Khim., 28, 613 (1954).

[10] L. A. Repa and V. Ya. Shtern, Zhur. Fiz. Khim., 28, 414 (1954).

[11] V. N. Kondrat'ev, Acta physicochim. URSS, 4, 556 (1936).

[12] N. J. Emeleus, J. Chem. Soc., 1733 (1929).

[13] B. V. Aĭzazov, M. B. Neĭman and I. I. Khanova, Izv. Akad. Nauk, SSSR, Ot. Khim. Nauk, 307 (1938).

[14] M. B. Neĭman, Usp. Khim., 7, 341 (1938).

[15] R. N. Pease, 'Equilibrium and Kinetics of Gas Reactions', Princeton, 1942.

[16] J. H. Knox and R. G. W. Norrish, Proc. Roy. Soc., A 221, 151 (1954); Trans. Far. Soc., 50, 928 (1954).

[17] N. S. Enikopolyan, S. S. Polyak and V. Ya. Shtern, Dok. Akad. Nauk SSSR, in press.

[18] V. V. Voevodskiĭ and V. I. Vedeneev, Dok. Akad. Nauk SSSR, 106, 679 (1956).

[19] P. S. Shantarovich, Zhur. Fiz. Khim., 10, 700 (1937).

[20] R. Spence, J. Chem. Soc., 686 (1932).

[21] D. Ya. Sadovnikov, Zhur. Fiz. Khim., 9, 575 (1937).

[22] R. G. W. Norrish and J. D. Reagh, Proc. Roy. Soc., A 176 429 (1940).

[23] G. R. Holy and K. O. Kutschke, Canad. J. Chem., 33, 496 (1955).

[24] K. U. Ingold and W. A. Bryce, J. Chem. Phys., 24, 360 (1956).

[25] D. E. Hoare and A. D. Walsh, Trans. Far. Soc., 53, 1102 (1957).

[26] L. A. Avramenko and R. V. Lorentso, Dok. Akad. Nauk SSSR, 67, 867 (1949).

[27] A. B. Nalbandyan and V. V. Voevodskiĭ, 'Mechanism of Oxdiation and Combustion of Hydrogen', Moscow, 1949.

[28] A. B. Nalbandyan, Dok. Ak. Nauk Arm. SSSR, 9, 101 (1948).

[29] N. S. Enikopolyan and L. V. Karmilova, Zhur. Fiz. Khim., (1958), in press.

[30] L. V. Karmilova, N. S. Enikopolyan and A. B. Nalbandyan, Zhur. Fiz. Khim., 31, 851 (1957).

[31] N. S. Enikopolyan and L. V. Kormilova, Zhur. Fiz. Khim., (1958), in press.

[32] N. S. Enikopolyan, G. V. Korolev and G. P. Savushkina, Zhur. Fiz. Khim., 31 865 (1957).

[33] R. G. W. Norrish and S. G. Foord, Proc. Roy. Soc., A 157, 503 (1936).

[34] D. E. Hoare and A. D. Walsh, Fifth Symposium on Combustion, p. 467, 1955.

[35] L. I. Avramenko and R. V. Lorentso, Dok. Akad. Nauk SSSR, 69, 205 (1949).

[36] N. M. Emanuel', Zhur. Fiz. Khim., 14, 863 (1940); in 'Kinetics of Chain Oxidation', Moscow, 1950, p. 79.

[37] N. N. Semenov and N. M. Emanuel', Dok. Akad. Nauk SSSR, 28, 220 (1940).

[38] N. M. Emanuel', Zhur. Fiz. Khim., 19, 15 (1945).

[39] Z. K. Maĭzus, V. M. Cherednichenko and N. M. Emanuel', Dok. Akad. Nauk SSSR, 70, 855 (1950).

[40] S. S. Polyak and V. Ya. Shtern, Zhur. Fiz. Khim., 27, 631 (1953).

[41] J. J. Batten and M. J. Ridge, Austr. J. Chem., 8, 370 (1955).

[42] V. Ya. Shtern, in 'Chain Oxidation of Gaseous Hydrocarbons', Moscow, 1955, p. 37.

[43] Z. K. Maĭzus and N. M. Emanuel', in 'Chain Oxidation of Gaseous Hydrocarbons', Moscow, 1955, p. 81.

[44] P. W. Schenk, Z. anorg. allg. Chem., 211, 150 (1933).

[45] D. I. Meschi and R. I. Myers, J. Am. Chem. Soc., 78, 6220 (1956).

[46] K. I. Ivanov, 'Intermediate Products and Reactions in Hydrocarbon Autoxidation', Moscow, 1949.

[47] A. Robertson and W. A. Waters, Trans. Far. Soc., 42, 201 (1946).

[48] E. H. Farmer and D. A. Sutton, J. Chem. Soc., c. 119 (1943).

[49] B. Andersen, Arkiv Kemi, 2, 33, 451 (1950).

[50] H. W. Melville and S. Richards, J. Chem. Soc., 944 (1954).

[51] K. K. Hargrave and A. L. Morris, Trans. Far. Soc., 52, 89 (1956).

[52] I. V. Berezin, L. Sinochkina, V. G. Dzantiev and N. F. Kazanskaya, Zhur. Fiz. Khim., 31, 554 (1957).

[53] L. S. Vartanyan, Z. K. Maĭzus and N. M. Emanuel', Zhur. Fiz. Khim., 30, 856 (1956).

[54] P. George, Proc. Roy. Soc., A 185, 337 (1946).

[55] J. L. Bolland and G. Gee, Trans. Far. Soc., 42, 236 (1946).

[56] V. L. Vaĭser, Dok. Akad. Nauk SSSR, 67, 839 (1949).

[57] J. L. Bolland and P. TenHave, Trans. Far. Soc., 45, 93 (1949).

[58] J. P. Wibaut and A. Strong, Proc. Kon. Neder. Akad. Wet., 54 B, 102 (1951).

[59] F. M. Lewis and M. S. Matheson, J. Am. Chem. Soc., 71, 747 (1949).

[60] L. F. Fieser, R. C. Clapp and W. H. Daubt, J. Am. Chem. Soc., 64, 2052 (1942).

[61] W. S. M. Grieve and D. H. Hey, J. Chem. Soc., 1797 (1934).

[62 W. A. Waters, J. Chem. Soc., 113 (1937).

[63] L. Bateman and G. Gee, Proc. Roy. Soc., A 195, 376 (1948).

[64] K. I. Ivanov and E. D. Vilenskaya, in 'Chemical Kinetics, Catalysis and Reactivity', Moscow, 1955, p. 260.

[65] C. H. Bamford and M. J. S. Dewar, Proc. Roy. Soc., A 198, 252 (1949).

[66] J. L. Bolland, Proc. Roy. Soc., A 186, 48 (1946).

[67] J. L. Bolland, Trans. Far. Soc., 44, 669 (1948).

[68] J. L. Bolland, Trans. Far. Soc., 46, 358 (1950).

[69] L. Bateman and G. Gee, Trans. Far. Soc., 47, 155 (1951).

[70] K. Ivanov, V. Savinova and V. Zhakhovskaya, Dok. Akad. Nauk SSSR, 59, 905 (1948).

[71] D. G. Knorre, Z. K. Maǐzus and N. M. Emanuel', Dok. Akad. Nauk SSSR, 112, 457 (1957).

[72] D. G. Knorre, Z. K. Maǐzus, L. K. Obukhova and N. M. Emanuel', Usp. Khim., 26, 416 (1957).

[73] L. Bateman and G. Gee, Proc. Roy. Soc., A 195, 391 (1948).

[74] C. H. Bamford and M. L. S. Dewar, Nature (London), 163, 215 (1949).

[75] W. Kern and H. Willersin, Makromol. Chem., 1, 1 (1956).

[76] L. Bateman, H. Hughes and A. L. Morris, Dics. Far. Soc., No. 14, 190 (1953).

[77] W. Kern and H. Willersin, Angew. Chem., 67, 573 (1955).

[78] L. Bateman, Quart. Rev., 8, 147 (1954).

[79] J. W. R. Fordham and H. L. Williams, Canad. J. Res. 27 B, 943 (1949).

[80] G. H. Twigg, Disc. Far. Soc., No. 14, 240 (1953).

[81] A. Farkas and E. Passaglia, Jour. Am. Chem. Soc., 72, 3333 (1950).

[82] V. Stanett and R. B. Mesrobian, Disc. Far. Soc., No. 14, 242 (1953).

[83] C. F. H. Tipper, J. Chem. Soc., 1675 (1953).

[84] E. R. Bell, J. H. Raley, F. F. Rust, F. H. Seubold and W. E. Vaughan, Disc. Far. Soc., No. 10, 242 (1951).

[85] V. Stannett and R. B. Mesrobian, J.A.C.S., 72, 4125 (1950).

[86] S. S. Medvedev in 'Problems in Kinetics and Catalysis', 1940, p. 23.

[87] D. G. Knorre, V. L. Pikaeva and N. M. Emanuel', Dok. Akad. Nauk SSSR, in press.

[88] D. K. Tolopko, Dok. Akad. Nauk SSSR, 104, 101 (1955).

[89] N. M. Emanuel', Zhur. Fiz. Khim., 30, 847 (1956).

[90] N. M. Emanuel', Dok. Akad. Nauk SSSR, 102, 559 (1955).

[91] E. A. Blyumberg and N. M. Emanuel', Izv. Akad. Nauk SSSR, Ot. Khim. Nauk, 274 (1957).

[92] N. M. Emanuel', Dok. Akad. Nauk SSSR, 110, 245 (1956).

[93] N. M. Emanuel', Dok. Akad. Nauk SSSR, 111, 1286 (1956).

[94] E. A. Blyumberg, D. M. Ziv, V. L. Pikaeva and N. M. Emanuel', Dok. Akad. Nauk SSSR, in press.

[95] N. M. Emanuel', Dok. Akad. Nauk SSSR, 95, 603 (1954).

[96] Z. K. Maǐzus and N. M. Emanuel', Dok. Akad. Nauk SSSR, 57, 271 (1947).

[97] Z. K. Maǐzus and N. M. Emanuel', Izv. Akad. Nauk SSSR, Ot. Khim. Nauk, 57 (1948).

[98] Z. K. Maǐzus and N. M. Emanuel', Izv. Akad. Nauk SSSR, Ot. Khim. Nauk, 182 (1948).

[99] A. A. Babaeva, Z. K. Maǐzus and N. M. Emanuel', Izv. Akad. Nauk SSSR, Ot. Khim. Nauk, in press.

[100] K. E. Kruglyakova and N. M. Emanuel', in press.

[101] M. F. Sedova and N. M. Emanuel', Izv. Akad. Nauk SSSR, Ot. Khim. Nauk, 658 (1956).

[102] V. Ya. Shtern and S. S. Polyak, Zhur. Fiz. Khim., 27, 341 (1953).

[103] G. Chavanne and G. Tock, Bull. Soc. Chim. Belg., 41, 630 (1932).

[104] A. Lemay and C. Ouellet, Canad. J. Chem., 33, 1316 (1955).

[105] J. J. Batten, H. J. Gardner and M. J. Ridge, J. Chem. Soc., 3029 (1955).

[106] W. Bone and R. E. Allum, Proc. Roy. Soc., A 134, 586 (1932).

[107] W. Bone and A. Haffner, Rance. Proc. Roy. Soc., A 143, 16 (1933).

[108] D. M. Newitt and L. S. Thornes, J. Chem. Soc., 1656 (1937).

[109] R. Pease, J. Am. Chem. Soc., 56, 2034 (1934).

[110] Z. K. Maĭzus and N. M. Emanuel', Dok. Akad. Nauk SSSR, 87, 241 (1952).

[111] Z. K. Maĭzus and N. M. Emanuel', Dok. Akad. Nauk SSSR, 87, 437 (1952).

[112] Z. K. Maĭzus and N. M. Emanuel', Dok. Akad. Nauk SSSR, 87, 801 (1952).

[113] Z. K. Maĭzus, A. M. Markevich and N. M. Emanuel', Dok. Akad. Nauk SSSR, 89, 1049 (1953).

[114] Z. K. Maĭzus and N. M. Emanuel', Dok. Akad. Nauk SSSR, 95, 1009 (1954).

[115] N. M. Emanuel', Izv. Akad. Nauk SSSR, Ot. Khim. Nauk, in press.

[116] V. I. Urizko and M. V. Polyakov, Dok. Akad. Nauk SSSR, 95, 1234 (1955).

APPENDIX I: THE METHOD OF THE ACTIVATED COMPLEX*

Let us suppose that we succeed in surmounting the purely numerical difficulties of determining theoretically the rate of chemical reactions and in completing such calculations even for the reactions that are the simplest from the viewpoint of the chemist. But these reactions are indeed complex from the viewpoint of the kineticist since they proceed by a chain mechanism. Suppose, for instance, that it becomes possible to calculate from theory the rate of the reaction:

$$2H_2 + O_2 \longrightarrow 2H_2O \quad . \qquad (1)$$

This calculation will consist of several distinct steps.

The overall reaction (1) is made of about ten separate simple processes such as:

$$H + O_2 \longrightarrow OH + O$$

$$O + H_2 \longrightarrow OH + H$$

$$OH + H_2 \longrightarrow H_2O + H$$

Similarly, every simple reaction consists of the sum of a collection of elementary steps. Each of these is a reaction between particles submitted to thermal agitation and they all differ somewhat by the relative position of the atoms at the moment of collision, the velocity and direction of motion of the particles themselves.

In this way we see that the theoretical calculation of reaction rates involves three types of problems:

1) A consideration of the elementary steps. What is needed here is to find how the energy of the system of reacting atoms depends on their relative position and their velocity of motion. These are problems dealt with in quantum chemistry.

2) A determination of the rates of the elementary steps. The

* This Appendix was written by M. I. Temkin.

rate is the sum of a tremendous number of individual reaction acts. This problem is statistical in nature and its solution is handled by the methods of statistical mechanics.

3) The determination of the rate of the complex process considered as the sum of its elementary steps. This problem is solved by the theory of chain reactions.

The solution of each problem presupposes in principle the solution of the preceding one. In fact, at the present time, methods of solution have been developed especially for the third problem. Methods of solution of the first problem, namely, the quantum-chemical one are in such a state that they are incapable of yielding quantitative results. However, approximate quantum-chemical methods give a correct qualitative picture of the elementary act, for the simplest cases [1]. These qualitative results will be used in what follows.

The statistical problem is in better condition. Here, the methods of the kinetic theory of gases have been applied successfully. They have been further elaborated in the so-called theory of the transition state or the activated complex, which uses statistical mechanics to calculate rates of reaction.

The fundamental basis of the activated complex method is the assumption that the reaction does not perturb essentially a Maxwell-Boltzmann distribution of energy among the molecules. This assumption will be discussed later and for the moment we will accept it.

According to the law of Maxwell-Boltzmann, the probability of each state is proportional to $\exp(-\epsilon/kT)$ where ϵ is the energy of that state. Thus, states of high energy are rare. Therefore, the overwhelming majority of elementary reaction acts will take place in practically the most favorable manner, i.e., following a path which differs energetically very little from the most favorable one. Consider the situation for the case of the simplest reaction of all:

$$H_2 + D \longrightarrow H + HD \quad . \qquad (2)$$

This reaction requires some small activation energy. Approximate quantum-chemical calculations show that a deuterium atom approaching a hydrogen molecule will suffer the least repulsion when the three atoms are collinear. If this is true, then reaction will be due almost always to an approach of the deuterium atom along the line of the hydrogen atoms.

A stretching of the H_2 molecule decreases the repulsion experienced by the D atom but this stretching requires the expenditure of energy. There will exist some optimum distance between the atoms for which the total energy required for reaction will be a minimum. On the basis of the law of Maxwell-Boltzmann, we may assert that the overwhelming majority of reaction acts will correspond to a spacing between atoms close to its optimum value.

This spacing is a result of thermal motion — the vibration of the atoms in the H_2 molecule along the molecular axis. Reaction will take place almost always when the D atom approaches a vibrating H_2 molecule possessing the necessary vibration amplitude and at the moment of its maximum stretching. At first glance, it appears quite improbable that the correct direction of approach, the required amplitude and phase of vibration and the required relative velocity of motion of a D atom and the center of gravity of an H_2 molecule, will all obtain at the same time. Still, this is considerably more probable than an alternative reaction path requiring a larger energy since the probability of states of higher energy decreases sharply with increasing energy according to the law $\exp(-\epsilon/kT)$.

On the other hand, the most favorable exact path cannot be realized since the probability of such an event is equal to zero.

In this way, we conclude that elementary reaction acts follow a path close to the most favorable one.

During an elementary reaction act, the total energy of the group of reacting atoms stays constant but its components — kinetic and potential energy — change. By potential energy we mean that fraction of the energy determined by the relative position of the atomic nuclei. It also includes, besides the true potential energy, the kinetic energy of the electrons. By kinetic energy we mean that fraction of the energy determined by the velocity of motion of the nuclei.

In each reaction act, the reacting system passes through a configuration corresponding to a maximum value of the potential energy (and consequently a minimum value of the kinetic energy), in other words, a configuration corresponding to the top of a potential barrier. This configuration is called the activated (or active) complex or the transition state.

The idea of the method consists then in selecting the configuration of the activated complex as a sort of observation point to determine the reaction rate. In order to calculate the number of reaction acts, it is sufficient to count the number of systems going through that configuration.

Since we accept the applicability of the law of Maxwell-Boltzmann, it is not necessary to consider all the steps separating the initial state from the activated complex. We may determine immediately the number of states of interest to us, independently of the mechanism by which statistical equilibrium is maintained.

To do this, we represent the dependence of the potential energy of the reacting system on atomic distances by means of some multidimensional potential surface. The potential energy corresponding to the configuration of the activated complex is a maximum with respect to atomic displacements leading to reaction. At the same time, since each deviation from the most favorable reaction path is associated with a higher potential energy, we deal with a point which is also a minimum with respect to displacements corresponding to deviations from the most favorable path. Thus the activated complex is represented by a minimax point or a saddle point on the potential surface. This distinguishes the activated complex from a stable molecule, the potential energy of which increases for arbitrary displacements of the atoms from their equilibrium position. This situation leads to a minimum point on the potential energy surface.

Let us now introduce normal coordinates measuring small deviations of the atomic system from the configuration corresponding to the extremum of the potential surface. For a stable molecule, these are the normal coordinates of vibration; to these we must add three coordinates to characterize the rotation of the molecule (or two if the molecule is linear) and three coordinates of the center of gravity of the molecule to represent the translational motion of the molecule as a whole. In a similar way, the activated complex may be described by three centers of gravity coordinates, three coordinates for rotation (or two if the activated complex is linear as for the case of three atoms considered above), a certain number of vibration amplitudes characterizing the degree of deviation from the most favorable path and finally one coordinate that has no counterpart in the case of a stable molecule — the 'reaction' coordinate representing the motion of the system along the most favorable path.

The words 'rotation' and 'vibration' as applied to the activated complex must not be taken literally. They express the fact that the coordinates and their conjugated momenta depend on potential and kinetic energy in a way formally similar to that prevailing for the normal coordinates of a stable molecule.

In fact, it is only this dependence that we need in order to make calculations based on a Maxwell-Boltzmann distribution. Of course, insofar as the activated complex has only a momentary existence, its rotations and vibrations cannot actually take place.

The number of vibrational degrees of freedom of the activated complex is easily obtained if one remembers that, in all, the number of degrees of freedom must be equal to $3n$ where n is the number of atoms. Thus, for instance, in the case of a linear activated complex consisting of three atoms, the total number of degrees of freedom is 9: 3 are translational, 2 rotational, 1 corresponds to the reaction coordinate and therefore there are 3 vibrational degrees of freedom.

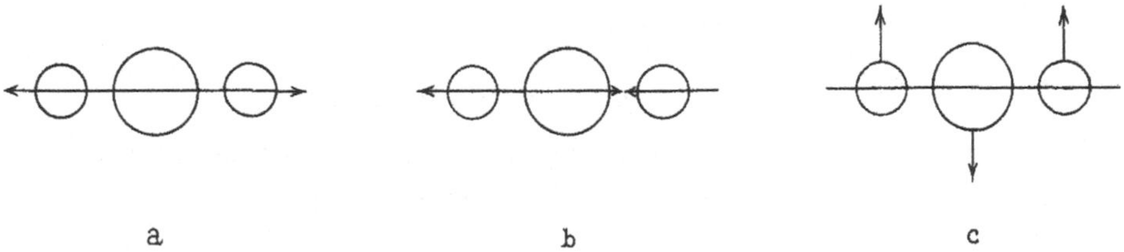

FIGURE 78. Normal vibrations of a CO_2 molecule

a: symmetric valence vibration
b: antisymmetric valence vibration
c: deformation vibration

To fix the ideas, let us compare the vibrations of a linear triatomic molecule to the vibrations of a triatomic linear activated complex.

Figure 78 represents the normal vibrations of a CO_2 molecule.

Deformation vibrations may take place in two mutually perpendicular planes: any other deformation vibration may be represented by the sum of these two. Thus, a CO_2 molecule possesses 4 independent normal vibrations: symmetric and antisymmetric valence modes and 2 deformation modes.

It is easy to see that in the case of a linear triatomic activated complex, for instance, that for the $H_2 + D$ reaction, the motion along the reaction path replaces the antisymmetric valence vibration. Thus, there remain three vibrational degrees of freedom characterizing deviation from the most favorable path. For instance, the deformation modes represent deviation from a collinear arrangement of the atom and the molecule.

Until now we have only talked about the configuration of the activated complex. Let us now note that to a given atomic configuration representing the activated complex, there correspond two opposite directions of motion along the reaction coordinate. One corresponds to the direct reaction, the other to its reverse, e.g., one to the reaction:

$$H_2 + D \longrightarrow H + HD$$

and the other to the reaction:

$$H + HD \longrightarrow H_2 + D \quad .$$

Therefore, the activated complex is completely specified only if its direction of motion is specified besides its configuration.

It is necessary to make a clear distinction between the activated complex and intermediate complexes, critical complexes and the like used in a number of earlier theories.

Let us assume that the change

$$A \longrightarrow B \tag{3}$$

takes place by means of an intermediate substance X that is short-lived, unstable, and transformed back into A. Then the mechanism of the change is:

$$A \rightleftarrows X \longrightarrow B \ . \tag{4}$$

Let the transformations $A \longrightarrow X$ and $X \longrightarrow A$ be very fast as compared to the conversion $X \longrightarrow B$. Then, in the reacting system, an equilibrium between A and X is established and it is only slightly perturbed by the process $X \longrightarrow B$. This type of equilibrium has been precisely postulated in earlier theories using the concept of intermediate complexes and the like.

The relation between the substances A and B, and the activated complex Z for reaction (3) is expressed by the mechanism:

$$A \longrightarrow Z \longrightarrow B \ . \tag{5}$$

The activated complex formed from A always goes into B and never goes back to A. (We assume here that the course of reaction is adiabatic. This limitation will be further discussed below.) The atoms that reach the activated complex configuration continue their motion by inertia. The representative point on the potential surface goes over the top of the barrier and goes down into the valley of products.

The total energy (kinetic and potential) of the system of reacting atoms will always be either smaller or larger than the height of the potential barrier. In the first case, the top of the potential barrier is not reached, in the second case it is crossed and reaction takes place. The probability that the system will possess an energy exactly equal to the height of the barrier, is equal to zero. Therefore there can be no stopping of the system on top of the barrier.

If reaction (3) is reversible, equilibrium may be established:

$$A \rightleftarrows B \ . \tag{6}$$

This equilibrium is realized by the mechanism:

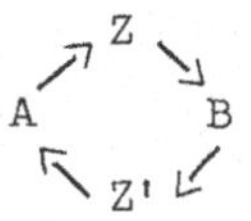

(7)

Here, Z' represents the activated complex for the reverse reaction. It has the same configuration as Z but the opposite direction of motion along the reaction coordinate.

In the case of an intermediate complex X, the equilibrium (6) is realized by means of an equilibrium between X and both A and B:

$$A \rightleftarrows X \rightleftarrows B \quad . \tag{8}$$

In a system in chemical equilibrium as shown by (6), all the possible molecular states are in statistical equilibrium. In particular, the frequency of occurrence of the configurations Z and Z' will be determined by the laws of statistical equilibrium.

Let us now assume that the concentration of A be the same as that corresponding to chemical equilibrium but that the concentration of B is suddenly made equal to zero. In this case, only the reaction $A \longrightarrow B$ will take place in the system, not its reverse $B \longrightarrow A$. The activated complexes Z' for the reverse reaction cannot form since they originate only in B and not in A. The basic assumption of the method of the activated complex may now be stated: it is assumed that the reaction rate $A \longrightarrow B$ does not depend on the circumstance that has just been discussed, in other words, Z will occur in the system just as frequently as if chemical equilibrium prevailed. We assume, therefore, that the direct and inverse processes are without mutual action on each other. This assumption is not new — indeed, it was already made in the first kinetic formulation of Guldberg and Waage's law of active masses for chemical equilibrium.

These are the main physical postulates of the activated complex method.

The calculation of reaction rate according to this method may be explained in the following manner: Let us select some small segment of length $\Delta\ell$ along the reaction coordinate in such a way that it includes the top of the potential barrier. Systems for which the value of the reaction coordinate falls within this interval and which move along the reaction coordinate in the right direction, we will call activated complexes for the direct reaction. Now we can talk about the concentration of activated complexes, since for a given value of $\Delta\ell$, their number per unit volume will be fully determined. This number, since $\Delta\ell$ is a small quantity, will be proportional to $\Delta\ell$. Therefore, the concentration of activated complexes is expressed by $c_a\Delta\ell$ where c_a is a coefficient of proportionality between the concentration of activated complexes and the quantity $\Delta\ell$ or the concentration of activated complexes per unit length of reaction coordinate.

Let $\bar{v}$ be the average velocity corresponding to the change of reaction coordinate with time. This quantity $\bar{v}$ is a rate of change of configuration and is essentially the relative velocity of motion of the reacting atomic system. The average life-time of the activated complex is equal to $\Delta\ell/\bar{v}$, i.e., it is the time required to move along a path $\Delta\ell$ with a velocity $\bar{v}$.

Each time an activated complex is formed, a reaction act takes place. Therefore, per unit volume, there take place $c_a\Delta\ell$ reaction acts in a time $\Delta\ell/\bar{v}$. Hence, the number of reaction acts per unit volume per unit time (i.e., the reaction rate) is equal to $c_a \cdot \bar{v}$.

Until now, we have assumed that reaching the activated complex configuration guarantees reaction. In some cases this is not so (see below); therefore we introduce an additional factor — the transmission coefficient κ that may be smaller or equal to unity. The reaction rate then becomes equal to $\kappa c_a \bar{v}$.

Since we assume that there is statistical equilibrium between reactants and activated complexes during the course of reaction, we may calculate $\bar{v}$ and c_a by methods of statistical mechanics. The calculation of c_a makes use of the generalized form of the law of Maxwell-Boltzmann (expression of equilibrium through partition functions) and $\bar{v}$ is calculated as is usually done for an average velocity of thermal motion.

In this way we arrive at the equation of Eyring [2] for a homogeneous gaseous reaction:

$$k_c = \kappa \frac{kT}{h} \frac{f_a'}{f} \cdot \exp(-\epsilon_0/kT) \quad . \tag{9}$$

In this equation: k_c is the reaction rate constant (the concentrations being expressed in number of molecules per unit volume; k is the Boltzmann constant; h the Planck constant; T the temperature; f the product of partition functions for reactants; f_a the partition function of the activated complex, not including a factor corresponding to the reaction coordinate; ϵ_0 is the height of the energy barrier for reaction, i.e., the difference between the minimum energies of activated complexes and reactants.

It must be noted that there are various determinations of the average velocity along the reaction coordinate. Eyring [2] expresses it as:

$$\bar{v} = \frac{\int_{0}^{\infty} |v| e^{-\frac{m^* v^2}{kT}} dv}{\int_{-\infty}^{+\infty} e^{-\frac{m^* v^2}{kT}} dv} \tag{10}$$

where m^* is the reduced mass of the activated complex. Evans and Polanyi use the following expression for $\bar{v}$:

$$|\bar{v}| = \frac{\int_{-\infty}^{+\infty} |v|\, e^{-\frac{m^* v^2}{kT}} dv}{\int_{-\infty}^{+\infty} e^{-\frac{m^* v^2}{kT}} dv} = \frac{\int_{0}^{\infty} |v|\, e^{-\frac{m^* v^2}{kT}} dv}{\int_{0}^{\infty} e^{-\frac{m^* v^2}{kT}} dv} \; . \tag{11}$$

This corresponds to a usual definition of average velocity. Obviously:

$$\bar{v} = \frac{1}{2} |\bar{v}| \; .$$

If c_a is the concentration of activated complexes per unit length of reaction coordinate, irrespective of the direction of motion, then at chemical equilibrium:

$$c_a^+ = \frac{1}{2} c_a$$

and

$$w^* = \frac{1}{2} \kappa c_a |\bar{v}| \; . \tag{12}$$

This is the equation of Evans and Polanyi [4]. Or also:

$$w^* = \kappa c_a \bar{v} \tag{13}$$

which is Eyring's equation. Unlike equations (12) and (13) equation (9) is valid for any values of [B] and not only for values of [B] corresponding to chemical equilibrium.

The transmission coefficient κ requires special consideration. To discuss it, we must introduce the concept of adiabatic course of reaction.

The term 'adiabatic' is used here with the meaning it first received in the theory of 'adiabatic invariants' of Ehrenfest. It has nothing to do with adiabatic processes known in thermodynamics.

If the parameters of a mechanical system change very slowly (at the limit: infinitely slowly), the quantum numbers characteristic of the motion may not change. Such a process is called adiabatic in quantum mechanics.

For the electronic motion in an atomic system, for instance in a system corresponding to any elementary reaction act such as $D + H_2$, the quantities that determine the relative position of the atomic nuclei, play the role of parameters. As the atoms approach each other (e.g., D approaching H_2), these parameters change. Since the heavy nuclei move much more slowly than the light electrons, it is natural to assume that the electrons succeed in 'adapting' themselves to every new nuclear configuration and to preserve almost exactly the character of motion they would have if the nuclei spent an infinitely long time in a given position. This assumption is precisely that on which the adiabatic course of an elementary act is based.

If this condition is fulfilled, the representative point of the system moves on a potential surface corresponding to a given quantum state of the system. If this point reaches the top of the potential barrier, the reaction necessarily takes place. Thus, for adiabatic processes $\kappa = 1$ (considering the approximate nature of activated complex theory, we must say: the coefficient κ is close to unity).

However, a quantum-mechanical treatment shows that the assumption of adiabaticity of the elementary reaction act may not be true. It turns out that the process may be non-adiabatic if the total electron spin changes or some other forbidden transition takes place. The theory of non-adiabatic processes, developed by L. D. Landau [3] shows that reaching the top of the energy barrier does not then guarantee reaction; in most cases the system returns to the initial state. As a result, typical values of κ for non-adiabatic processes are of the order of 10^{-5}.

To assess the relative importance of adiabatic and non-adiabatic processes, one may turn to experimental data concerning unimolecular reactions. Equation (9) shows that when $\kappa = 1$, the pre-exponential factor of unimolecular rate constants is of the order of $10^{13} - 10^{14} sec^{-1}$. Chain mechanisms that are first order affect only the magnitude of the activation energy, while the pre-exponential factor keeps a value of the same order of magnitude as true unimolecular decompositions.

Therefore, whatever the mechanism may be, if a reaction is first order, if the pre-exponential factor is close to $10^{13} sec^{-1}$ or somewhat larger, there is indication of the adiabaticity of the process. A much smaller value of the pre-exponential factor, say $10^{8} sec^{-1}$ indicates that the reaction path is non-adiabatic. In this way it can be shown that about 90% of reactions are adiabatic.

Turning our attention to the basic starting point of the activated complex method, we must remark that the assumption of a Maxwell-Boltzmann distribution is not always justified. The best known illustration is shown by gaseous unimolecular reactions at low pressures when the number of activating collisions is not sufficient to maintain an equilibrium concentration of active molecules.[1] In this case, the reaction rate is smaller than that calculated by means of equation (9).

At one of the stages of development of chain theory, deviations from a Maxwell-Boltzmann distribution as a result of reaction, were emphasized and this led to the concept of energy chains. These concepts are now out of fashion. A large amount of experimental material shows that chains are actually propagated by carriers and that there are no energy chains. Thus, deviations from a Maxwell-Boltzmann distribution do not play the important role formerly attributed to them. One may believe that individual simple processes participating in a chain reaction obey quite well the equations derived, without taking into account any deviation from a Maxwell-Boltzmann distribution.

Let us remark that the range of applicability of the main assumption behind the activated complex method, coincides with that of the laws of mass action and of Arrhenius. Indeed, deviations from a Maxwell-Boltzmann distribution should perturb the law of mass action and this actually occurs for the case of unimolecular gaseous reactions at low pressures, when these reactions cease to be first order. The law of Arrhenius is thereby also invalidated.

Besides neglecting deviations from a Maxwell-Boltzmann distribution and thus introducing some errors (small in the majority of cases), the activated complex method, in the form normally used, also involves a series of unprecise premises: the curvature of the reaction path is neglected, the quantification of the vibrations of the activated complex is not taken into account, etc. Nevertheless, it must be emphasized that the

---

[1] In unimolecular rate theory, we call active molecules those which possess among their vibrational degrees of freedom, a total energy larger than the height of the potential barrier. One must make a clear distinction between active molecules and activated complexes.

method represents a successful approximation, quite applicable at the present stage of development of chemical kinetics. The activated complex method can be applied successfully to the calculation of rates of adsorption of gases on solid surfaces and of heterogeneous catalytic processes [5].

The realm of application of the activated complex method is severely limited by our ignorance of potential surfaces corresponding to elementary reaction acts. But several results may be obtained by using the method without any knowledge of potential surfaces: the general form of the law of mass action in non-ideal systems, the dependence of the rate on the action of external forces on a reacting system, for instance hydrostatic pressure for reactions in solutions, an approximate calculation of pre-exponential factors in the Arrhenius expression, etc.

Further progress in the theoretical calculation of rates of simple reactions does not at first require an improvement of the statistical-mechanical approach. What it most requires is a better treatment of the quantum-mechanical part of the problem.

In conclusion, let us draw attention to some mistakes that are made concerning the concepts of the activated complex method.

It is frequently stated that the method of the activated complex postulates the existence of some sort of equilibrium between activated complexes and original reactants, as in chemical equilibrium. Then the inverse of the factor kT/h in equation (9) (which has the dimension of a frequency) is treated like the life-time of the activated complex. It is then said that the activated complex has a life-time of about 10^{-13}sec. Since this time is very short, it is concluded that establishment of equilibrium between activated complexes and original molecules is, in fact, impossible.

All these arguments are based on a misunderstanding. The 'life-time' of the activated complex (and also the concentration of activated complexes) is an undetermined quantity unless we specify some small interval along the reaction path including the top of the barrier, within the limits of which there exists a configuration which we call the activated complex. The law of Maxwell-Boltzmann determines the number of particles having coordinates within the limits $q_1 + \Delta q_1$, $q_2 + \Delta q_2$, etc. and momenta within the limits $p_1 + \Delta p_1$, $p_2 + \Delta p_2$, etc. where Δq_1, Δp_1 etc. are small quantities. How much time does the activated complex spend in such a state? This depends on the choice of values of Δp_1, Δq_1 etc. Similarly, we may say that the life-time of the system in a state corresponding to an interval of reaction path between ℓ and $\ell + \Delta\ell$, depends on the choice of $\Delta\ell$. Since $\Delta\ell$ is a small quantity, it is proportional to $\Delta\ell$.

Therefore, to say that the activated complex exists during a time of the order of 10^{-13} sec, is, strictly speaking, devoid of meaning. The life-time of the activated complex is a quantity having a differential character, just as the life-time of a state as defined by the law of Maxwell-Boltzmann. This circumstance, of course, does not invalidate the applicability of the law of Maxwell-Boltzmann in calculations concerning a reacting system. In such a calculation, no other equilibrium is assumed than a Maxwell-Boltzmann equilibrium between states.

It is hoped that these remarks will foster a better understanding of the method of the activated complex and its limits of applicability.

APPENDIX II: QUANTUM MECHANICAL CALCULATIONS OF THE ACTIVATION ENERGY*

One of the first attempts to calculate the activation energy of a bimolecular reaction is due to London [5]. He made a calculation based on the approximation that the dissociation energy of the molecule H_2 can be expressed by the simple formula:

$$w_0 = Q_{H-H} = A_0 + \alpha_0 \quad . \tag{1}$$

Here A_0 is the so-called Coulomb integral numerically equal to the electrostatic interaction between the two unexcited electron clouds and the two hydrogen atom nuclei; α_0 is the so-called exchange integral which is obtained because of the fact that the wave function of the molecule, made of one-electron wave functions of each separate atom, must satisfy a particular requirement expressing the symmetry of the electronic system. This integral expresses a characteristic of interactions between atoms; in particular the very existence of the exchange integral provides an explanation to the saturation of chemical bonds. The expression under the integral sign in both α_0 and the Coulomb integral A_0 contain, besides products of the corresponding wave functions, the energy of interaction between the electrons and the hydrogen nuclei. In the approximation considered, at distances between H atoms and close to the equilibrium value, the exchange integral consists of 80 to 95% of the total bond energy.

Formula (1) is obtained in the first approximation of perturbation theory, on the assumption that the wave function of a system of two separated hydrogen atoms stays the same when both atoms approach each other. Moreover, in the derivation of (1), it is assumed that the square of the so-called non-orthogonality integral S characterizing the degree of overlap of the two atomic wave functions in the molecule, is small compared to unity. If this quantity is not taken into account, normalization of the wave function of the molecule is not accurately satisfied.

Because of these simplifications, formula (1) is not suitable for quantitative calculations. If the quantity S^2 is not neglected, the energy of the bond H_2 is given by the expression:

---

* This Appendix was written by N. D. Sokolov in collaboration with the author.

$$w_0 = \frac{A_0 + \alpha_0}{1 + S^2} \quad . \tag{1'}$$

The value of w_0 calculated by means of this formula amounts to 66% of the true energy of the H_2 bond. However, if S^2 is neglected [formula (1)], as a result of a chance compensation of errors, the value of w_0 is close to the experimental value. This, of course, does not justify the approximations that have been made.

Consider a reaction of the type:

$$XY + ZW \longrightarrow XZ + YW \tag{2}$$

where X, Y, Z and W are univalent atoms.

A calculation of the four-electron problem at the same level of approximation as that leading to (1), easily gives a formula (due to London) for the energy of the system of four atoms X, Y, Z, W:

$$W = \Sigma - \frac{1}{2}\left[(\alpha - \beta)^2 + (\beta - \gamma)^2 + (\gamma - \alpha)^2\right]^{1/2} \quad . \tag{3}$$

Here Σ is the sum of the Coulomb integrals for all possible pairs of atoms of the system (Figure 79):

$$\Sigma = A_1 + A_2 + B_1 + B_2 + C_1 + C_2 \quad . \tag{4}$$

The quantities α, β, γ are the sums of the corresponding exchange integrals:

$$\alpha = \alpha_1 + \alpha_2; \quad \beta = \beta_1 + \beta_2; \quad \gamma = \gamma_1 + \gamma_2 \quad . \tag{5}$$

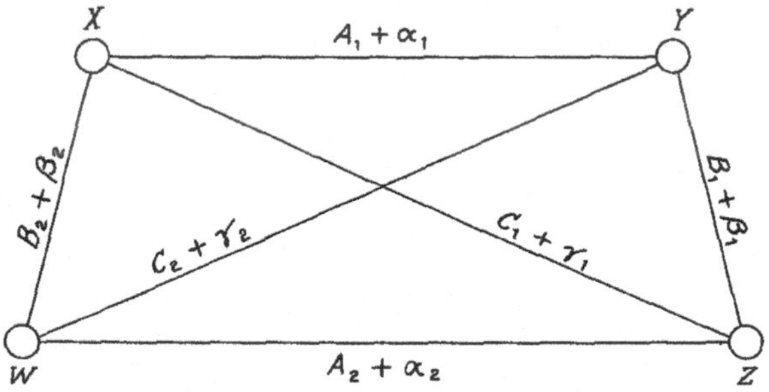

FIGURE 79. Activated complex for a reaction of the type: $XY + ZW = XW + YZ$

The integrals α_i, β_i, γ_i $(i = 1, 2)$ differ from the integral α_0 in equations (1) and (1') because they contain under the integral sign, not only terms expressing the electrostatic interaction between a given pair of atoms (electrons and nuclei), but also terms corresponding to the electrostatic interaction of these atoms with all the others and to the mutual interaction of the latter.

The derivation of formula (3) rests on a number of stringent assumptions. First of all, as in the derivation of formula (1), it is assumed that the wave functions of the separated atoms remain unexcited as the interatomic distance is reduced, and that the quantity S^2 is small as compared to unity. Second, formula (3) assumes that each atom possesses only one electron and it is therefore not strictly applicable to the case of atoms with many electrons. Third, superpositions of more than two electrons are completely neglected and this means neglecting additional quantities of the order of S^2. Moreover, formula (3) does not take bond polarity into account so that it is not applicable to ionic reactions. Finally, formula (3) does not consider excitation of atomic valence states during reaction.

In spite of these simplifications, formula (3) describes in a qualitatively correct manner the main features of the interaction between two molecules XY and ZW or between a molecule XY and an atom Z (at least when X, Y, Z and W are hydrogen atoms). This is easily shown by applying formula (3) to a three-atom system. Putting $A_2 = B_2 = C_2 = 0$ and $\alpha_2 = \beta_2 = \gamma_2 = 0$, we find for the energy of the system XYZ:

$$w' = A_1 + B_1 + C_1 - \frac{1}{2}\left[(\alpha_1 - \beta_1)^2 + (\beta_1 - \gamma_1)^2 + (\gamma_1 - \alpha_1)^2\right]^{1/2} . \quad (6)$$

At the limit, when atom Z is sufficiently far from the molecule XY and thus $B_1 = C_1 = \beta_1 = \gamma_1 = 0$, formula (6) gives:

$$w_0 = A_{01} - |\alpha_{01}| = A_{01} + \alpha_{01}$$

$$(\alpha_{01} < 0) .$$

This expression coincides with formula (1) i.e., it represents the binding energy of the isolated molecule XY. If atom Z starts approaching molecule XY, at distances sufficiently large as compared to the XY spacing, it may be written that $\beta_1/\alpha_1 \ll 1$ and $\gamma_1/\alpha_1 \ll 1$. Then, expanding (6) into a series with respect to these small quantities, we find:

$$w' = (A_1 + \alpha_1) + B_1 + C_1 - \frac{1}{2}\beta_1 - \frac{1}{2}\gamma_1 . \quad (7)$$

This expression tells us that as the atom approaches the saturated molecule, the energy of the system increases, since, as can be calculated for the case of s-electrons:

$$|B_1| < \frac{1}{2}|\beta_1|; \quad |C_1| < \frac{1}{2}|\gamma_1|; \quad \beta_1 < 0, \quad \gamma_1 < 0 \ .$$

Therefore, the atom and the molecule repel each other. If the atom is far from the molecule at first and starts approaching the molecule, then, as before, the energy of the system also increases from an initial value $W_0 = B_{01} + \beta_{01}$ to the value:

$$W' = (B_1 + \beta_1) + A_1 + C_1 - \frac{1}{2}\alpha_1 - \frac{1}{2}\gamma_1 \ . \tag{8}$$

It follows that for a reaction of the type

$$XY + Z \longrightarrow Z + YZ \tag{9}$$

there must exist some activation energy that can be calculated by means of formula (6) after numerical evaluation of all the integrals involved at all interatomic distances.

But even without a numerical solution, formula (6) illustrates the main features of the dependence of the energy of the XYZ system on interatomic distances.

To fix the ideas, consider a linear arrangement between three atoms each having one s-electron (Figure 80). Then the energy of the system is a function of two coordinates r_1 and r_2 and may be represented by means of a 'potential energy surface'. The coordinate axes represent the distances r_1 and r_2; energy values are represented by means of equipotential curves (Figure 81). Then at $r_2 = \infty$ and $r_1 = r_{01}$ (r_{01}: equilibrium interatomic distance in the isolated molecule XY), we will find ourselves in the upper left part of the diagram and the energy of the system will be equal to the energy of interaction of the atoms in their equilibrium position in the molecule XY. As the atom Z gets nearer (r_2 decreases), the energy of the system increases and if r_1 stays constant, it will increase indefinitely. The energy changes in a similar way if $r_2 = r_{02}$ stays constant and one starts at $r_1 = \infty$.

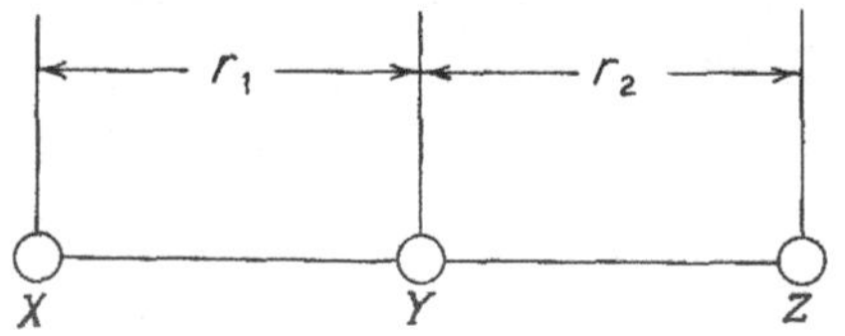

FIGURE 80. Activated complex for the reation XY + Z ⟶ X + YZ.

The state of the three isolated atoms is represented on the upper right corner of the potential surface ($r_1 = r_2 = \infty$), where the energy of the system is equal to zero. If we start from this state and at any time maintain $r_1 = r_2$, we will move along the diagonal PO (Figure 81). At some value of the interatomic distance $r_1 = r_2$, the energy will reach a minimum which, as shown on Figure 81 corresponds to a saddle point indicated by a cross. The transition from state XY + Z to state X + YZ passes through this point which obviously represents the least expenditure of energy. Figure 81 shows that it is energetically advantageous for the distance XY to increase somewhat as XY is approached by Z and that the maximum energy is reached at the saddle point.[1] After that, the distance r_2 will continuously decrease while r_1 increases until the final state is reached when $r_1 = \infty$ and $r_2 = r_{02}$. In this manner, there exists for reaction (9) a 'most favorable path' (represented by a dashed line on Figure 81) corresponding to the least expenditure of energy (the activation energy).

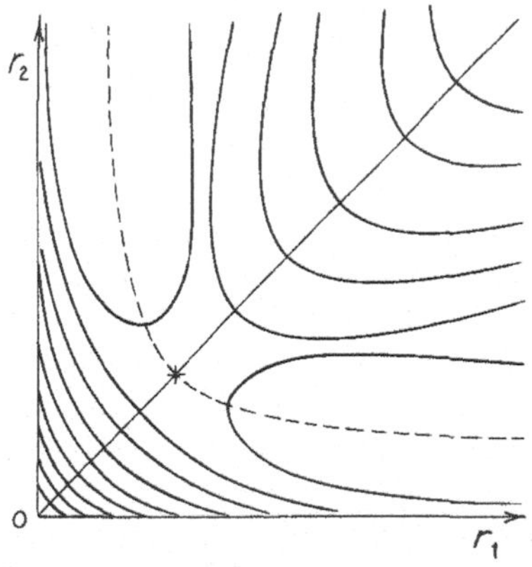

FIGURE 81. Potential energy surface.

By means of formula (3), it is possible to estimate approximately relative activation energies for different configurations of the reacting particles. As an example, consider the reaction:

$$H_2 + D \longrightarrow HD + H \qquad (10)$$

Assuming that $r_1 = r_2$, it is easily shown that the activation energy of reaction (10) is less for a linear configuration of the atoms than for an angular arrangements (Figure 82).

If $r_1 = r_2$, formula (6) may be rewritten in the form ($A_1 = B_1$; $\alpha_1 = \beta_1$):

$$w' = 2A_1 + C_1 + \alpha_1 - \gamma_1 \qquad (11)$$

---

[1] Following Eyring and Polanyi, for the reaction $H_2 + D \longrightarrow HD + H$, the minimum on the diagonal PO is the bottom of a shallow basin, about 2.5 kcal deep [4]. As a result the saddle point is somewhat displaced. But there is no experimental support and little theoretical foundation in favor of this basin.

The difference between this energy and the energy of the initial state $w_0 = A_{01} + \alpha_{01}$ [1] will be equal to:

$$\epsilon = \epsilon_0 = 2A_1 + C_1 - A_{01} + (\alpha_1 - \alpha_{01}) - \gamma_1 \quad (12)$$

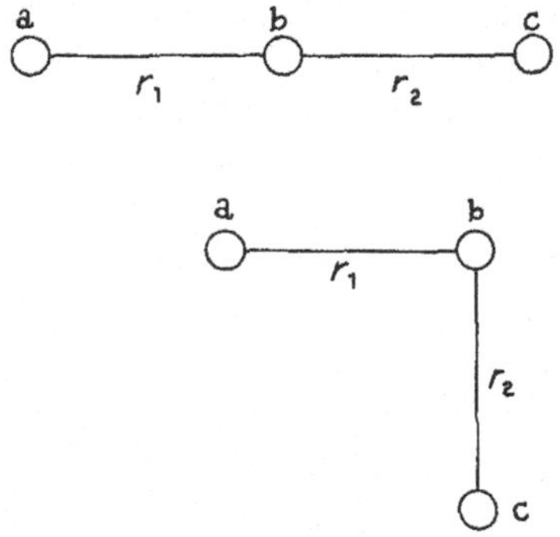

FIGURE 82. Linear and triangular arrangement of three atoms in the activated complex.

If the distance $r_1 = r_2$ corresponds to the saddle point of the potential surface, equation (12) gives the activation energy of reaction (10). From (12) it can be seen that the difference in energy between the linear and triangular configurations is due to the fact that the distance between atoms a and c (Figure 82) is smaller in the first case than in the second. Correspondingly, the repulsion between these atoms (the quantity $-\gamma_1$) is larger for the triangular arrangement than for the linear one. Since the exchange integral in the case of hydrogen is always considerably larger than the Coulomb integral, this increase in repulsion $(-\gamma_1)$ cannot be compensated at the same time by an increase in Coulombic attraction (C_1). Therefore it is expected that the activation energy will be larger in the triangular case. By an approximate estimation called the semi-empirical method (see below), it is found that the difference may reach 20 to 30 kcal.

Using formula (3), we can show easily that the activation energy of the elementary reaction between two hydrogen molecules

$$H_2 + D_2 \longrightarrow HD + HD \quad (13)$$

calculated on the assumption of a rectangular configuration of the activated complex (Figure 83), is twice as large as the energy required for the formation of a right-angle complex between a molecule H_2 and an atom H.

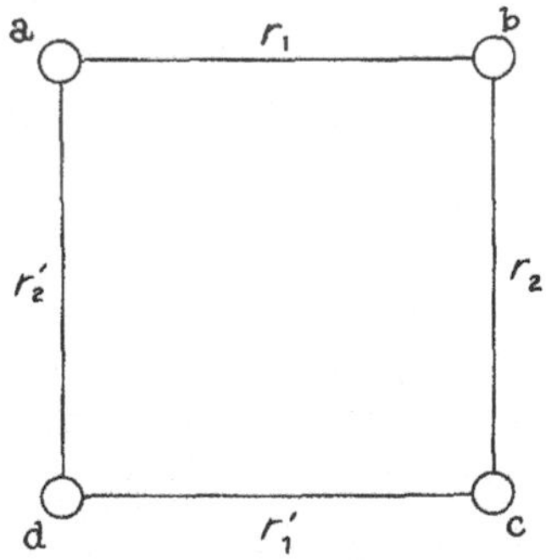

FIGURE 83. Square activated complex.

Indeed, putting $A_1 = A_2 = B_1 = B_2$; $\alpha_1 = \alpha_2 = \beta_1 = \beta_2$ and

$\gamma_1 = \gamma_2$ in formula (3), we get:

$$w = 4A_1 + 2C_1 + 2\alpha_1 - 2\gamma_1 \quad . \tag{14}$$

Subtracting the energy of the two isolated molecules, namely, $2w_0 = 2(A_{01} + \alpha_{01})$, we find for the activation energy of reaction (13):

$$E' = w - 2w_0 = 2[2A_1 + C_1 - A_{01} + (\alpha_1 - \alpha_{01}) - \gamma_1] \quad . \tag{15}$$

Comparison with (12) shows that E' is twice as large as ϵ.

Similarly, by means of equation (6), it can be shown that the activation energy of the reaction

$$H_2 + D_2 \longrightarrow H + HD + D \tag{16}$$

calculated for a linear arrangement of all four atoms (Figure 84) must be approximately equal to the energy Q_{H-H} of the bond in H_2. Indeed, assuming that the distances between neighboring atoms are equal and the same as for the three-atom problem, we have $A_1 = B_1 = B_2$; $C_1 = C_2$; $\alpha_1 = \beta_1 = \beta_2$; $\gamma_1 = \gamma_2$.

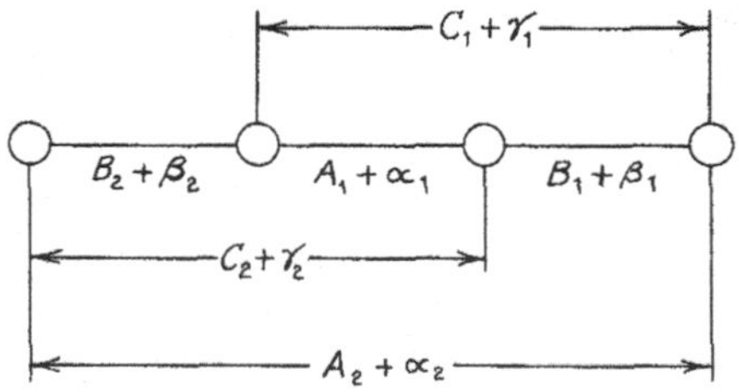

FIGURE 84. Linear arrangement of four atoms in the activated complex.

With this, if the quantity

$$\frac{(\gamma_1-\alpha_2)^2}{6(\alpha_1-\gamma_1)^2}$$

can be neglected as being considerably smaller than unity, the following expression is obtained for the activation energy of reaction (16):

$$E' = Q_{H-H} + \epsilon_0 + [C_1 + A_1 + A_2 + 0.73(\alpha_1 - \gamma_1)] \tag{17}$$

where ϵ_0 is the activation energy of reaction (10), [see equation (12)]. It can easily be seen that the quantity between brackets is negative, since $(\alpha_1 - \gamma_1) < 0$. Hence, the activation energy of reaction (16) is such that: $\epsilon_0 + Q > E' \geq Q_{H-H}$. The activation energy E_0 of the reverse exothermic process is such that: $\epsilon_0 > E_0 \geq 0$.*

* That the activation energy of reaction (16) for a linear configuration of four atoms is small was recently confirmed by a quantum-mechanical calculation of the linear complex H_4. This was done by the method of molecular orbitals with antisymmetrization and electronic interactions as well as use of a self-consistent field [12, 13].

A numerical evaluation of the activation energy by means of the London formula, is a difficult mathematical problem. An approximation was therefore developed, the so-called semi-empirical method. It is based on the use of potential curves for diatomic molecules. Because of the fact that the calculation uses experimental data based on the results of formulae (6) and (3), errors compensate each other to a certain extent. Nevertheless, the semi-empirical method itself introduces new assumptions that introduce new errors and a large degree of arbitrariness in the calculation.

In order to calculate, as a function of the interatomic distances, the integrals of (3), it is first assumed that the energy of interactions of the atoms of a molecule XY, expressed by equation (1), may be approximated by a Morse function of the form:

$$w_0 = A + \alpha = D\{\exp[-2a(r - r_0)] - 2\exp[-a(r - r_0)]\} \quad , \qquad (18)$$

where D is the bond energy (without the zero-point energy) r_0 is the equilibrium interatomic distance $X - Y$, a is a constant determined by spectroscopy. Then, it is assumed that the Coulomb integral A for each molecule is a constant, fixed fraction n of the total bond energy ϵ_0. Consequently, the integrals A and α may be calculated by means of the relations:

$$A = nD\{\exp[-2a(r - r_0)] - 2\exp[-a(r - r_0)]\} \quad , \qquad (19)$$

$$\alpha = (1 - n)D\{\exp[-2a(r - r_0)] - 2\exp[-a(r - r_0)]\} \quad . \qquad (19')$$

Similar formulae are written for each pair of atoms $(X - Y) + (Z - W)$. Substituting the values for the Coulomb and exchange integrals obtained in this manner, into equations (3) or (6), one may calculate the energy of the system for any interatomic distance and therefore determine the activation energy.

There is no difficulty in estimating, for instance, the energy of the system H ... H ... H that represents the transition state of the elementary reaction (10). The experimental activation energy of that process amounts to 6 - 7 kcal.

Let us use equation (12), assuming a linear arrangement of the atoms and putting $r_1 = r_2$. The calculation shows that when $r_1 = r_2$, the energy is minimum when $r_1 \simeq 0.78$A. Taking $n = 0.14$, $D = 108.5$ kcal and $a = 1.94A^{-1}$, we find, from (19) and (19'): $A_{01} \approx -15$ kcal, $A_1 \approx -15$ kcal, $C_1 \approx -6$ kcal, $\alpha_{01} \approx -93$ kcal, $\alpha_1 \approx -93$ kcal, $\gamma_1 \approx -34$ kcal. Putting these values into (12), we obtain $\epsilon_0 \cong 13$ kcal. Approximately the same value for the activation energy of reaction (10) has been obtained by Eyring and Polanyi [6] who used practically the same procedure. The calculated value is about twice as high as the experimental figure. As is shown by the authors of the book 'Absolute Rate Theory' [2], agreement with experiment can be obtained if a somewhat higher value of n is taken. Indeed, taking $n = 0.17$ and multiplying the first three terms of equation (12) by $0.17/0.14$ and the last three by $0.83/0.86$, we find $\epsilon_0 \cong 7$ kcal for the same value of $r_1 = r_2 = 0.78$A.

In this method of calculation, two errors are introduced.

The first is due to the fact that the fraction n (ratio of the Coulomb energy to the total bond energy) is not at all constant but depends strongly on interatomic distance. For the case of two H atoms, it can easily be shown that, at large interatomic distances, the Coulomb integral A decreases as $\exp(-2r)$ and the exchange integral α as $\exp(-r)$. Similar differences exist also at smaller distances. Thus, for instance, in the case of the H_2 molecule, the fraction of the total energy represented by the Coulomb integral is 17% at $r = 1$A but about 5% at $r = 0.78$A, the distance corresponding to the activated complex H ... H ... H [2]. If the latter value $n = 0.05$ were used in the calculation, the activation energy of reaction (10) would be about 30 kcal.

The second error introduced in the calculation consists in the fact that the integral α for the three- or four-atom problem is put equal to the value of α for the bond between two isolated atoms. In reality, they may be quite different. Indeed, the operator in the wave equation contains, in the four-atom problem, not only terms corresponding to the two-atom problem, but also additional terms describing the mutual interaction between all atoms of the problem. Taking the same value of α as for the two-atom case introduces errors probably of the order of the quantity α itself.

Nevertheless, in removing these sources of error by calculation of all the integrals appearing in equation (6), with proper normalization, we do not obtain a result in agreement with the experimental value: the calculated activation energy becomes 19 kcal [7]. This means that the starting point of the London formula represents too gross an approximation. There is no satisfactory calculation of the activation energy available at the present time. Attempts to refine the calculation for the case of reaction (10) by taking into consideration changes ('deformations') of the atomic wave functions due to interaction between atoms, have not yielded satisfactory results. Thus, introduction into the wave function of the system of two variable parameters, has little effect on the energy of three atoms H ... H ... H, although a similar variation improves the calculated energy of the H_2 molecule. Therefore, the calculated activation energy does not show any improvement: it is now 25 kcal [7].

The energy of the system of three H atoms for non-linear

configurations is also quite different when calculated by the semi-empirical method or by a variation technique. Thus, for instance, the energy difference between linear and triangular configuration of the system H - H - H, for $r_1 = r_2 = 1.05A$ is calculated to be 20 kcal by the semi-empirical method and more than 50 kcal by a variation technique [8].

A numerical calculation of the activation energy for reaction between two hydrogen molecules has also been performed with the help of the semi-empirical method. Of course it suffers from the same basic defects as the calculation for the reaction between an atom and a molecule. In the book 'Absolute Rate Theory' [2], the value of 90 kcal for reaction (13) is given. In view of what has been said, this figure must be viewed as a rough estimate.

It must be strongly emphasized that the semi-empirical method represents quite an arbitrary way of calculating activation energies.

Indeed, the fraction representing the contribution of the Coulomb integral to the total energy, remains practically arbitrary in the framework of the semi-empirical method. As a result, there is plenty of lee-way to adjust the calculated figure to the experimental activation energy. We have already seen that, for reaction (10), a value of $n = 0.14$ was arbitrarily selected and that for a more correct choice of $n = 0.05$, the calculated activation energy is quite different from the observed value. Even more arbitrary is the choice of n for reactions involving atoms and molecules with p-electrons. In this case, a direct calculation of Coulomb and exchange integrals shows that they have values close to each other [9]; however, in a calculation based on the London formula, it is assumed, by analogy with the hydrogen problem, that the Coulomb integral is substantially smaller than the exchange integral. For instance, in the book of Glasstone et al. [2], concerning the activation energy for the reaction between H_2 and ICl, it is assumed at one place that $n = 0.14$ and at another place that $n = 0.17$ and 0.20. Let us note that even a small change in the value of n has a large effect on the activation energy. Thus, changing n from 0.17 to 0.20 modifies the activation energy by 8 kcal. This means a difference in reaction rates at normal temperatures, expressed by the factor $\exp(\Delta E/RT) \sim 10^6$.

As a result of the errors and simplifications introduced in a calculation of activation energies from the London formula by means of the semi-empirical method, it follows that the error in the calculated value of E may reach several hundreds of a percent. Moreover, the magnitude of the error cannot be controlled at all theoretically.

In this fashion, we can be convinced that the formula of London is quite unsuitable to a quantitative calculation of activation energies and may be used only for qualitative estimates. We have indicated earlier what kind of qualitative information can be obtained for the case of various relative orientations of an H_2 molecule and an H atom.

Let us note that these conclusions cannot be transferred to the case of atoms and molecules with many electrons, especially if the elementary reaction act involves p-electrons. Indeed, as is shown by calculations concerning even quite simple diatomic molecules such as Li_2, LiH etc., results are strongly affected by the participation of internal electrons. This is even more important for a molecule containing electrons that are not forming intra-molecular bonds (unfilled electron pairs). The picture is further complicated by the specific configuration of p-electron clouds. As shown on Figure 85, the interaction between three atoms with p-electrons is much different for different relative orientations than in the case of atoms with s-electrons. Interaction between two molecules (ab) and (cd) having p - p bonds (Figure 86) is again quite different from the interaction between three atoms with p-electrons and arranged in perpendicular fashion.

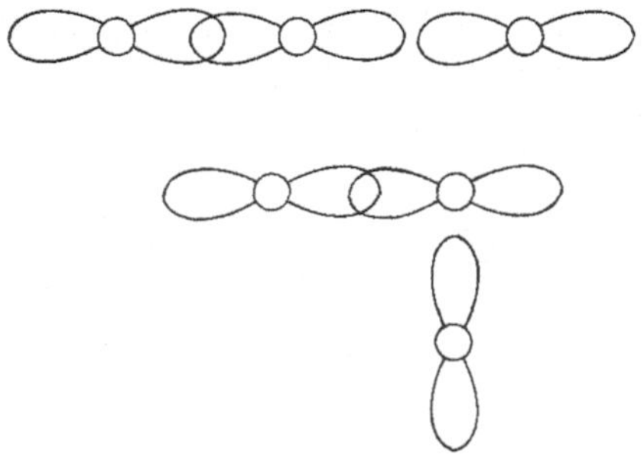

FIGURE 85. Linear (above) and triangular (below) arrangement of three atoms with p-electrons in the activated complex (schematic).

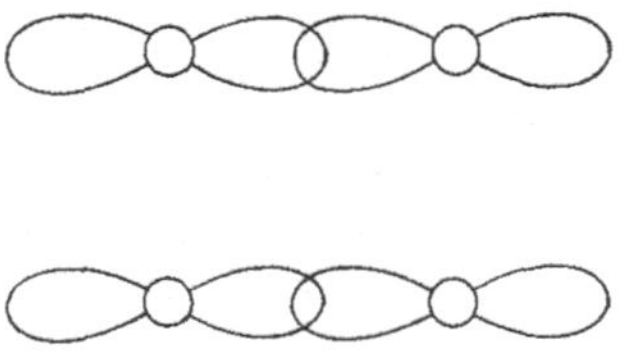

FIGURE 86. Arrangement of four atoms with p-electrons in the activated complex (schematic).

Here, in contrast with the case involving s-electrons, it is difficult to expect any simple relation between the corresponding activation energies. The situation is even more complex if there are unpaired electrons with the axis of their clouds in a p-state lying perpendicularly to the bond axis. This probably increases repulsion between the particles. In such cases, nothing at all can be said a priory concerning the activation energy. It is not ruled out that future calculations might reveal that the nature of the activation energy is quite different than in the case of hydrogen atoms and molecules.

To sum up, we conclude that the problem of the theoretical determination of activation energies of gaseous bimolecular reactions remains unsolved from a quantitative standpoint. The existing qualitative treatment is, strictly speaking, applicable only to hydrogen atoms and molecules. It seems that further developments must follow some alternative approach.

REFERENCES

[1] See for instance, N. N. Semenov, Usp. Khim., 1, 19 (1934).

[2] S. Glasstone, K. Laidler and H. Eyring, Absolute Rate Theory, McGraw-Hill, New York, 1941.

[3] L. D. Landau, Phys. Z. Sowietunion 1, 88 (1932); 2, 46 (1932).

[4] M. I. Temkin, Zhur. Fiz. Khim., 11, 169 (1938).

[5] F. London, "Probleme der modernen Physik (Sommerfeld Festschrift)", c. 104, 1928; Zs. Elektrochem. 35, 552 (1929).

[6] H. Eyring and M. Polanyi, Zs. phys. Chem., B 12, 279 (1931).

[7] J. Hirschfelder, H. Eyring and N. Rosen, J. Chem. Phys., 4, 121 (1936).

[8] J. Hirschfelder, J. Chem. Phys., 6, 795 (1938).

[9] M. F. Mamotenko, Papers in a Collection of Physico-Chemical Studies, Acad. Sci. USSR, p. 1 (1947).

[10] Virginia Griffing, Jour. Chem. Phys., 23, 1015 (1955).

[11] Virginia Griffing and J. T. Vanderslice, Jour. Chem. Phys., 23, 1035 (1955).

APPENDIX III: ADDITIONS TO VOLUME I*

1. On Radical Isomerization (Chapter I Section 9).

A. N. Nesmeyanov, R. Kh. Freĭdlina and co-workers [1] have discovered the isomerization in solution of neutral aliphatic radicals of type A into radicals of type B:

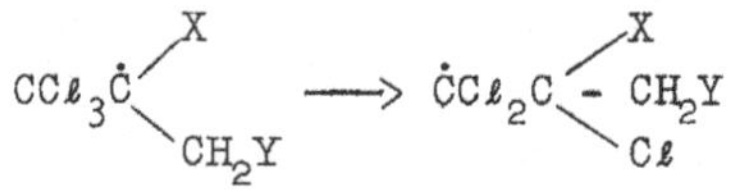

with:

$$X = H\ ,\ H\ ,\ H\ ,\ CH_3,\ CH_3,\ C\ell,\ C\ell,\ Br$$

$$Y = Br,\ C\ell,\ CC\ell_3,\ Br\ ,\ C\ell\ ,\ Br,\ C\ell,\ Br$$

Radicals of type A are made by a chain addition of hydrogen bromine, bromotrichloromethane and bromine to unsaturated compounds of the type $CC\ell_3CX = CH_2$ in the presence of benzoyl peroxide. The reaction is carried out under mild conditions in carbon tetrachloride at 50 - 60°C. If isomerization did not take place during reaction, then, for instance, in the case of hydrogen bromide, reaction should proceed following mechanism I:

$$CC\ell_3CX = CH_2 + HBr \xrightarrow{Bz_2O_2} CC\ell_3CXHCH_2Br \quad . \qquad \text{I}$$

In fact, in all investigated cases ($X = H$, CH_3, $C\ell$) compounds of type (1) are not formed but rather compounds (2), (3) and (4):

$$CC\ell_3CX = CH_2 + HBr \xrightarrow{Bz_2O_2} \begin{cases} HCC\ell_2CXC\ell CH_2Br & (2) \\ CC\ell_2 = CXCH_2Br & (3) \\ HCC\ell_2CXC\ell CH_2C\ell & (4) \end{cases} \quad . \qquad \text{II}$$

---

* Translator's Note: This additional material was received from the author after publication of Volume I and is included at the author's request.

These products can be explained if it is assumed that a radical of type A: $CCl_3\dot{C}XCH_2Br$ isomerizes to give a radical of type B: $\dot{C}Cl_2CXClCH_2Br$. The latter reacts with HBr, giving $HCCl_2CXClCH_2Br$ and regenerating a bromine atom. The role of benzoyl peroxide is to provide bromine atoms.

Thus, A. N. Nesmeyanov and R. Kh. Freĭdlina assume a mechanism similar to that accepted for HBr additions to olefins in the presence of peroxides (mechanism III), with inclusion of steps (3) and (6) where radicals A and A' are isomerized into B and B':

$$\begin{aligned}
&0)\quad Bz_2O_2 \longrightarrow \dot{C}_6H_5\\
&1)\quad \dot{C}_6H_5 + HBr \longrightarrow C_6H_6 + \dot{B}r\\
&2)\quad CCl_3CX = CH_2 + \dot{B}r \longrightarrow \underset{A}{CCl_3\dot{C}X\,CH_2Br}\\
&3)\quad \underset{A}{CCl_3\dot{C}XCH_2Br} \xrightarrow{\text{isomerization}} \underset{B}{\dot{C}Cl_2CXClCH_2Br}\\
&4)\quad \dot{C}Cl_2CXClCH_2Br + HBr \longrightarrow \underset{(2)}{HCCl_2CXClCH_2Br} + \dot{B}r\\
&5)\quad CCl_3\dot{C}XCH_2Br + CCl_3CX = CH_2 \longrightarrow \underset{(3)}{CCl_2 = CXCH_2Br}\\
&\qquad\qquad + \underset{A'}{CCl_3\dot{C}XCH_2Cl}\\
&6)\quad \underset{A'}{CCl_3\dot{C}XCH_2Cl} \xrightarrow{\text{isomerization}} \underset{B'}{\dot{C}Cl_2CXClCH_2Cl}\\
&7)\quad \dot{C}Cl_2CXClCH_2Cl + HBr \longrightarrow \underset{(4)}{HCCl_2CXClCH_2Cl} + \dot{B}r\ .
\end{aligned}\qquad \text{III}$$

Products (2) to (4) were isolated and their structure identified by the usual techniques of organic chemistry.

In the reaction of 1,1,1-trichloropropene with bromine under the same conditions, two products were formed: (5) corresponding to normal bromine addition and (6) resulting from isomerization:

$$CCl_3CH = CH_2 + Br_2 \xrightarrow{Bz_2O_2} \begin{cases} \longrightarrow CCl_3CHBrCH_2Br & (5)\\ \longrightarrow BrCCl_2CHClCH_2Br & (6)\end{cases} \qquad \text{IV}$$

The formation of (5) may be due to a simultaneous electrophilic addition of bromine to the double bond or possibly to the formation of a radical A reacting with bromine with or without preliminary isomerization:

$$\underset{A}{CCl_3\dot{C}HCH_2Br} \xrightarrow{\text{isomerization}} \dot{C}Cl_2CHClCH_2Br \xrightarrow{Br_2}$$

$$\longrightarrow BrCCl_2CHClCH_2Br + Br$$

V

$$Br_2 \longrightarrow CCl_3CHBrCH_2Br + Br \quad .$$

When bromine is added to 1,1,1-trichloropropene under the action of light, compounds (5) and (6) are also formed. When the bromination is carried out in a medium of dilute acetic acid, without light and without peroxides, compound (5) is the unique reaction product.

In all the examples discussed, isomerization of radicals takes place. These radicals are formed from the reactants by some chemical transformation.

It would be quite interesting to study a purely homolytic (molecular) isomerization in solution, the isomerizing compound being transformed into an isomer through a radical chain process. All the numerous molecular rearrangements in solutions studied until now belong to the class of heterolytic reactions.

The investigations just discussed contain the discovery of such a radical chain (molecular) isomerization. A. N. Nesmeyanov, R. Kh. Freǐdlina and V. N. Kost have found that 1,1,1-trichloro-2-bromopropene isomerizes completely to give 1,1,2-trichloro-3-bromopropene-1 at room temperature after an induction period of one to two days:

$$CCl_3CBr = CH_2 \xrightarrow{h\nu} CCl_2 = CClCH_2Br \quad .$$

This isomerization is initiatied by ultra-violet light and is inhibited by addition of small quantities of hydroquinone or dimethylaniline.

The mechanism of this isomerization is apparently very similar to that of the rearrangement of chlorinated radicals just discussed. The authors propose the scheme:

$$C_3Cl_3BrH_2 \xrightarrow{h\nu} \dot{B}r + \dot{C}_3Cl_3H_2$$

$$CCl_3CBr = CH_2 + \dot{B}r \longrightarrow CCl_3\dot{C}BrCH_2Br$$

$$CCl_3\dot{C}BrCH_2Br \xrightarrow{\text{isomerization}} \dot{C}Cl_2CBrClCH_2Br$$

$$\dot{C}Cl_2CBrClCH_2Br + CCl_3CBr = CH_2 \longrightarrow$$

$$CCl_2 = CClCH_2Br + CCl_3\dot{C}BrCH_2Br \quad .$$

VI

REFERENCES

[1] A. N. Nesmeyanov, R. Kh. Freĭdlina and L. I. Zakharkin, Dok. Akad. Nauk SSSR, 81, 199 (1951); Usp. Khim., 25, 665 (1956); Quart. Rev., 10, 330 (1956).

A. N. Nesmeyanov, R. Kh. Freĭdlina and V. N. Kost, Dok. Akad. Nauk SSSR, 113, 828 (1957); Tetrahedron, 1, 241 (1957); R. Kh. Freĭdlina, A. B. Belyavskiĭ and A. N. Nesmeyanov, Izv. Akad. Nauk SSSR, Ot. Khim. Nauk, (in the press).

2. On Polar Factors in Organic Reactions (Chapter I, Section 11)

Connection Between the Relations of Hammett and of Polanyi

The arguments in favor of the role of heats of reaction in the elementary steps of heterolytic processes find additional support, as it seems to us, in the rule of Hammett [1, 2]. This rule has been verified for a very large number of heterolytic reactions (solvolysis, alkylation, acylation, esterification, etc.) in the side chain of a benzene ring containing substituents in meta and para positions.

An example of such reactions is the base hydrolysis of complex esters, involving as rate determining step the addition of OH^- to the carbonyl group of the side chain:

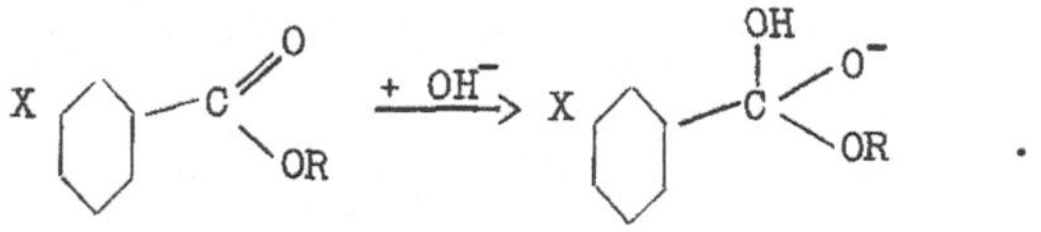

The simplest example is the dissociation of substituted benzoic acids:

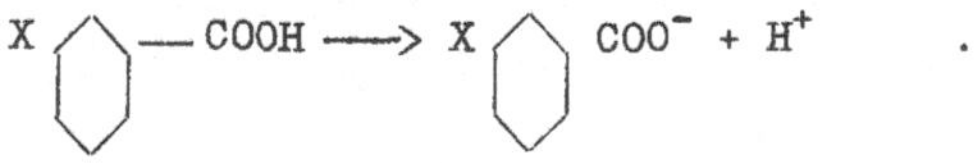

Hammett proposes the following relation:

$$\log_e \frac{k_{ij}}{k_{oj}} = \sigma_i \rho_j \quad . \qquad (1)$$

Here k_{ij} is the experimental rate constant for different reactions (subscript j) in the side chain* for various substituents (subscript i) in

* The same heterolytic reaction in different solvents has different values of ρ because of different degrees of solvation.

the benzene ring. The subscript $i = 0$ corresponds to a reaction with no substituent in the ring. The quantity σ_i depends only on the nature and the position of the substituent. The quantity ρ_j characterizes each given reaction. From general considerations, all that could be said is that $\log_e(k_{ij}/k_{0j})$ is a certain function f_{ij} of both substituent and reaction type. Hammett's rule says that for many reactions f_{ij} is the product of two functions σ_i and ρ_j. Its experimental verification can be made as follows. For an arbitrary reaction of a given type, $j = m$, let us put $\rho_m = 1$ and let us find experimentally for various substituents the quantity $\sigma_i = \log_e(k_{im}/k_{0m})$. Then let us plot in ordinates the values $\log_e(k_{ij}/k_{0j})$, for other reactions and in abscissae the values σ_i for various substituents. It turns out that the points corresponding to each one of these other reactions, lie on straight lines passing through the origin. The slopes of these lines (i.e., the values ρ_j) are different for the different reactions and may be positive or negative. The expression for the rate constants is as usual:

$$k_{ij} = C_{ij} \exp(- E_{ij}/RT) \quad .$$

It is quite natural to assume that in many cases, the pre-exponential factor will be a constant C_j for each type of reaction with various ring substituents. Then $k_{ij} = C_j \exp(- E_{ij}/RT)$ and:

$$\log_e \frac{k_{ij}}{k_{0j}} = - \frac{E_{ij} - E_{0j}}{RT} = - \frac{\Delta E_{ij}}{RT} = \sigma_i \rho_j \quad .$$

For any other reaction $j = j_1$:

$$- \frac{\Delta E_{ij_1}}{RT} = \sigma_i \rho_{j_1} \quad .$$

Hence:

$$\frac{\Delta E_{ij}}{\Delta E_{ij_1}} = \frac{\rho_j}{\rho_{j_1}} \tag{2}$$

or

$$\Delta E_{ij} = \frac{\rho_j}{\rho_{j_1}} \Delta E_{ij_1} \quad . \tag{3}$$

Therefore, if for a given substituent i_1 we compare the differences in activation energies ΔE_{ij} for various reactions j and find a certain proportionality, the same proportionality remains valid for any other substituent i_n. In other words, the differences ΔE_{ij} remain proportional to each other when the substituents are changed. This is a suitable form for Hammett's rule when the steric factors determining the pre-exponential factors of the rate constants retain a constant value for a given type of reaction.*

Hammett's rule applies not only to rate constants but also to equilibrium constants of various reactions in the side chain of the benzene ring. This is shown by the many data on dissociation of acids and bases. Since the equilibrium constant is a ratio of rate constants for forward and backward processes, both the rate constants for heterolytic dissociation and the reverse association of ions also obey Hammett's rule.

Indeed, if Hammett's rule obtains for the forward and backward processes, then:

$$\log_e \frac{k_{ij_1}}{k_{oj_1}} = \sigma_i \rho_{j_1} \quad \text{and} \quad \log_e \frac{k_{ij_2}}{k_{oj_2}} = \sigma_i \rho_{j_2} \quad .$$

Hence:

$$\log_e \frac{k_{ij_1}}{k_{ij_2}} \cdot \frac{k_{oj_2}}{k_{oj_1}} = \log_e \frac{K_{ij(1 \longrightarrow 2)}}{K_{oj(1 \longrightarrow 2)}} = \sigma_i(\rho_{j_1} - \rho_{j_2}) = \sigma_i \rho_{j(1 \longrightarrow 2)} \tag{4}$$

where the K's are equilibrium constants with and without substiuents. The equilibrium constant is:

$$K_{ij(1 \longrightarrow 2)} = D_{ij(1 \longrightarrow 2)} \exp(Q_{ij(1 \longrightarrow 2)}/RT)$$

where $Q_{ij(1 \longrightarrow 2)}$ is the heat of the reaction $1 \longrightarrow 2$.

Assuming that in the majority of cases the pre-exponential factor D does not depend on the nature of the substituent but is determined solely by the reaction type, we get $D_{ij(1 \longrightarrow 2)} = D_{j(1 \longrightarrow 2)}$. Comparing different equilibria, we get then:

$$\frac{Q_{ij(1 \longrightarrow 2)} - Q_{oj(1 \longrightarrow 2)}}{Q_{ij(3 \longrightarrow 4)} - Q_{oj(3 \longrightarrow 4)}} = \frac{\Delta Q_{ij(1 \longrightarrow 2)}}{\Delta Q_{ij(3 \longrightarrow 4)}} = \frac{\rho'_{j(1 \longrightarrow 2)}}{\rho'_{j(3 \longrightarrow 4)}} \ . \tag{5}$$

---

* We have already indicated that in the case of steric hindrance, Hammett's rule does not imply proportionality between ΔE's. But it is then also found experimentally that the rule of Hammett is not valid either in its original form. It may be concluded that the proportionality between differences in activation energies is the basic rule and that Hammett's rule in its original form is a consequence of this basic rule.

Consequently there exists a relation between differences in heats of reaction similar to that found with the activation energies.* Moreover, we also have a direct proportionality between the difference in activation energies ΔE and the difference in heats of reaction ΔQ for each reaction. Indeed:

$$-\Delta E_{ij_1} = \sigma_i \rho_{j_1} \quad \text{and} \quad \Delta Q_{ij(1 \longrightarrow 2)} = \sigma_i \rho'_{j(1 \longrightarrow 2)} \quad .$$

Hence:

$$-\frac{\Delta E_{ij_1}}{\Delta Q_{ij(1 \longrightarrow 2)}} = \frac{\rho_{j_1}}{\rho'_{j(1 \longrightarrow 2)}} \quad . \tag{6}$$

Consequently, it follows from Hammett's rule that changes in activation energies are proportional to changes in heats of reaction and the coefficient of proportionality for each given reaction keeps a constant value when changes are made in the nature of the substituent:

$$-\Delta E_{ij_1} = \alpha_j \Delta Q_{ij(1 \longrightarrow 2)} \quad .$$

But this relation coincides with that of Polanyi. Indeed, it says that for reactions of a given type:

$$\begin{aligned} E_{ij} &= A_j - \alpha_j Q_{ij(1 \longrightarrow 2)}; \quad E_{oj} = A_j - \alpha_j Q_{oj(1 \longrightarrow 2)} \quad ; \\ \Delta E_{ij} &= E_{ij} - E_{oj} = -\alpha_j (Q_{ij(1 \longrightarrow 2)} - Q_{oj(1 \longrightarrow 2)}) = -\alpha_j \Delta Q_{ij} \end{aligned} \tag{7}$$

which is what we set out to prove.

If then the activation energies and heats of heterolytic reactions (or, which is the same, the rate constants and the equilibrium constants) obey Hammett's rule, the same reactions, just like radical reactions obey Polanyi's rule which says that for a given series of reactions changes in activation energies are determined by changes in heats of reaction.** In other words, the activation energy is largely determined by the heat of reaction. Hammett's rule for the rate constants is a direct consequence

* If the heterolytic reaction proceeds through a certain number of elementary steps, the overall rate of reaction is determined by that of the slow step. The rate constant of the slow step obeys Hammett's rule. It is natural to assume that the reverse of that slow step also obeys the rule. But in equilibrium, there is equilibrium for all individual steps. Thus the equilibrium constant and therefore the difference in heats of reaction ΔQ will also obey the rule.

** This point was brought to my attention by A. E. Shilov.

of Polanyi's rule and of Hammett's rule for the heats of reaction. However, if Hammett's rule is not observed by the rate constants, this does not mean that Polanyi's rule is not obeyed. Indeed, if Polanyi's rule is observed but not Hammett's rule for the heats of reaction, then Hammett's rule for the rate constants will not be obeyed.

To sum up, there are good reasons to believe that the relation of Polanyi is observed by heterolytic reactions, at any rate as well as by radical reactions.

As basic relations we shall take Planyi's rule and Hammett's rule for heats of reaction (i.e., for equilibrium constants). Polanyi's rule (as shown on page 32) and Hammett's rule for the heats of reaction (see below) have a theoretical basis, qualitative as it may be. From these two relations, Hammett's rule for the reaction rates can be derived mathematically. A direct theoretical proof of the latter would be difficult. The relation between heats of reaction and molecular structure is much more direct than that between structure and activation energies.

Consider any reaction, e.g., the elementary step of hydrolysis of a complex esther:

$$\bigcirc\!\!-P \;+\; OH^- \;\longrightarrow\; \bigcirc\!\!-P'$$

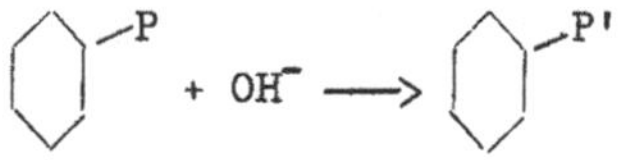

where   P is the group

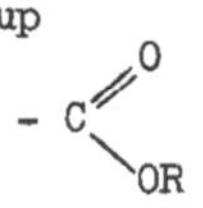

$$- C\!\!\begin{array}{c}{}^{\displaystyle O}\\ {}_{\displaystyle OR}\end{array} \quad \text{and} \quad P' \text{ is } \; - C - O^-\!\!\begin{array}{c}{}^{\displaystyle OH}\\ {}_{\displaystyle OR}\end{array} \quad .$$

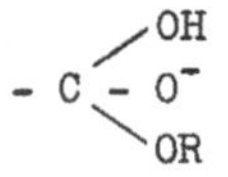

The heat of reaction is the difference between heat contents of final and initial states: $Q_0 = H_0 + H_{OH} - H_0'$. After introduction of a substituent at any position in the ring, the heat content is changed and thus also the heat of reaction: $Q_x = H_x + H_{OH} - H_x'$.

The difference ΔQ_x is:

$$\Delta Q_x = Q_x - Q_0 = (H_x - H_0) - (H_x' - H_0') \quad ,$$

where $H_x - H_0$ and $H_x' - H_0'$ are the differences in heat content of

$$\bigcirc\!\!-P \qquad \text{and} \qquad \bigcirc\!\!-P'$$

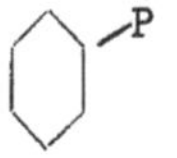

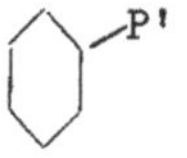

with and without substituent.

Let us express H_X and H_0 as the sum of three terms:

$$H_X = \epsilon_X^\Phi + D_X^P + \epsilon_P \quad \text{and} \quad H_0 = \epsilon_0^\Phi + D_0^P + \epsilon_P$$

where

ϵ_X^Φ and ϵ_0^Φ are the heat contents of substituted and unsubstituted phenyl radicals with a free valence instead of group P

ϵ_P is the heat content of radical P

D_X^P and D_0^P are the heats of recombination of the phenyl radical with radicals P and P'.

The heat content of the reaction product is correspondingly:

$$H_X' = \epsilon_X^\Phi + D_X^{P'} + \epsilon_{P'} \quad \text{and} \quad H_0' = \epsilon_0^\Phi + D_0^{P'} + \epsilon_{P'} \quad .$$

Hence:

$$\begin{aligned} \Delta Q = Q_X - Q_0 &= \epsilon_X^\Phi + D_X^P - \epsilon_0^\Phi - D_0^P - (\epsilon_X^\Phi + D_X^{P'} - \epsilon_0^\Phi - D_0^{P'}) \\ &= D_X^P - D_0^P - (D_X^{P'} - D_0^{P'}) = \Delta D_X - \Delta D_0 \quad . \end{aligned} \tag{8}$$

Thus ΔQ is determined only by the differences in heats of recombination of substituted and unsubstituted phenyl radicals with radicals P and P'. In the majority of cases, these differences are small as compared to the quantities D themselves. Variations in the heats of recombination, i.e., in the bond dissociation energy of C – R are determined by the effect of substituents and that due to the bonds broken and made during reaction.* If the effect is small and may be characterized by a single parameter σ_1 depending only on the nature and position of the substituent, we have, in first approximation:

$$D_X^P = D_0^P + \left(\frac{\partial D^P}{\partial \sigma}\right)_0 \sigma_1 \quad \text{and} \quad D_X^P - D_0^P = \left(\frac{\partial D^P}{\partial \sigma}\right)_0 \sigma_1 = b_j^{(P)} \sigma_1 \quad ,$$

where $b_j^{(P)}$ is determined only by the type of group P and the reaction type, does not depend on the substituent, while σ_1 depends only on nature and position of the substituent and practically does not depend on the

---

* The first effect may be due, for example, to the accumulation of an electric charge at the location of group P as a result of the maldistribution of electron density in the ring caused by the substituent. The second effect is then related to the polarity of the bonds in group P.

reaction type. Hence:

$$\Delta Q = (b_j^{(P)} - b_j^{(P')})\sigma_1 = \rho_j^{(P \longrightarrow P')}\sigma_1 \quad . \tag{9}$$

We have therefore obtained in first approximation from general considerations, an expression identical to Hammett's rule for the heats of reaction. There is some doubt concerning the assumed independence of σ_1 from the nature of the radical P. Yet, for small values of σ, this may be considered as approximately correct.

To see this more clearly, consider that σ is due to the change in ring electronic density caused by the substitution of X for H. Then obviously σ does not depend on the nature of group P. When the phenyl radical and the P radical recombine, the electronic density and the difference in electronic densities change in the unsubstituted and substituted rings. Yet, if these variations are small, like the parameters σ themselves, ΔQ can be expanded in a series the first term in σ being retained only:

$$\Delta Q = \left(\frac{d\Delta Q}{d\sigma}\right)_0 \sigma_1 \tag{10}$$

where

$$\left(\frac{d\Delta Q}{d\sigma}\right)_0$$

is the derivative at $\sigma \longrightarrow 0$, i.e., a quantity that does not depend on the nature of the substituent.*

This result may be illustrated in a different way. Let us imagine that we attach a 'stopcock' to the C - P bond in the unsubstituted compound. This 'stopcock' is already there before X is substituted for H. The closed 'stopcock' prevents the flow of electrons from the ring to the side-chain and vice-versa. Then a substituent is introduced that changes the electronic density in the ring. Then the 'stopcock' is opened and electrons flow in one direction or another till the electronic capacity of the group P is satisfied. If this capacity is small, the variation in electronic density will be proportional to the density in the ring (depending only on the substituent and to the capacity of group P which does not depend on the substituent $(\sigma_1\rho_j)$.

It must be noted that Hammett's rule is not always obeyed even

---

* The qualitative character of all these arguments is not due to the fact that higher order terms are neglected but to the assumption that the heats of reaction D_X depend only on σ. This assumption is based on a crude model.

for benzene derivatives (the ortho substituted compounds). It is not obeyed by the majority of aliphatic compounds substituted one way or another. This is probably due to the determining role of steric factors after introduction of polar substituents so that the assumed equality of entropy terms is no more valid. This is particularly clear in the case of ortho-substituted compounds where the substituent is adjacent to the position of reaction. Hammett's rule is not obeyed [1]. Steric effects act not only on the activation entropy but also on the activation energy. Sometimes this effect can be estimated so that steric factors can be separated from polar factors. This has been done by Taft [3] in a variety of cases. It was shown that the variations in rate constants due to the purely polar factors of substituents, obey Hammett's rule even in these cases. It therefore appears that the linear relation between activation energies and heats of reaction due to polar substituents is a very wide-spread phenomenon in chemistry.

It is interesting to note that M. I. Kabachnik [3] has shown recently that Hammett's rule can be applied to the equilibrium dissociation constants of substituted phosphoric and thiophosphoric acids. The values of σ for this series were smaller than (but of the same order of magnitude as) the σ values determined by Hammett. It may be assumed that Hammett's rule will also apply to the rate constants of the reactions of phosphoric and thiophosphoric acids. Thus a phosphorus atom, like a benzene ring, transmits the effect of substituents to other groups bound to that phosphorus atom.

The transmission of the effect of substituents is made more difficult by the separation of the reacting group from the conjugated system by means of several CH_2 groups. For example, in the series C_6H_5COOH, $C_6H_5CH_2COOH$, $C_6H_5CH_2CH_2COOH$, the dissociation constants of these acids are characterized by a decrease in the value of ρ from unity to 0.212 [2] as the number of intermediate CH_2 groups increases from 0 to 2. This is quite natural since non-conjugated aliphatic chains prevent an easy transmission of the polar effects.

REFERENCES

[1] L. P. Hammett, Physical Organic Chemistry, McGraw Hill, New York (1940).

[2] H. H. Jaffe, Chem. Rev., 53, 191 (1953).

[3] R. W. Taft, Jour. Am. Chem. Soc., 74, 2729, 3121 (1952); M. I. Kabachnik, Dok. Akad. Nauk SSSR, 110, 393 (1956).

3. On the Nature of Intermediates in the Oxidation of Hydrocarbons (Chapter II, Section 2)

Recently, at the Institute of Chemical Physics, E. A. Blyumberg and Yu. D. Norikov have proved conclusively that aldehydes and acids do not come for the most part from peroxides in the gas phase oxidation of hydrocarbons at temperatures above 200°C. They studied the oxidation of normal butane at 250°C and 700 mm Hg. It was shown by means of the kinetic isotope technique that the total rate of formation of stable reaction products (aldehydes and acids) exceeds by a factor of more than two the rate of conversion of normal butyl hydroperoxide. Thus more than 50% of stable products are formed by a path that bypasses the hydroperoxide stage.

Also recently, N. A. Kleĭmenov and A. B. Nalbandyan have used the kinetic method of tracers to study the mercury photosensitized oxidation of methane. The temperature was 300°C, the pressure 1 atm., the mixture 90% CH_4 + 10% O_2 to which tagged methyl hydroperoxide was added (~ 0.05%). The results show that ~ 2/3 of the formaldehyde formed in the reaction does not come from the peroxide but directly from the peroxidic radical (Dok. Akad. Nauk SSSR, in the press).

4. On the Mechanism of Propane Oxidation (Chapter II, Section 3)

Norrish [1] postulates a mechanism for decomposition of n - RO_2, namely $RO_2 \longrightarrow OH + R'CHO$ different from that of Shtern. The latter used his analytical data [2] to show that the mechanism of Norrish leads to a quantity of water three times in excess of the experimental value. It seems that the mode of decomposition of RO_2 proposed by Norrish does not take place significantly in the oxidation of propane.

REFERENCES

[1] J. H. Knox and R. G. W. Norrish, Proc. Roy. Soc., A 222, 151 (1954).

[2] N. Ya. Chernyak, V. L. Antonovskiĭ, A. F. Revzin and V. Ya. Shtern, Zhur. Fiz. Khim., 28, 240 (1954).

5. On the Kinetic Tracer Method* (Chapter II, Section 3, page 118)

Besides the rates of formation and disappearance of intermediate products, the elucidation of reaction mechanisms also requires a knowledge

* This is a correction. It replaces the corresponding paragraphs on pages 118 and 121 (Translator's Note).

of the order in which the various products appear. This problem can also be solved by the kinetic method of M. Neĭman that determines the rate of formation of any product B from its various precursors. It is then necessary to tag one of the precursors, say A and to follow with time the change in concentration of B and the specific activity α and β of the products A and B.

Let us call w_1 the rate of formation of B from A, w_1' that from other precursors and w_2 the rate of disappearance of B. Then:

$$\frac{d(B)}{dt} = w_1 + w_1' - w_2 \quad .$$

Since $w_1^* = \alpha w_1$ and $w_2^* = \beta w_2$, the rate of increase of the specific activity is:

$$\frac{d\beta}{dt} = \frac{1}{(B)} \cdot \frac{d(B^*)}{dt} - \frac{\beta}{(B)} \cdot \frac{d(B)}{dt} = \frac{w_1}{(B)} \left(\alpha - \beta \frac{w_1 + w_1'}{w_1} \right). \qquad (1)$$

If at the start of reaction, a small quantity of untagged product B is added, at $t = 0$, $\beta = 0$. In this case, β first increases to a maximum value and then decreases since α decreases. The maximum value of β is determined by the condition:

$$\frac{d\beta}{dt} = 0 \quad \text{i.e.,} \quad \left(\alpha - \beta \frac{w_1 + w_1'}{w_1} \right) = 0 \quad .$$

Hence:

$$\frac{\alpha}{\beta} = \frac{w_1 + w_1'}{w_1} \quad . \qquad (2)$$

Also, if B is formed only from A, $w_1' = 0$ and

$$\frac{\alpha}{\beta} = 1$$

or $\alpha = \beta$. The curves of activity must then cross each other. But if $w_1' \neq 0$, they do not cross and from the value of $\frac{\alpha}{\beta}$ at the maximum, the ratio of the rates w_1 and w_1' can be determined.

This method was used by M. B. Neĭman, A. B. Nalbandyan et al [35] to decide whether in the course of methane oxidation, carbon monoxide is formed only from HCHO or also from some other substances. In order to prove this point, they added 0.07% tagged $C^{14}H_2O$ and 0.5% CO to a mixture of 33% CH_4, 66% air and 0.1% NO. The reaction was carried out at 670°C. During the course of the process, the specific activities α and β of the products HCHO and CO were measured. At the point of maximum value of β (see Figure 16), $\alpha = \beta$. This may be the case only if $w_1' = 0$. Therefore, carbon monoxide is produced only from HCHO by the overall scheme $CH_4 \longrightarrow HCHO \longrightarrow CO$.

6. On the Mechanism of Butene Cracking.* (Chapter II, Section 3, page 132)

The chain decomposition of butene-1 may be represented by the sequence:

1) $C_4H_8 \longrightarrow CH_2 = CH - \dot{C}H_2 + CH_3$ chain initiation
2) $CH_3 + C_4H_8 \longrightarrow CH_4 + \dot{C}_4H_7$
3) $CH_2 = CH - \dot{C}HCH_3 + CH_2 = CH - CH_2CH_3 \longrightarrow$
$CH_2 = CH - CH = CH_2 + \dot{C}H_2CH_2CH_2CH_3$ or $CH_3\dot{C}HCH_2CH_3$
4) $\dot{C}H_2CH_2CH_2CH_3 \longrightarrow C_2H_4 + \dot{C}_2H_5$
5) $\dot{C}_2H_5 + C_4H_8 \longrightarrow C_2H_6 + \dot{C}_4H_7$
6) $\searrow C_2H_4 + \dot{C}_4H_9$
7) $\dot{C}H_2CH_2CH_2CH_3 \longrightarrow CH_3\dot{C}HCH_2CH_3$
8) $CH_3\dot{C}HCH_2CH_3 \longrightarrow CH_3 - CH = CH_2 + \dot{C}H_3$
9) $2\ C_4H_7 \longrightarrow$ chain termination.

According to this mechanism, the radical C_4H_7 formed from butene-1 may follow two reaction paths. It may form a diene by reaction with the original material, giving an alkyl radical, normal or secondary (reaction 3).

It follows from this scheme that the main gaseous products of butene-1 cracking should be methane, ethane, propylene and ethylene. These products are also those found experimentally.

V. V. Voevodskiĭ, from the distribution of gaseous products of pentene-1 cracking, as determined by Yu. A. Abruzov [55] estimated the relative rates of the elementary steps entering in the cracking mechanism. The olefinic radical C_5H_9 may react in three ways, with probabilities α, β and γ, the sum of which is equal to unity (for sufficiently long chains). The third path is the decomposition of C_5H_9 along the C - C bond to give a CH_3 radical and butadiene. The alkyl radicals that are formed may react in two ways: the radical $CH_2CH_2CH_2CH_2CH_3$ may either isomerize by a reaction analogous to 7) or decompose in a reaction similar to 4) with a probability δ. The radical C_3H_7, following a reaction similar to 6) gives propylene or propane following a reaction similar to 5) with a probability φ. V. V. Voevodskiĭ has written the kinetic equations corresponding to the formation of the main products of cracking of pentene-1. He

* This is a slightly different mechanism of butene cracking replacing the mechanism given on pages 132 and 133 (Translator's Note).

calculates probabilities α, β, γ, δ and φ respectively equal to 0.1, 0.38, 0.51, 0.58 and 0.34. As can be seen, the least probable step is the scission of the radical C_5H_9 along a C - C bond (α = 0.1).

7. On the Inhibition of Reactions in the Liquid Phase (Chapter IV, Section 5, page 188)

Several compounds, especially polyphenols, secondary aromatic amines, quinones and nitro-compounds are well known inhibitors of polymerization processes. During a long time, inhibition by polyphenols and amines has been explained by their ability to react with the free radicals propagating the polymerization chain [17]. But special studies by Dolgoplosk et al. [38-40] on the inhibition mechanism of thermal and initiated polymerization have revealed that no inhibition takes place in the total absence of oxygen. Then even in the case of such monomers as methylacrylate and vinylacetate which give very active free radicals, no inhibition can be observed. Addition of polyphenols or amines in the absence of oxygen changes only very slightly the kinetics of the process. It is only when some oxygen is present in the system that an induction period is observed during which polymerization stops altogether. The authors believe that the inhibiting action of polyphenols in the presence of oxygen is due to the formation of RO_2 radicals which are particuarly active in hydrogen abstraction so that favorable circumstances exist for the oxidation of phenols and aromatic amines to corresponding semiquinone radicals. The final quinone products are effective acceptors of free radicals. The proposed interpretation of the phenomenon, follows from the experimental data and explains the so-called 'inhibiting effect' of small quantities of oxygen in polymerization. This effect is due to the sequence of reactions just described:

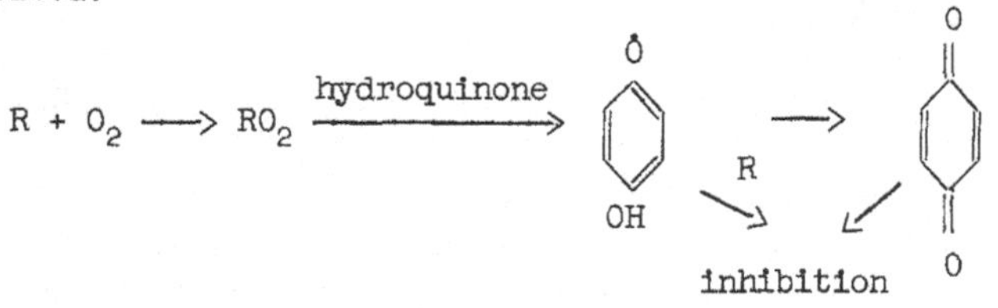

where R is a polymer radical. In this way, oxygen in systems containing traces of polyphenols and aromatic amines acts as an inhibitor of polymerization processes. A similar effect exists when an oxygen free system contains peroxides (e.g., benzoyl peroxide, cumene hydroperoxide) and salts of heavy metals (Fe, Cu, Mn). Then the following cycle of transformations takes place:

$$Me^{n} \underset{\text{hydroquinone}}{\overset{\text{peroxide}}{\rightleftharpoons}} Me^{n+1}$$

hydroquinone ↙ quinone

These observations indicate that inhibition of polymerization is not due to polyphenols and aromatic amines but to the quinone compounds derived from them. It is known that quinones and especially benzoquinone inhibit polymerization effectively whether there is oxygen in the system or not [39-42]. The mechanism of inhibition of polymerization by quinone compounds is explained by their ability to react with polymer radicals to give semiquinone radicals. The latter are not active enough to continue the polymerization chain. In particular, the reaction between benzoquinone and free radicals gives the corresponding ether of hydroquinone [40, 43]:

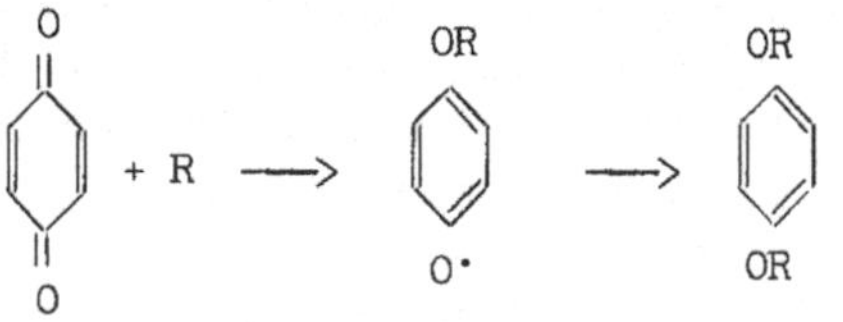

To the same type of inhibitors belong aromatic nitro-compounds which have been investigated by several investigators [44-48]. The inhibiting action of these compounds in polymerization is also reinforced markedly by oxygen [48]. The ability of quinones and nitro-compounds to react with free radicals is illustrated by the data collected in Table 66 which show the relation between methane yield and the nature of the inhibitor when methyl radicals react with hydrocarbons. As a source of methyl radicals, methylphenyltriazene was used. This compound decomposes in solutions when heated:

$$CH_3 - N = N - NHC_6H_5 \longrightarrow CH_3 + N_2 + C_6H_5NH \quad .$$

The inhibitors react with methyl radicals in just the same order of inhibition activity in polymerization. Obviously, inhibition in this case is due to the alkylating action of free radicals and not to hydrogen abstraction from the inhibitor since then no decrease in methane yields would be observed.

This explanation relative to the quinone mechanism of inhibition cannot be transposed completely to oxidation processes. In oxidation, polyphenols and aromatic amines are much more effective inhibitors than quinones.

TABLE 66 [48]

Effect of Inhibitors on the Yield of Methane Formed by Thermal Decomposition of a 1% Solution of Methylphenyltriazene

Inhibitor	T°	Solvent	Ratio: inhibitor/triazene	CH_4 content % in gas
Without inhibitor	125	Ethylbenzene	-	31.0
M-dinitrobenzene	125	"	5:1	21.0
M-dinitrobenzene	125	"	8:1	18.1
2,4,dinitrochlorobenzene	125	"	8:1	16.7
Trinitrobenzene	125	"	1:1	10.0
Trinitrobenzene	125	"	5:1	2.5
Naphthoquinone	100	"	2:1	1.5
Benzoquinone	100	"	2:1	0

REFERENCES

[1] I. W. Breitenbach and K. Horeishy, Ber. 74, 1386 (1941).
[2] B. A. Dolgoplosk and D. S. Korotkina, Trudy V.N.I.I.S.K., 198 (1951).
[3] B. A. Dolgoplosk and G. A. Parfenova, Trudy V.N.I.I.S.K., 224 (1951).
[4] H. W. Melville and W. F. Watson, Trans. Far. Soc., 44, 886, (1948).
[5] S. G. Cohen, J. Am. Chem. Soc., 69, 1057, (1947).
[6] A. F. Bickel and W. A. Waters, J. Chem. Soc., 1764, (1950).
[7] S. G. Foord, J. Chem. Soc., 48, (1940).
[8] Ch. T. Price, R. W. Bell and E. Krebs, J. Am. Chem. Soc., 64, 1103, (1942).
[9] R. L. Franck and C. E. Adams, J. Am. Chem. Soc., 68, 908, (1946).
[10] G. Schulz, Markromolek. Chem., 1, 94, (1947).
[11] B. A. Dolgoplosk, B. L. Erusalimskiĭ and E. B. Milovskaya, Trudy V.N.I.I.S.K., 163 (1953).
[12] E. E. Nikitin, Dok. Akad. Nauk SSSR, 116, 584 (1952).
[13] E. E. Nikitin, Dok. Akad. Nauk SSSR, 1958 (in press).

8. Heterogeneous Catalysis in Biology (Chapter VI Section 8)

In living organisms, various types of chemical transformations take place at the surface of highly specialized catalysts: the enzymes. All enzymes that have been investigated chemically until now are either pure proteins or complex proteins with non-protein 'prosthetic' groups. In this way, biocatalysis is essentially heterogeneous catalysis at the surface of protein macromolecules. Consider, for instance, the mechanism of the processes of cellular respiration, representing the main path of oxidation in cells and the principal source of energy required for sustaining life. The overall reaction is ultimately the transfer of hydrogen from the oxidized substrate to oxygen of the air, with formation of water. But it takes place in a multitude of individual steps in which the potential of the system is lowered gradually. Some of these steps involve the transfer of hydrogen between adjacent components of the main oxidation chain. Other steps involve electron transfer. Many experimental facts have led to the conclusion that these reactions may take place in the cell without immediate contact between reacting molecules. Moreover, it must be kept in mind that the energy released in these exothermic processes is not dissipated into heat but is used to bring about a series of endothermic processes e.g., the binding of orthophosphoric acid to form adenosin triphosphoric acid (ATP) required in muscular contraction. The efficiency of this energy transfer is very high and may reach 85%. In order to realize directly this energy transfer, simultaneous collision between at least six reacting molecules would be required, two of these being high polymers. These facts and many others relative to muscular contraction and photosynthesis indicate migration of energy in biological systems.

In 1941, A. Szent-Györgyi [31] proposed an electronic mechanism of energy transfer, based on an analogy between proteins and metals. However, elementary calculations as well as data on the semi-conductivity and phosphoresence spectra of proteins show that the corresponding 'conduction bands' of protein systems, if they exist, must lie at least 3 or 4 electron-volts above the valence band. These conduction bands should then be empty at the temperature of living organisms.

In 1954, Commoner, Townsend and Pake [32] using the technique of electron paramagnetic resonance (EMR) studied animal and plant tissues and observed relatively large concentrations of unpaired electrons: 10^{-6} to 10^{-8} mole/grams. Later, Blyumenfel'd [33] showed that these unpaired electrons are not associated with low molecular weight free radicals or ions with variable valence but to protein enzymes. But they are found only during the course of enzymatic processes. The absence of a hyperfine structure and the small width of the EMR lines indicate a large degree of delocalization of these electrons in the protein structure. It was proposed that the conductivity paths follow the hydrogen bonds perpendicular to the main polypeptide chains, as shown on the following page.

In enzyme catalysis, the substrate forms a temporary complex with the enzyme. Sometimes, the non-protein prosthetic group plays the role of impurity. The electronic levels of this impurity lie near the conduction or valence bands of the protein structure. These bands, unlike those of usual semi-conductors are very narrow (a few tenths of e.v.). Then, the

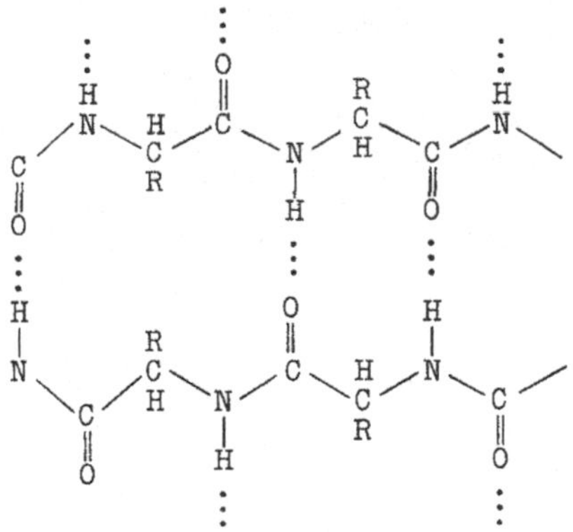

complex substrate-enzyme may be regarded as an impurity semi-conductor with electron or hole conduction. The electron of the conduction band may not fall back to the completely full valence band and migrates along the chain of hydrogen bonds in the polypeptive structure, till it falls into the potential trap of an acceptor forming a complex with the enzyme. This is how an oxidation-reduction process may take place at a distance without immediate contact between reacting molecules. It is believed that the endothermic processes make use of these conduction bands which are populated by unpaired electrons in excess of equilibrium amounts as a result of the exothermic oxidation processes.

Blyumenfel'd and Kolmanson [34,35] have studied the EMR spectra of free radicals formed after γ irradiation of aminoacids, peptides, native and denatured proteins. They confirmed the existence of delocalized unpaired electrons in protein structures and their relation to the regularity of the network of hydrogen bonds.

Quite recently, Commoner [36] in a study of typical oxidation enzyme processes, succeeded in observing two types of EMR spectra. One was due to low molecular intermediate compounds of the semiquinone type. The other was due to high molecular weight materials acting like semi-conductors.

To sum up, there is direct proof of the role of semi-conductor properties in heterogeneous catalytic processes in biological systems. It is very likely that a similar approach and similar techniques will be fruitful in studies of usual catalytic processes at surfaces.

REFERENCES

[1] A. Szent-Györgyi, Science, 93, 609 (1941).

[2] B. Commoner, J. Townsend and G. Pake, Nature, 174, 689 (1954).

[3] L. A. Blyumfel'd, Izv. Akad. Nauk SSSR, ser. biol., 285 (1957).

[4] L. A. Blyumenfel'd and A. E. Kalmanson, Biofizika, 2, 552 (1957).

[5] L. A. Blyumenfel'd and A. E. Kalmanson, Dok. Akad. Nauk SSSR, 117, 72 (1957).

[6] D. Commoner, 'Free radicals in biological systems', Symposium Program. Symposium on the Formation and Stabilization of Free Radicals, September, 1957.

AUTHOR'S ACKNOWLEDGEMENTS

I want to express my gratitude to my collaborators at the Institute of Chemical Physics of the Academy of Science USSR: S. S. Polyak, V. I. Vedeenev, V. V. Voevodskiĭ, N. S. Enikopolyan, D. G. Knorre, R. V. Kolesnikova, A. B. Nalbandyan, M. B. Neĭman, N. D. Sokolov, A. E. Shilov, N. Ya. Shlyapintokh, N. M. Emanuel' and also to I. V. Berezin, L. A. Blyumenfel'd, A. N. Pravednikov, G. B. Sergeev, M. I. Temkin and D. A. Frank-Kamenetskiĭ. With them I have discussed many problems treated in the book and they have contributed to the writing of several sections and chapters of the first and second editions.

N. Semenov

Lightning Source UK Ltd.
Milton Keynes UK
UKOW01f1045311017
311936UK00004B/218/P

9 780691 626277